Lecture Notes in Computer Science

W0079382

Lecture Notes in Computer Science

Vol. 1: GI-Gesellschaft für Informatik e.V. 3. Jahrestagung, Hamburg, 8.–10. Oktober 1973. Herausgegeben im Auftrag der Gesellschaft für Informatik von W. Brauer. XI, 508 Seiten. 1973.

Vol. 2: GI-Gesellschaft für Informatik e.V. 1. Fachtagung über Automatentheorie und Formale Sprachen, Bonn, 9.–12. Juli 1973. Herausgegeben im Auftrag der Gesellschaft für Informatik von K.-H. Böhling und K. Indermark. VII, 322 Seiten. 1973.

Vol. 3: 5th Conference on Optimization Techniques, Part I. (Series: I.F.I.P. TC7 Optimization Conferences.) Edited by R. Conti and A. Ruberti. XIII, 565 pages. 1973.

Vol. 4: 5th Conference on Optimization Techniques, Part II. (Series: I.F.I.P. TC7 Optimization Conferences.) Edited by R. Conti and A. Ruberti. XIII, 389 pages. 1973.

Vol. 5: International Symposium on Theoretical Programming. Edited by A. Ershov and V. A. Nepomniaschy. VI, 407 pages. 1974.

Vol. 6: B. T. Smith, J. M. Boyle, J. J. Dongarra, B. S. Garbow, Y. Ikebe, V. C. Klema, and C. B. Moler, Matrix Eigensystem Routines – EISPACK Guide. XI, 551 pages. 2nd Edition 1974. 1976.

Vol. 7: 3. Fachtagung über Programmiersprachen, Kiel, 5.–7. März 1974. Herausgegeben von B. Schlender und W. Frielinghaus. VI, 225 Seiten. 1974.

Vol. 8: GI-NTG Fachtagung über Struktur und Betrieb von Rechensystemen, Braunschweig, 20.–22. März 1974. Herausgegeben im Auftrag der GI und der NTG von H.-O. Leilich. VI, 340 Seiten. 1974.

Vol. 9: GI-BIFOA Internationale Fachtagung: Informationszentren in Wirtschaft und Verwaltung. Köln, 17./18. Sept. 1973. Herausgegeben im Auftrag der GI und dem BIFOA von P. Schmitz. VI, 259 Seiten. 1974.

Vol. 10: Computing Methods in Applied Sciences and Engineering, Part 1. International Symposium, Versailles, December 17–21, 1973. Edited by R. Glowinski and J. L. Lions. X, 497 pages. 1974.

Vol. 11: Computing Methods in Applied Sciences and Engineering, Part 2. International Symposium, Versailles, December 17–21, 1973. Edited by R. Glowinski and J. L. Lions. X, 434 pages. 1974.

Vol. 12: GFK-GI-GMR Fachtagung Prozessrechner 1974. Karlsruhe, 10.–11. Juni 1974. Herausgegeben von G. Krüger und R. Friehmelt. XI, 620 Seiten. 1974.

Vol. 13: Rechnerstrukturen und Betriebsprogrammierung, Erlangen, 1970. (GI-Gesellschaft für Informatik e.V.) Herausgegeben von W. Händler und P. P. Spies. VII, 333 Seiten. 1974.

Vol. 14: Automata, Languages and Programming – 2nd Colloquium, University of Saarbrücken, July 29–August 2, 1974. Edited by J. Loeckx. VIII, 611 pages. 1974.

Vol. 15: L Systems. Edited by A. Salomaa and G. Rozenberg. VI, 338 pages. 1974.

Vol. 16: Operating Systems, International Symposium, Rocquencourt 1974. Edited by E. Gelenbe and C. Kaiser. VIII, 310 pages. 1974.

Vol. 17: Rechner-Gestützter Unterricht RGU '74, Fachtagung, Hamburg, 12.–14. August 1974, ACU-Arbeitskreis Computer-Unterstützter Unterricht. Herausgegeben im Auftrag der GI von K. Brunnstein, K. Haefner und W. Händler. X, 417 Seiten. 1974.

Vol. 18: K. Jensen and N. E. Wirth, PASCAL – User Manual and Report. VII, 170 pages. Corrected Reprint of the 2nd Edition 1976.

Vol. 19: Programming Symposium. Proceedings 1974. V, 425 pages. 1974.

Vol. 20: J. Engelfriet, Simple Program Schemes and Formal Languages. VII, 254 pages. 1974.

Vol. 21: Compiler Construction, An Advanced Course. Edited by F. L. Bauer and J. Eickel. XIV. 621 pages. 1974.

Vol. 22: Formal Aspects of Cognitive Processes. Proceedings 1972. Edited by T. Storer and D. Winter. V, 214 pages. 1975.

Vol. 23: Programming Methodology. 4th Informatik Symposium, IBM Germany Wildbad, September 25–27, 1974. Edited by C. E. Hackl. VI, 501 pages. 1975.

Vol. 24: Parallel Processing. Proceedings 1974. Edited by T. Feng. VI, 433 pages. 1975.

Vol. 25: Category Theory Applied to Computation and Control. Proceedings 1974. Edited by E. G. Manes. X, 245 pages. 1975.

Vol. 26: GI-4. Jahrestagung, Berlin, 9.–12. Oktober 1974. Herausgegeben im Auftrag der GI von D. Siefkes. IX, 748 Seiten. 1975.

Vol. 27: Optimization Techniques. IFIP Technical Conference. Novosibirsk, July 1–7, 1974. (Series: I.F.I.P. TC7 Optimization Conferences.) Edited by G. I. Marchuk. VIII, 507 pages. 1975.

Vol. 28: Mathematical Foundations of Computer Science. 3rd Symposium at Jadwisin near Warsaw, June 17–22, 1974. Edited by A. Blikle. VII, 484 pages. 1975.

Vol. 29: Interval Mathematics. Procedings 1975. Edited by K. Nickel. VI, 331 pages. 1975.

Vol. 30: Software Engineering. An Advanced Course. Edited by F. L. Bauer. (Formerly published 1973 as Lecture Notes in Economics and Mathematical Systems, Vol. 81) XII, 545 pages. 1975.

Vol. 31: S. H. Fuller, Analysis of Drum and Disk Storage Units. IX, 283 pages. 1975.

Vol. 32: Mathematical Foundations of Computer Science 1975. Proceedings 1975. Edited by J. Bečvář. X, 476 pages. 1975.

Vol. 33: Automata Theory and Formal Languages, Kaiserslautern, May 20–23, 1975. Edited by H. Brakhage on behalf of GI. VIII, 292 Seiten. 1975.

Vol. 34: GI – 5. Jahrestagung, Dortmund 8.–10. Oktober 1975. Herausgegeben im Auftrag der GI von J. Mühlbacher. X, 755 Seiten. 1975.

Vol. 35: W. Everling, Exercises in Computer Systems Analysis. (Formerly published 1972 as Lecture Notes in Economics and Mathematical Systems, Vol. 65) VIII, 184 pages. 1975.

Vol. 36: S. A. Greibach, Theory of Program Structures: Schemes, Semantics, Verification. XV, 364 pages. 1975.

Vol. 37: C. Böhm, λ-Calculus and Computer Science Theory. Proceedings 1975. XII, 370 pages. 1975.

Vol. 38: P. Branquart, J.-P. Cardinael, J. Lewi, J.-P. Delescaille, M. Vanbegin. An Optimized Translation Process and Its Application to ALGOL 68. IX, 334 pages. 1976.

Vol. 39: Data Base Systems. Proceedings 1975. Edited by H. Hasselmeier and W. Spruth. VI, 386 pages. 1976.

Vol. 40: Optimization Techniques. Modeling and Optimization in the Service of Man. Part 1. Proceedings 1975. Edited by J. Cea. XIV, 854 pages. 1976.

Vol. 41: Optimization Techniques. Modeling and Optimization in the Service of Man. Part 2. Proceedings 1975. Edited by J. Cea. XIII, 852 pages. 1976.

Vol. 42: James E. Donahue, Complementary Definitions of Programming Language Semantics. VII, 172 pages. 1976.

Vol. 43: E. Specker und V. Strassen, Komplexität von Entscheidungsproblemen. Ein Seminar. V, 217 Seiten. 1976.

Vol. 44: ECI Conference 1976. Proceedings 1976. Edited by K. Samelson. VIII, 322 pages. 1976.

Vol. 45: Mathematical Foundations of Computer Science 1976. Proceedings 1976. Edited by A. Mazurkiewicz. XI, 601 pages. 1976.

Vol. 46: Language Hierarchies and Interfaces. Edited by F. L. Bauer and K. Samelson. X, 428 pages. 1976.

Vol. 47: Methods of Algorithmic Language Implementation. Edited by A. Ershov and C. H. A. Koster. VIII, 351 pages. 1977.

Vol. 48: Theoretical Computer Science, Darmstadt, March 1977. Edited by H. Tzschach, H. Waldschmidt and H. K.-G. Walter on behalf of GI. VIII, 418 pages. 1977.

Lecture Notes in
Computer Science

Edited by G. Goos and J. Hartmanis

87

5th Conference on Automated Deduction

Les Arcs, France, July 8-11, 1980

Edited by W. Bibel and R. Kowalski

Springer-Verlag
Berlin Heidelberg New York 1980

Editorial Board

W. Brauer P. Brinch Hansen D. Gries C. Moler G. Seegmüller
J. Stoer N. Wirth

Editors

Wolfgang Bibel
Institut für Informatik, Technische Universität München, Postfach 20 24 20
8000 München 2/Germany

Robert Kowalski
Department of Computing and Control, Imperial College,
180 Queen's Gate
London, SW7 2 BZ/England

AMS Subject Classifications (1979): 68 A 40, 68 A 45
CR Subject Classifications (1974): 5.21, 3.60

ISBN 3-540-10009-1 Springer-Verlag Berlin Heidelberg New York
ISBN 0-387-10009-1 Springer-Verlag New York Heidelberg Berlin

This work is subject to copyright. All rights are reserved, whether the whole or part of the material is concerned, specifically those of translation, reprinting, re-use of illustrations, broadcasting, reproduction by photocopying machine or similar means, and storage in data banks. Under § 54 of the German Copyright Law where copies are made for other than private use, a fee is payable to the publisher, the amount of the fee to be determined by agreement with the publisher.

© by Springer-Verlag Berlin Heidelberg 1980
Printed in Germany

Printing and binding: Beltz Offsetdruck, Hemsbach/Bergstr.
2145/3140-543210

F O R E W O R D

This FIFTH CONFERENCE ON AUTOMATED DEDUCTION, held at Les Arcs, Savoie, France, July 8 - 11, 1980, was preceded by earlier meetings at Argonne, Illinois (1974), Oberwolfach, West Germany (1976), Cambridge, Massachusetts (1977), and Austin, Texas (1979).

This volume contains the papers which were selected by the program committee from the 62 papers submitted to the conference. These papers range over most of the main approaches to the automation of deductive reasoning. They describe both theoretical status and practical investigations of implementations and their applications, especially in computing science.

PROGRAM COMMITTEE

W. Bibel, München (program co-chairman);
W.W. Bledsoe, Austin;
A. Colmerauer, Marseille;
M. Davis, New York;
L. Henschen, Evanston;
R. Kowalski, London (program co-chairman);
D. Mc Dermott, New Haven;
R. Milner, Edinburgh;
U. Montanari, Pisa;
J S. Moore, Menlo Park;
D. Oppen, Stanford;
V. Pratt, Cambridge;
J.A. Robinson, Syracuse;
S. Sickel, Santa Cruz (past program chairman).

ORGANISER

G. Huet, INRIA, Rocquencourt.

CONTENTS

TUESDAY MORNING

TUESDAY AFTERNOON

WEDNESDAY MORNING

FRIDAY AFTERNOON

USING META-THEORETIC REASONING TO DO ALGEBRA

Luigia Aiello (†,‡)
Richard W. Weyhrauch (‡)

(†) Istituto di Elaborazione della Informazione, CNR, Pisa, Italy
(‡) Artificial Intelligence Laboratory, Stanford University, Stanford, USA

ABSTRACT

We report on an experiment in interactive reasoning with FOL. The subject of the reasoning is elementary algebra. The main point of the paper is to show how the use of meta-theoretic knowledge results in improving the *quality* of the resulting proofs in that, in this environment, they are both easier to find and easier to understand.

1. INTRODUCTION

In this paper we report on an experiment in interactive reasoning, namely reasoning with FOL about elementary algebra.

FOL is a conversational system designed by Richard Weyhrauch at the Stanford Artificial Intelligence Laboratory. It runs in LISP on a KL10. It implements First Order Logic using the natural deduction formalism of Prawitz [Pra65] enriched in many ways. In this paper we make no attempt to be self-contained, mostly because of space limitations. We refer to the literature [Wey77,78,79;Aie80;Fil78,79;Tal80] for an introduction to FOL, its many features and some examples of its use.

We chose elementary algebra as the subject of our reasoning because it is a commonly known field and has an established axiomatic mathematical presentation. Another reason for this choice is that future applications of FOL will require knowledge of the algebraic facts of arithmetic. Hence, it is important to provide FOL with the ability to reason about arithmetic and to discover what arithmetic and meta-arithmetic facts are used in ordinary conversations about numbers.

The goal of our experiment was epistemological: to verify the adequacy of FOL as a framework for knowledge representation. We were interested in verifying the ability of FOL to "be told" about elementary mathematics. In order to understand the spirit of the experiment it is important to explain our criterion for determining when a reasoning system is adequate. It is *not* simply its ability to carry out proofs that is relevant. In the case of arithmetic, a special purpose theorem prover designed for performing algebraic manipulations can certainly attain the same theorems as our treatment in FOL, sometimes in a faster way (in terms of cpu time). The central issue is how rich a mode of expression is allowed by the system. That is, what facts are implicit by being part of the code of the system and what facts can be explicitly explained to it. It is the *quality* of the conversation that matters. From this viewpoint not many existing theorem provers can be considered adequate. The treatment of elementary algebra and algebraic manipulations presented in this paper uses modes of expression (namely, meta-theoretic) that are simply not explicitly available to most theorem proving systems.

FOL provides a rich enough conversational facility so that an agreement with the user is made about the subject of the conversation and a domain of consensus is established in which to carry on the reasoning. The first thing to be agreed with FOL is which language to speak, i.e. which tokens we are going to use in our conversation and what syntactic role they will play. FOL then expects to know what facts we assume as basic truths (axioms) of our subject domain. At this point it is ready to do reasoning with us.

Making the above kind of conversation explicit is part of the epistemological flexiblity of FOL. Another aspect of the flexibility of FOL is that it is capable of being told to relativize things to the right context. In fact, it can deal with many theories at the same time. One of them, named META, generally contains meta-theoretic knowledge. It plays a special role because of the way it can communicate with other theories. The possibility of representing meta-theoretic knowledge at the right level, along with the possibility of using it intermixedly with knowledge at the theory level, is a major feature of FOL.

In ordinary math books the distinction between statements at the theory level and those at the meta-theory level is often blurred, if not completely absent. Conversely, if the reasoning system you are dealing with has no capability of explicitly representing meta-theoretic knowledge, many of the statements in elementary math books cannot even be expressed. This might not be interesting if such meta-statements never appeared in practice. On the contrary, they arise *very* often in mathematics books (as well as in ordinary conversation, but this is not the point of the present paper).

In order to have a reasonable conversation with a system about algebra, you want it to have an *understanding* of algebra. In other words, you require that the ability of the system to perform manipulations and answer questions is (at least) as good as yours. We have chosen to bring FOL to this understanding by following a *foundational* approach. It consists in starting with some axioms and incrementally build a theory using in new proofs only facts that are either axioms or have already been proved. Math books are frequently written from a foundational point of view, with the intent of producing an understanding in the reader.

From an epistemological point of view we consider this experiment a success as an examination of our experience with FOL shows that: (1) Proofs expressed to FOL closely resemble the informal proofs of math books, both for their length and for the kind of knowledge they use. (2) Proofs become shorter and shorter along the way, i.e. the more facts you have already proved, the simpler it becomes to prove new ones. (3) What in the book is left as an "easy exercise for the reader", usually is an easy exercise for FOL too - it can be proved with a single line proof.

Our developement of algebra in FOL follows the presentation in [BML65], paying particular attention to consider representing the content of each sentence in the book. We have proved theorems about integers considered both as an integral domain and as an ordered integral domain. We omit the presentation of order and concentrate on the first part. This gives us chances to speak about the use of meta-theoretic knowledge in building proofs, which is one the main purposes of this paper.

The use of meta-theoretic knowledge has been prompted by the observation that in rewriting (i.e. "simplifying") arithmetic expressions we have to deal with the commutativity of the operators plus and times. It is known that commutativity cannot be used as a rewrite rule, since it would cause the rewritings to loop.

This observation has led some authors to deal with equivalences for equational theories involving associative and commutative operators in special ways. For example, finding complete sets of equations that could be used as rewriting rules, or special code to deal with particular cases. The approach we have followed to perform algebraic manipulations on arithmetc expressions is different. The fact that the operators plus and times are commutative, instead of being used at the theory level (by using the relevant axioms), has been embedded into FOL at the meta-theory level. This has been done by specifying what kind of manipulations are allowed on the symbols occurring in an arithmetic expression. Namely, arguments to the functions plus and times can be reordered (for instance to check that **x*y** and **y*x** are the same, hence that **x*y=y*x** holds). As a result, we have devised a "simplification" algorithm that manipulates arithmetic *expressions* by using both theoretical and meta-theoretical knowledge.

The paper is organized as follows. We first explain in detail what has been done at the theory level, and make remarks about proof building within the theory. Then we explain what has been described at the meta-theory level, and finally, how the the simplification algorithm for arithmetic expressions works, in particular, how it uses at the same time knowledge represented in both the theory and the meta-theory.

2. REASONING AT THE THEORY LEVEL

FOL [Wey77,78,79] interacts with the user in a sorted first order language. One of the characterizing features of FOL is that knowledge is represented in the form of *L/S structures* (or *L/S pairs*). These can be explained as a pair of descriptions: a syntactic one (a sorted language and a set of axioms) and a semantic one (a domain of interpretation and information about the interpretation of some of the symbols of the language). The specification of the semantics is done by attaching LISP objects and LISP code to syntactic entities, via what is called *semantic attachment*. The semantic description is also called a *simulation structure*. It functions in FOL as its internal mechanizable analogue of a model for the theory specified by the given syntax. Note that, with an abuse of language, we frequently call an L/S structure a theory.

In order to build a theory for algebra we start by telling FOL that we want to build a new L/S structure, name it ARITH, and direct our attention to it. Then, by means of the following declarations:

```
DECLARE SORT INTEGER,NATNUM,NEGNUM;
MG INTEGER ≥ {NATNUM,NEGNUM};
DECLARE INDVAR u v w x y z ϵ INTEGER;
DECLARE OPCONST -(INTEGER)=INTEGER;
DECLARE OPCONST + * - (INTEGER,INTEGER)=INTEGER;
```

we tell FOL that we are going to speak about integers (i.e. INTEGER is a sort), natural and negative numbers. We specify that natural and negative numbers are integers, i.e.

the sort INTEGER is more general (MG) then both the sort NATNUM and the sort NEGNUM. We then introduce some individual variables ranging over the integers and some operator constants along with their arity and their domains and ranges.

After the language has been established, we can tell FOL the axioms for integral domains.

```
AXIOM COMMLAW:   ∀ u v.(u+v)=(v+u),
                 ∀ u v.(u*v)=(v*u);
      ASSOLAW:   ∀ u v w.u+(v+w)=(u+v)+w,
                 ∀ u v w.u*(v*w)=(u*v)*w;
      DISTLAW:   ∀ u v w.u*(v+w)=(u*v+u*w);
      ZERO:      ∀ u.u+0=u;
      UNITY:     ∀ u.u*1=u;
      NONTRIV:   ¬ 0=1;
      ADDINV:    ∀ u.∃ v.u+v=0;
      CANCEL:    ∀ u v w.(¬ u=0 ∧ u*v=u*w ⊃ v=w);;
```

As for the simulation structure, the objects of sort NATNUM are attached to the LISP natural numbers. There is no need for an explicit attachment to the negative numbers because, as an example, the expression "-3" in FOL is interpreted to be the operator unary minus applied to the natural number three. The operators plus and times are attached to the LISP functions PLUS and TIMES, respectively.

As already noted, we adopt the foundational approach, hence no use is made of the operators unary and binary minus in the axioms. After the theorem (named UNINV)

```
∀ u v w.(u+v=0 ∧ u+w=0 ⊃ v=w)
```

stating the unicity of the inverse for the operator plus has been proved, unary minus is introduced by giving FOL the implicit definition (named INVAX)

```
∀ u.u+(-u)=0
```

and attaching the symbol "-" to the LISP function MINUS. We can then introduce the binary minus by the definition

```
∀ x y.(x-y)=(x+-y)
```

In our interaction with FOL, there is no attempt to derive each proof by starting with a minimal set of notions: we introduce and use new notions when they contribute to making proofs more natural. The axiom ADDINV, together with the theorem UNINV may seem to convey the same information as the axiom INVAX, hence they may appear to be of the same use in proof building. Actually this is false: when performing deductions, the axiom INVAX can be used directly but the use of ADDINV involves the application of the deduction rule for existential elimination.

The goal of building proofs that are as natural as possible has been pursued not only by introducing the relevant notions as soon as they are available, but also by exploiting many built-in features of FOL. No attempt has been made to build proofs using the bare logic, i.e., the deductive apparatus of Prawitz's natural deduction alone. We have freely used both the tautology tester and the rewrite/eval commands of FOL. The tautology tester checks whether or not a well formed formula follows as a tautological consequence from a given set of formulas.

The use of the rewrite/eval commands has played a central role in many proofs. The command REWRITE expects a term (or a wff) and a simpset, which is a set of rewrite (simplification) rules and rewrites the term (wff) according to the given simpset. The command EVAL is similar to REWRITE, but it makes use of both the syntactic and the semantic knowledge it is provided with, in the form of a simpset and of semantic attachments, respectively. More details about the workings of REWRITE and EVAL can be found in [We77,78,79;Aie80].

To provide an example of use of REWRITE, a proof of the statement:

$$\forall x.((x*x=x) \supset (x=0 \vee x=1))$$

can be produced starting from the instantiation of the cancellation law on the terms x, x and 1, i.e., $\neg x=0 \wedge x*x=x*1 \supset x=1$. This can be rewritten by a simpset containing $\forall x.x*1=x$ and yields $\neg x=0 \wedge x*x=x \supset x=1$. The theorem follows as a tautological consequence.

The example just presented is very simple. The situation is more complicated when large simpsets and/or many rewritings are involved. This prompted us to construct a "standard" simpset which allowed us to perform most of the standard algebraic manipulations by a single REWRITE, or EVAL.

The following simpset, named SS for Simplification Set, contains either axioms or formulas which were proved using FOL.

```
UNIT   ∀ u.(u*1)=u
       ∀ u.(1*u)=u
NULL   ∀ u.(u+0)=u
       ∀ u.(0+u)=u
       ∀ u.(0*u)=0
       ∀ u.(u*0)=0
INVX   ∀ u.(u+-u)=0
MINS   ∀ u v.(u-v)=(u+-v)
       ∀ u v.-(u+v)=(-u+-v)
       ∀ u.--u=u
        -0=0
       ∀ u v.(u*-v)=-(u*v)
       ∀ u v.(-u*v)=-(u*v)
ASSO   ∀ u v w.((u+v)+w)=(u+(v+w))
       ∀ u v w.((u*v)*w)=(u*(v*w))
DIST   ∀ u v w.(u*(v+w))=((u*v)+(u*w))
       ∀ u v w.((u+v)*w)=((u*w)+(v*w))
```

The evaluation of an arithmetic expression by means of the above simpset (i.e. EVAL ... BY SS;) puts it in the form of a sum of monomials, with no redundant occurrences of zeroes, ones and minus signs, and where a minus sign possibly prefixes some of the monomials.

It should be noticed that the commutativity laws for plus and times do not *explicitly* appear in the above simpset. Hence, whenever commutativity has to be used in a proof, the user has the choice of either instantiating the relevant axiom (which often results in a pretty long and "unnatural" proof) or better, to use the meta-theoretic knowledge (explained in the following) to further process the arithmetic expression(s) he is dealing with.

3. REASONING AT THE META-THEORY LEVEL

In order to speak with FOL about facts in the meta-theory of arithmetic expressions we have to set up the language (actually, the meta-language) of arithmetic. We start by telling FOL, within the L/S structure named META, that AREX (ARithmetic EXpressions) is a sort and that objects of sort AREX are TERMs. The sort TERM, as well as the sorts INDCONST, OPCONST, etc., appearing in the following are part of the description in META of the language objects (which has been developed by R. Weyhrauch and C. Talcott).

In META we give names to the objects of the theory. We start by introducing individual constants for the operator symbols of the theory, and attaching them to the corresponding objects.

```
DECLARE INDCONST Minus, Sum, Prod ϵ OPCONST;
MATTACH Sum   ⇔  ARITH:OPCONST: + ;
MATTACH Prod  ⇔  ARITH:OPCONST: * ;
MATTACH Minus ⇔  ARITH:OPCONST: - ;
```

The command MATTACH binds names of individual constants in META to objects in a particular theory. In this case the theory is ARITH and the objects are OPCONSTs.

The sort Integer-INDVAR is then introduced in META in order to speak about the individual variables of sort INTEGER in the theory. We specify that objects of sort Integer-INDVAR are particular arithmetic expressions and also particular individual variables. We also give names in META to the variables of sort INTEGER introduced in the theory (i.e. T-u is MATTACHed to u, T-v to v, etc.).

```
DECLARE PREDCONST Integer-INDVAR(AREX);
MG AREX ≥ {Integer-INDVAR};
MG INDVAR ≥ {Integer-INDVAR};
DECLARE INDCONST T-u,T-v,T-w,T-x,T-y,T-z ϵ Integer-INDVAR;
MATTACH T-u ⇔ ARITH:INDVAR: u;
...
```

In order to speak in META about arithmetic expressions in their most general form (i.e. expressions in ARITH that are built out of both individual variables and constants and the operators plus, times and minus, as well as function symbols applied to some arguments), we have to describe in META function application at the theory level. This is done by extending the language in META with the new sort ARGS (ARGumentS).

```
DECLARE SORT ARGS;
MG ARGS ≥ {AREX};
```

We have specified that the sort ARGS is more general than the sort AREX. This implies that an arithmetic expression can be an argument to a function. Conversely, no requirement is imposed on arguments to consist of one or more arithmetic expressions. Hence, in an arithmetic expression, function symbols can occur that have been declared to map objects of any sort into objects of any other sort, as far as they eventually yield to an arithmetic expression. In other words, an arithmetic expression in ARITH is any term that is hereditarily well sorted and whose sort is INTEGER.

The constructors and selectors needed to handle arguments are also introduced, as well as those needed to handle the application of function symbols to arguments. These constructors and selectors are then attached to the relevant LISP code.

4. COMPUTING IN META

To provide an example of how computations are performed in META and how the relevant information is retrieved from the theory ARITH, whenever needed, let's discuss how the predicate MONOMIAL is defined and how its evaluation goes.

```
AXIOM MONOMIAL: ∀ ae.(MONOMIAL(ae) ≡ INDEL(ae) ∨ ¬(funof(ae)=Sum));;
```

where

```
AXIOM INDEL:  ∀ ae.(INDEL(ae)   ≡  INDSYM(ae) ∨ NEGNUMRAL(ae));;

AXIOM INDSYM: ∀ o.(INDSYM(o)    ≡  INDCONST(o) ∨ INDVAR(o) ∨ INDPAR(o)),
              ∀ o.(INDCONST(o)  ≡  syntype(o)=Indconst),
              ∀ o.(INDVAR(o)    ≡  syntype(o)=Indvar),
              ∀ o.(INDPAR(o)    ≡  syntype(o)=Indpar);;

AXIOM NEGNUM: ∀ ae.(NEGNUMRAL(ae) ≡
                        mainsym(ae)=Minus ∧ NATNUMRAL(arg(1,ae)));;
```

Note that the predicate MONOMIAL is supposed to be applied to arithmetic expressions (aes) that have the form of a sum of monomials (namely, to arithmetic expressions that have already been simplified at the theory level by SS). Hence, to check if an arithmetic expression in this form consists of only one monomial we have to check whether it consists of an individual element or if its function part is not a plus.

Individual elements are defined to be either individual symbols (i.e., individual constants, variables or parameters) or negative numerals. In FOL negative "numerals" aren't individual symbols: they are natural numerals prefixed by the unary minus sign.

To check whether or not an object is an INDCONST (or an INDVAR, or an INDPAR) its syntactic type has to be determined in ARITH. This is done by the function **syntype** which is attached to the following LISP code.

```
ATTACH syntype ↔  (LAMBDA(X)(SYNT-DN X 'ARITH));
```

The LISP function SYNT-DN applied to X and ARITH switches attention from the L/S structure META to ARITH, computes the syntactic type of X in ARITH, then switches back to META and brings back its result.

As an example, MONOMIAL(T-u) will evaluate to true. This is because syntype(T-u) evaluates to Indvar since u has been declared to be an INDVAR in ARITH. On the other hand, if AE has be declared (in META) to be of sort AREX and has been MATTACHed to the term x+y (in ARITH), then MONOMIAL(AE) will evaluate to false. Its syntactic type (in

ARITH) is neither INDCONST, nor INDVAR, nor INDPAR, and its function part is the operator plus (i.e., funof(AE)=Sum is true). Note also that the evaluation of funof(AE) involves an L/S structure switching, to check that the main symbol of the expression attached to AE actually is an operator, i.e., its syntactic type in ARITH is either OPCONST or OPVAR.

5. THE SIMPLIFICATION ALGORITHM

Due to space limitations we cannot report the complete set of meta-axioms that implements in FOL the simplification algorithm for arithmetic expressions. We can only sketch it.

In order to simplify an arithmetic expression, it is first evaluated at the theory level by means of the simpset SS. This process, as already noted, puts it in the form of a sum of monomials. Then, the manipulations that are performed on the resulting expression at the meta-theory level consist in a reordering and merging of the monomials constituting it. The reordering of a monomial is performed by matching the variable (and function) symbols occurring in it with a lexicon, which is the list of all the variable and function symbols occurring in it, in the order they first occur. Thus, in META, given an arithmetic expression, its lexicon is built first, then it is used to reorder the monomials and merge them. The choice to build a lexicon for each arithmetic expression in the way just described has been done on the assumption that there is no pre-extablished order among the symbols of an arithmetic expression. On the contrary, the order is intrinsic in the expression itself. In other words, there is no reason for considering the expression y*x "less ordered" then the expression x*y, per se, while the expression x*y+y*x has to be transformed into 2*x*y and not into 2*y*x (i.e., the reordering is to be done according to the lexicon (x y), because this is the order in which the variables x and y occur).

After the reordering and merging is done, some further simplification of the arithmetic expression can still be possible (for instance if a monomial ends up with 0 or 1 as a coefficient). Thus, the arithmetic expression is processed again at the theory level, evaluating it by means of the following simplification set, named SSS (Small Simplification Set).

```
∀ u.  (1*u)  = u
∀ u.  (0*u)  = 0
∀ u. (-1*u)  =-u
∀ u.  (0+u)  = u
∀ u.  (u+0)  = u
```

The behaviour of the entire algorithm that simplifies arithmetic expressions can then be summarized as follows:

At the theory level: evaluate the arithmetic expression by means of the simpset SS.

At the meta-theory level: build the lexicon; simplify all the arguments of functions (if any) that are arithmetic expressions; reorder each monomial according to the lexicon and merge similar monomials.

At the theory level: evaluate the arithmetic expression by means of the simpset SSS.

This entire process is invoked at the meta-theory level, by the following statement.

```
AXIOM SIMPL: ∀ ae. simpl(ae)= eval-dn(arrange(eval-dn(ae,Ss,ARITH),
                                              lexarex(ae,Emptylex)),
                                      Sss,
                                      ARITH)
```

Note that **ae** has been declared to be an individual variable ranging over arithmetic expressions, **eval-dn(Term,Simp,LSname)** is the function that allows to activate from **META** the FOL evaluator in the L/S structure named **LSname** on the term whose name (in **META**) is **Term** and with the simpset whose name (in **META**) is **Simp**. In the case of the axiom above, the simpset **SS** has been named **Ss**, the simpset **SSS** has been named **Sss**. The function **lexarex** builds the lexicon for an arithmetic expression, starting from the empty one (**Emptylex**), while **arrange** is the function that, given an arithmetic expression (in the form of a sum of monomials) and a lexicon does the reordering and merging.

Note that, the evaluation of the function **simpl** causes many L/S structure switchings, besides the two explicitly invoked by **eval-dn**. They are performed in order to compute (in **ARITH**) information that is needed (in **META**) about the arithmetic expression being manipulated (as in the case of the evaluation of the meta-predicate **MONOMIAL** shown in Section 3.1). It is worth noting that this doesn't result in an unaffordable loss of efficiency of the entire process.

6. REFLECTING META-THEORETIC KNOWLEDGE INTO THE THEORY

FOL allows the reflection into a theory of knowledge represented in its meta-theory through the so called *reflection principle* (implemented by the **REFLECT** command). We point to [Wey79] for a throughful exposition of the workings of **REFLECT**. Here we limit ourselves to note that **REFLECT**, if given as arguments the name of a fact in **META** and a list of objects in the theory, instantiates and evaluates (in **META**) that fact on the given objects, and gives back a result as an object at the theory level.

In order to reflect into **ARITH** the action of the function **simpl**, we have given **META** the two following axioms.

```
AXIOM SIMPL1:
        ∀ t1 .THEOREM(mkapplw2(Equal,t1,simpl(t1)));;
AXIOM SIMPL2:
        ∀ t1 t2 .(simpl(mkapplt2(Sum,t1,mkapplt1(Minus,t2)))=ZERO ⊃
                    THEOREM(mkapplw2(Equal,t1,t2)));;
```

where **t1** and **t2** are terms, **mkapplw2** and **mkapplt2** are the constructors that build wffs and terms (respectively) out of a binary symbol and two terms. The constructor **mkapplt1** is analogous, for unary symbols. The individual constant **ZERO** is the name in **META** for the numeral 0. The predicate **THEOREM(wff)** tests whether **wff** is a theorem or not.

As examples of use of reflection, the following facts (taken from [BML65]) have been proved by a reflection (of the axiom **SIMPL2**), followed by a generalization. For example, for the first fact the reflection is performed by typing **REFLECT SIMPL2 (u+(v+w)) ((w+u)+v);**.

```
∀ u v w.(u+(v+w))=((w+u)+v)
∀ u v w.(u*(v*w))=((w*u)*v)
∀ u v x y.((u-v)+(x-y)=(u-y)+(x-v))
∀ u v x y.((u-v)-(x-y))=((u+y)-(v+x))
∀ u v x y.(u-v)*(x-y)=((u*x)-(u*y))-((v*x)-(v*y))
∀ u v.-(v-u)=(u-v)
∀ u v.(u-v)=-(v-u)
∀ x y v.((x+y)+-x)+v=v+y
```

By reflecting the meta-axiom SIMPL1, the terms (containing uninterpreted function symbols) appearing in the following at the left hand side of the two arrows have been simplified to the terms appearing at the right hand side.

```
                    x*y+f(y*x,x)    ↠    x*y+f(x*y,x)
           x*(y+z*f(y*x,g(z*x)))    ↠    x*y+x*z*f(x*y,g(x*z))

 x*y*f(y*x,z)*g(x)-y*x*g(x)*f(x*y,z)    ↠    0
                      x*(-y)-(-y)*x    ↠    0

            x*3*g(4-y)-g(-y+4)*3*x    ↠    0
                      x*0*f(4-y,y)    ↠    0
x*3*(4-y)+(z-u)*f((x+y)*w,w*u) -
  12*x-(u-z)*f(w*x+w*y+0,u*1*w)+3*x*y    ↠    0
```

7. CONCLUSIONS

The examples at the end of the paper show that a certain level of algebraic manipulations have been reached within the FOL context. It is important to realize that this has been done without any additional special-purpose FOL system code. Furthermore it is worth pointing out once again the advantages of explicitly using meta-theoretic knowledge. They can be summarized as follows:

(1) The use of meta-theoretic knowledge has made proofs shorter at the theory level;

(2) It has provided an alternative solution to the problem of commutativity;

(3) It has shown that the interaction between theory and meta-theory is a practical tool for expanding the power and efficiency of the theorem prover.

ACKNOWLEDGEMENTS

This research was supported by the Advanced Research Projects Agency of the Defense Departement (contract MDA903-76-0206), by the National Science Foundation (contract MCS79-05998), by a NATO grant, and by the Italian National Research Council (CNR).

REFERENCES

[Aie80] Aiello, L. **Evaluating Functions Defined in First Order Logic or In Defense of Semantic Attachment in FOL**, Stanford Artificial Intelligence Laboratory Memo, in preparation (1980).

[BM79] Boyer, R.S. and Moore, J.S. **Metafunctions: Proving them correct and using them efficiently as new proof procedures**, Computer Science Laboratory, SRI International (1979).

[BML65] Birkoff, G. and McLane, S. **Brief Survey of Modern Algebra**,The McMillan Company (1965).

[Fil78] Filman, R.E. **The Interaction of Obseravation and Inference**, Ph.D. Thesis, Stanford University, Stanford (1978).

[Fil79] Filman, R.E. **Observation and Inference applied in a Formal Representation System**, *Proc. of the 4-th Workskop on Automated Deduction*, Austin, Texas (1979)

[Pra65] Prawitz, D. **Natural Deduction – a Proof-Theoretical Study**, Almqvist & Wiksell, Stockholm (1965).

[Tal80] Talcott, C. **FOLISP: a System for Reasoning about LISP Programs**, Stanford Artificial Intelligence Laboratory Memo, in preparation (1980).

[Wey77] Weyhrauch, Richard W. **FOL: A Proof Checker for First-order Logic**, Stanford Artificial Intelligence Laboratory Memo AIM-235.1 (1977).

[Wey78] Weyhrauch, Richard W. **The Uses of Logic in Artificial Intelligence**, Lecture Notes prepared for the Summer School on the Foundations of Artificial Intelligence and Computer Science (FAICS '78), Pisa (1978).

[Wey79] Weyhrauch, Richard W. **Prolegomena to a Mechanized Theory of Formal Reasoning**, Stanford Artificial Intelligence Laboratory Memo AIM-315 (1979); to appear in *Artificial Intelligence Journal* (1980)

APPENDIX

We include here two referee's comments, because we think they represent legitimate and frequent questions about the entire FOL effort. Rather than address them by local changes in the paper we have decided to answer them directly.

REFEREE 1: [omissis] In particular, the statement about the need for "metamathematical" reasoning in ordinary mathematics are only true in one interpretation (in which a modest amount of set theory is not part of the underlying theory). [omissis]...the distinction between theory and meta-theory (the main topic of the paper) does not seem to play a crucial role. [omissis] If there is any significance to the theory – meta-theory distinction, the authors did not make it clear.

ANSWER: While it is true that if we had chosen to axiomatize algebra in set theory the explicit use of meta-theory could have been avoided, the point of this example is to examine how we can represent the ordinary elementary understanding of algebra. In this sense our axiomatization is a substantially more accurate account of actual practice. That is, in elementary algebra we are taught how to manipulate equations, and not taught how to represent algebra in set theory. This manipulation of equations (i.e. syntatic expressions) is straightforwardly meta-mathematical. It is the ability to do this *directly* that makes our formalization attractive. The claim that a certain reduction to a different formalization could be done, is analogous to the argument that COBOL might as well be used for writing all programs. One of the issues that the FOL project addresses is the question of which natural representation of reasoning facilitates automatic derivations.

We think that a reasoning system has to provide the user with the ability of *explicitly* representing and using meta-theoretic knowledge for at least three reasons.

Uniformity: The attempt to finesse the explicit use of meta-theory by axiomatizing algebra in set theory is not satisfactory in general because it is easy to formulate meta-theoretical questions we can ask about this theory. Since this is *always* the case we need to recognize it and deal with it directly.

Universality: The system should be universal in the sense that everything that we want to say should be expressible in the system itself. Only general purpose code should be part of the system – special purpose code should be added to the system only as part of one or more L/S structures. For this reason the algebraic simplifier was introduced by adding it to the axiomatization of META instead of writing a special purpose theorem prover. Thus, using meta-theory, we were able to describe the result of the *activity* of reasoning about numbers and this description resulted in FOL having the ability to carry out this reasoning. The way in which these descriptions are related to the corresponding programs is worked out in detail in [Aie80].

Epistemology: This is strongly connected to the above reason. FOL is a tool for knowledge representation. Representing knowledge *within* a system does not mean "hardwiring" it as a part of the system. In the case of the example presented in this paper, we stress once again that the aim was the *development in* FOL of elementary algebra, not the *addition to* FOL of the ability of doing some algebraic manipulations.

REFEREE 2: Attachment begs the question of whether the attached function code behaves properly (obeys the axioms stated for the symbol to which it is attached). How can this be verified within FOL, or how can you reduce what has to be verified to a minimum? In the illustration it seems too crude to introduce SIMPL1 and SIMPL2 as meta-*axioms*; one would hope to have them as meta-*theorems*, proved from much simpler meta-axioms.

ANSWER: These remarks about semantic attachment in FOL are common.

The answer to the first remark is simple. If the user sets up in FOL a specification that is inconsistent (or that doesn't "model" what he has in mind) he is the only one to be blamed. An inconsistency can be introduced in a L/S structure at the syntax level (by giving an inconsistent set of axioms) as well as at the semantic level (by making crazy attachments to function and predicate symbols). There is no reason for considering one of these kinds of potential inconsistency more perilous than the other.

The second question seems more legitimate. If the user of FOL wants to prove some properties of the specifications he is putting together, he should be able to do it within FOL itself. In this respect, the addition to FOL of the "compiling algorithm" described in [Aie80], which allows systems of (mutually) recursive function definitions in FOL to be automatically transformed into LISP code (and made into semantic attachments) represents a step in that direction. In fact, the FOL user can now introduce all the notions as FOL axioms (or derive them as theorems), and then transform them into semantic information by a suitable compilation. Similiar to [BM79] the correctness of this procedure can be verified once and for all. In addition, FOL can be used to derive properties of these functions before doing the translation.

The remark that SIMPL1 and SIMPL2 should be theorems rather than axioms is answered in the same way: we didn't care very much about the "correctness" of those facts (except that we believe in our intuition as axiomatizers ...or as programmers...). Certainly a proof that SIMPL1 and SIMPL2 are consequences of more elementary axioms about proofs can be shown in META, but we were more interested in setting up an axiomatization and using it than in making it minimal.

GENERATING CONTOURS OF INTEGRATION: AN APPLICATION OF PROLOG IN SYMBOLIC COMPUTING

Gábor Belovári
Alkotás u. 25.II.VII.53, 1123 Budapest, Hungary

J.A. Campbell
Department of Computer Science, University of Exeter, Exeter EX4 4QL, England

ABSTRACT

One standard technique of evaluation of real definite integrals is transformation to a complex variable, and integration over a closed contour in the complex plane by means of Cauchy's Theorem. Textbook presentation of the technique tends to rely on examples, and to state no general principles governing generation of appropriate contours or applicability of the technique. This note states some general principles and a computation strategy, and outlines their implementation in a PROLOG program for automatic deduction of appropriate contours. The program complements the functions of existing systems of symbolic computing programs for integration and for manipulations in complex analysis.

Complex Analysis and Symbolic Computing.

The largest symbolic computing systems, like MACSYMA (1), contain sub-systems of programs, usually in LISP, for most standard symbolic mathematical manipulations. Until recently, only the very large systems included non-trivial packages for symbolic integration, but now improvements in general algorithms for indefinite integration are making it possible for this facility to be added to small systems.

Because of the concentration on the problem of indefinite integration in the past, and the fact that evaluation of many definite integrals merely involves substitutions in results obtained by consideration of that problem, awkward aspects of definite real-variable integration have received comparatively little attention. Among them is the observation that some classes of definite integral require evaluation by transformation to integrals around closed contours in the complex plane.

Apart from operations involving contours, it is possible to implement standard manipulations of complex analysis in programs having no flavour of artificial intelligence. Campbell, Kent and Moore (2) summarise the results of one such experiment in LISP. Features explicitly excluded from that experiment, such as conformal mappings, are easy to build into the larger systems, which have the necessary facilities to deal with general and space-consuming computations on rational functions.

The standard manipulations are repeated in examples in many textbooks, though the presentations do not often give an impression of their origins. Systematic presentations, which therefore hint at methods of generalation, are available in books by Curtiss (4) and Ahlfors (3).

The object of examining real definite integration with the help of logic programming is to explore the question of whether or not useful generalisations exist.

If they do, they should be capable of expression as broad principles which permit automatic deduction of a suitable contour and suitable complex-variable transformations for an appropriate real definite integral. Moreover, they should help a deductive procedure to identify integrals for which transformations to contour integrals are not appropriate.

We report here an approach to this problem in PROLOG (5) which starts from our cnoice of general principles that are consistent with all or some of the textbook examples of transformations. We conclude that logic programming is a helpful method of attacking the problem of definite integration and may be used as an adjunct to conventional symbolic computing systems in this area.

Programming Preliminaries.

We follow the approach of (2), in which the basic quantities of use for the computation of contours (integrands, points in the complex plane which need naming for repeated reference, constraints on constants etc.) are described in some detail. The individual declarations of data used by the program for integration in (2) translate easily into PROLOG, e.g. the association of the names p_1, p_2 and p_3 with the points a+ib, 0 and 3i respectively ($i^2=-1$) becomes

```
points ([p1, p2, p3], [(a.b),(0.0),(0.3)] ).
```

Administrative functions are provided to extract relevant parts of such data, e.g. the imaginary coordinate C of a distinguished point P is found with the help of

```
coord(P,C):- points(K,L), select(P,K,L,Q), cdr(Q,C).
select(_,[],_,_):-fail.
select(P,[P,.._],[Q,.._],Q).
select(P,[_,..L],[_,..M],R):- select(P,L,M,R).
cdr((..X),X).
```

Following the initial declarations, we start the computation of the contour with a functor (contour) of arity 6 , having arguments LOW, HIGH (lower and upper limits of integration), INT (the integrand), POLELIST (a list of the names of distinguished points which are poles of INT), CONSTR (a list of constraints on symbolic constants, with individual elements expressed as (<atom>. <relational operator, e.g. gt,le>. <atom>)), and eventual result PATH. In (2), POLELIST and other expressions obtainable from INT by operations of mathematical analysis are not requested separately from the user, but they are left in place here because the programming required to imitate those operations (which are fully deterministic) in PROLOG is substantial.

The goal contour(LOW,HIGH,INT,POLELIST,CONSTR,PATH) has four subgoals: completion of an administrative step, selection of a list DPOLIST of poles which should ideally be located inside the contour to be computed, success of contest (HIGH, LOW,INT,DPOLIST,DPOINTS,ALREADY), which carries out the bulk of the computation (where the result is unified with ALREADY, and DPOINTS is a list of all the distinguished points which may serve as end-points of (or targets for) elements of the contour), and a "cleaning up" goal cleanup(ALREADY,PATH).

The preliminary step involves transformation of INT to a form in which HIGH stands for (∞.0) and LOW for (-∞.0) or (0.0), arrangement of the symbolic items in CONSTR into a list with monotonic ordering (e.g. if CONSTR contains (a.LT.10), (a.EQ.c) and c.GT-10), the resulting list has a form [...., -10,....,c,a,..10,...]), assignment of random integer values to the symbols in that list, respecting the constraints, association of resulting numerical coordinates with all distinguished points, and sorting of POLELIST into a new order, with the pole which is closest to the line from LOW to HIGH, in this coordinate system, in first place. Subsequently, all geometrical operations which assist in construction of a contour are performed on the numerical coordinates. The justification for this approach is given in (2).

Starting with individual poles, all combinations of poles are tried in DPOLIST, a list which helps to guide the subsequent computation, until success is achieved or until no more possibilities exist.

Integrals and Principles.

In (2), contours were built out of elements of form l(p,q), a line from p to q, c(p,r,s), all or part of a circle with centre p, radius r and sense s, ce(p,q,s), a circular arc from p to q, and a "keyhole" element k(p,q). Here, the goal contest (HIGH,LOW,INT,DPOLIST,DPOINTS,ALREADY) may be described purely in terms of l and ce elements, with no loss of generality. Therefore contest has only to try an l-element or a ce-element for the next element of a partly-constructed contour which has reached the point BEGIN, and to stop if the end-point of the most recently generated element is equal to LOW, the overall starting-point. Suitable clauses are:

```
contest(_,LOW,_,_,_,ALREADY):- final(ALREADY,Y), Y==LOW.
contest(BEGIN,LOW,INT,_,_,[(l.BEGIN.LOW),..X]):-
          highpoint(U),BEGIN ==U,makeline(BEGIN,LOW,V),
          !,usable(V,INT,X).
contest(BEGIN,LOW,INT,DPOLIST,DPOINTS,[Y,..X]):-
          final(X,BEGIN),penult(X,U),!,
          contest(U,LOW,INT,DPOLIST,DPOINTS,X),!,
          tryline(BEGIN,INT,LOW,DPOLIST,DPOINTS,X,Y);
          tryarc(BEGIN,INT,LOW,DPOLIST,DPOINTS,X,Y)).
final([(_._.X),.._],X).
penult([(_.X._),.._],X).
```

The second clause allows for the special case that (provided BEGIN is not the same as the original upper limit of integration) the contour can be completed immediately with a "usable" line from BEGIN to LOW. Usability is a concept which arises naturally from textbook considerations: the goal usable(V,INT,X) above can be specified in terms of the details below..

The chief implicit principle at the textbook level is that a real definite integral should allow the possibility of transformation to a "usable closed contour.

Because only one of the pieces of this contour will be the portion A of the real axis marked off by the limits of the original integral, the contributions of the remaining elements of the contour in the complex plane should be usable in the sense that they add no new details to the integration: they should vanish individually, or cancel in pairs, or be linearly related to the contribution from A. The administration of these tests, which involve only simple operations of arithmetic and geometry, is assigned to the goal usable(_,_,_). or, in suitable contexts, to some of its subgoals such as period(_,_,_), which is used in the definition of tryangle in the Appendix. In all cases, the sum of the contributions from all pieces of the contour, taken anticlockwise, equals $2\pi i$ times the sum of the residues of the integrand at those of its poles which are inside the contour. This is the content of Cauchy's Theorem: details are given in textbooks (3, 4). An example of the ideas above is shown in Fig. 1.

Fig. 1

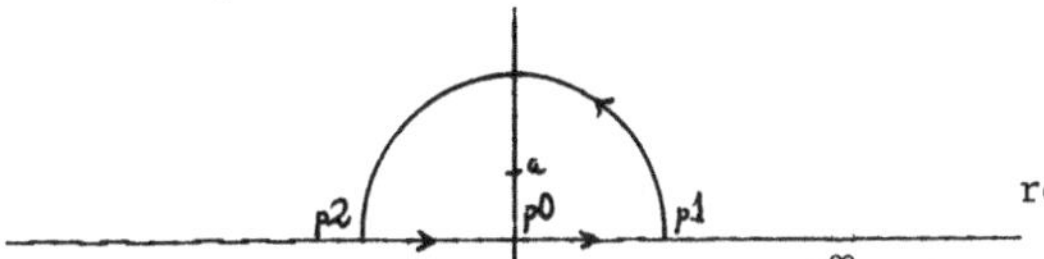

where we suppose that it is necessary to evaluate $I=\int_0^\infty f(x)dx$, and (for the sake of argument) that $f(x)=f(-x)$. If the contour of integration in Fig. 1 is used, the I is specified by the integral along the positive part of the real axis, from 0 to R, the integral along the negative real axis is also I, by symmetry of f, and 2I plus the integral on the semicircular arc in the upper half of the complex z-plane equals $2\pi i$ times the residue of f(z) at z=ia, if it is supposed that the only relevant singularity of f, as a complex function, is at that point. I matches the original definition only when $R\to\infty$, and in that limit the integrand must tend to zero fast enough with R on the semicircular arc to avoid divergence of the integral. It turns out that this happens when f(z) behaves like R^{-2-p}, for non-negative p, when $|z|=R$, and that the contribution of the arc to the integral is then zero. By this method, it is possible to check that $I=\pi/2a$ when $f(x)=1/(x^2+a^2)$.

The same result would follow if the contours based on the real axis were completed in the upper half-plane by, say, a hexagonal figure enclosing z=ia. However, nobody chooses this alternative, because it involves more calculation for the same reward. It is a general principle consistent with textbook examples of contours that one tries to <u>select the "usable" contour with the smallest number of pieces</u>, although we have never seen this recommendation written explicitly in a textbook. We implement the principle inside the PROLOG program by making immediate closure of a contour to be the first priority of contest: if tryline or tryarc has just produced an element ending at P≠LOW, the possibility of a direct connection between P and LOW is investigated first.

When the program can find a closed contour for a given integral, it is specified in the notation of (2). For example, the contour in Fig. 1 is the three-piece contour [(1.p0.p1),(ce.p1.p2.1),(1.p2.p0)], where p1 and p2 are the points at positive and negative infinity on the real axis, and p0 is the origin. For the integral I, there is no usable contour enclosing z=ia with less than these three pieces.

When integrands, regarded as functions of a complex variable, have several poles in the complex plane away from the real axis, alternative contours enclosing different selections of poles may be proposed. Cauchy's Theorem applies to any of the alternatives, so that any one may be justified. In order to avoid the syndrome of "more calculation for the same reward", we suggest the third principle that the preferred contour is one which encloses the smallest non-zero number of poles. This, too, is consistent with many textbook examples, notably the case where f(x) is a sine or cosine of kx^n (n integer constant, k constant), and the recommendation is to integrate $\exp(ikz^n)$ around the contour [(1.p0.p1),(ce.p1.p2.1),(1.p2.p0)] , with p1 as in Fig. 1 and p2=p1 exp ($i\pi/2n$). We do not insist that this third principle be followed rigorously in all cases: we treat it permissively, by manipulating the order of choices for DPOLIST so that all individual poles are considered before any pair of poles, all pairs before any triples,and so on. Of course, this "advice" may be bypassed by the action of a principle of more importance: if it is possible to close a contour immediately from some point P to LOW, any side-effects on the number of poles enclosed are not investigated.

One joint property of the second and third principles above which makes them into interesting objects for investigation in logic programming is that they may be contradictory in certain cases, e.g. where poles are so disposed that a contour enclosing only one of them has more pieces than a contour enclosing a greater number. Since the amount of computing in deduction or symbolic manipulation is governed more by the number of pieces of a contour than the number of poles, the second principle takes precedence over the third. Similarly, the first principle takes precedence over both: if no plausible closed contour made out of 1- and ce-elements can be found by contest, for a given definite integral, it is likely that contour integration is an inappropriate method for its evaluation, and the original goal of the program should fail to be achieved.

In the previous Section, we mentioned a cleaning-up phase for any successful result obtained by contest. At present, this merely reverses the list of contour-elements (thereby putting the range of the original definite real integral in first place) and makes two trivial types of addition. Firstly, a user is permitted to say whether he wants specified poles p to be inside or outside the eventual contour. If the contour found by contest locates any p on the wrong side, then the appropriate "correction" to apply is to locate a suitable point q on the contour, and insert in the contour at q a keyhole element (k.q.p), which breaks the contour some small distance δ away from p and carries it by two parallel lines a distance 2δ apart plus a semicircle of radius δ around q, back to a distance δ from p on the opposite side. Secondly, contours for complex integrals often bypass certain points (usually poles) by arcs of a circle of infinitesimal radius, with centre p, which give finite or zero contributions to the overall integral when that radius goes to zero. We find it simpler, in construction of a contour, to include any such p (which we call "passable", and test for this property via the subgoal passable(_,_) which is mentioned in the Appendix) on the path during the operation of contest, and then to mark it during the cleaning-up phase by the insertion of a new type of element, an "avoiding" element (a.p.s) with sense s (s=1 for anticlockwise sense, otherwise -1), at p. A user can easily substitute the correct circular arc, by inspection in context, on reading the complete specification of the contour.

The main duty of contest is to build up a contour, element by element, by trying to find a new l-element first, and then a ce-element if there are no suitable l-elements. The details of both types of trials, apart from the administration of the principles set out above, are geometrical, and unlikely to be of general interest. The l-element goal, tryline, is described in an Appendix, with commentary, but without any detailed listing of the internal properties of the geometrical subgoals. The structure of the ce-element goal, tryarc, is of similar complexity.

Examples of Integrals.

Firstly, we identify an example normally presented in textbooks as a special case. If the limits of integration are 0 and 2π, and the integrand is found to be a rational function containing trigonometric arguments, then the transformation from real x to complex z is $\sin x=(z^2-1)/2iz$, $\cos x=(z^2+1)/2z$, and the contour is the circle with centre at z=0 and radius 1. In terms of our second principle, this contour is the simplest possible; no other closed contour has only one piece. Therefore integrals in this class are accommodated in our general framework.

All other integrals presented as examples (3,4) have infinite or semi-infinite ranges of integration. If the range is from some a to infinity, we change it to run from zero to infinity, and move all singularities of the integrand, regarded as a function of complex variable z, from their former positions z=p to z=p-a. If the integrand is an even function of z, as in the example associated with Fig.1, the range is further extended to cover the entire(positive and negative) real axis. A similar treatment is possible if the initial range is from minus infinity to some a. Therefore, without losing generality, we have to deal (after the case in the paragraph above) only with integrals where the range is $[-\infty,0]$, $[0,\infty]$ or $[-\infty,\infty]$.

The two examples given in the previous Section exploit the description of usable contours (our first principle) as contours on whose pieces the contributions to a complex integral are either zero or linearly related to the desired real integral. Another example of a relevant problem is $\int_{-\infty}^{\infty} e^{ax}(1+e^x)^{-1}dx$, where the contour is given in Fig. 2.

Fig. 2.

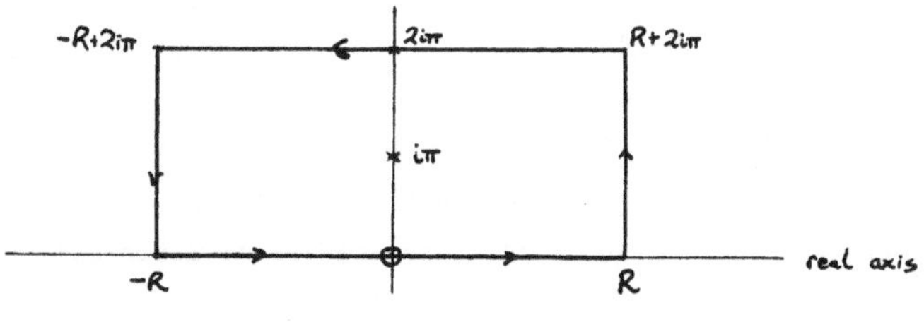

The integrand has poles at $\pm$ ($i\pi$, $3i\pi$, $5i\pi$...), but the contour shown (which our program derives) respects the third principle by enclosing just one pole. As R-> ∞, the contributions from the two vertical pieces of the contour cancel, and the contribution from the horizontal piece through the point $2i\pi$ is $-\exp(2i\pi a)$ times the original desired integral along the real axis.

We have tested the program on, for example, the integral problems listed in Section 4.4. of (2), and (to illustrate use of the "avoiding" element) on $\int_0^\infty x^{-1} \sin x \, dx$, where the relevant complex expression is $z^{-1}\exp(iz)$, and the contour is as in Fig. 1, avoiding the origin: [(1.p0.p1),(ce.p1.p2.1),(1.p2.p0),(a.p0,-1)] .

Discussion.

We began this study by considering the computation of complex substitutions for x on each piece of a contour to be as important as the computation of the contour itself. However, for the examples which we have considered, where substitutions address (in passing) the question of "usability", it is possible to answer the same question by other means. For example, on the semicircular arc of Fig. 1, the substitution is x -> R exp ($i\theta$), with θ as the new variable, but it is sufficient to verify that the integrand tends to zero faster than R^{-2} for R>∞ to learn all that needs to be known about the contribution of the arc to the integral. Assuming that a student or a program is to use the computed contour in an application of Cauchy's Theorem, it should be easy to recover the substitution, or a simple procedure to find the substitution for each piece of the contour, by access to a static table, using the integrand and the piece as keys. The characteristics of logic programming are needed only in the determination of the contour.

We can report two general successes in our work on the contour-generation project. Firstly, we have demonstrated an application of logic programming, in PROLOG, which is of significance to symbolic computing: PROLOG provides a vehicle for a computation which fills a gap in previous work. Therefore interfaces or overlays between PROLOG and symbolic computing systems are worth further investigation. Secondly, by logic programming, we have tested the reliability of some general principles and strategy for the determination of contours for the evaluation of real definite integrals, and thus have arrived at a somewhat more systematic view of the topic than that given by the ad hoc presentations in most textbooks.

We have found the main drawback of programming this exercise in PROLOG to be the need to write large numbers of clauses to express information about simple geometry and standard manipulations of symbolic computing, if the program is to be general, independent, and capable of relieving the user of non-trivial preliminary preparation of his data. The Appendix gives some impression of this difficulty! One way out is certainly to attempt to build overlays between PROLOG and symbolic systems. We admit to being undecided at present on whether the necessary work will be adequately repaid by the number of new applications in prospect. One area of possible application currently under study is field theory of the twistor type (6), in which some key calculations involve placement and transformation of contours for complex integrals of expressions with quite complicated pole-structure. We shall welcome suggestions of other areas in which programs such as ours may appear to be helpful.

REFERENCES

1. For a general survey of properties and applications of MACSYMA, see Proc.1977 MACSYMA Users' Conf., NASA publication CP-2012, NASA, Washington, D.C.(1977)
2. J.A.Campbell, J.G. Kent and R.J. Moore, B.I.T. 16, 241 (1976).
3. L.V. Ahlfors, "Complex Analysis", McGraw-Hill, N.Y.(1966)
4. J.H.Curtiss, "Introduction to Functions of a Complex Variable" M. Dekker Inc., N.Y. (1978)
5. P. Roussel, "PROLOG: Manuel de reference et d'utilisation", internal report, Groupe d'Intelligence Artificielle, Université d'Aix-Marseille (1975); L.M. Pereira, F.C.N. Pereira and D.H.D. Warren, User's Guide to DECsystem-10 PROLOG", Department of Artificial Intelligence, University of Edinburgh (1978.)
6. L.P. Hughston and R.S. Ward (eds.), "Advances in Twistor Theory". Pitman, London (1979)

APPENDIX:

Description of the program segment which searches for line-elements of a contour.

If tryline finds a suitable line from the point BEGIN, the end-point of the line is X. Information carried to assist the computation: integrand INT, starting-point LOW for the contour, list DPOLIST of the poles for which enclosure within the contour is being attempted, a list of the distinguished points in the complex plane, partial contour ALREADY computed.

```
tryline(BEGIN,INT,LOW,DPOLIST,_,ALREADY,X):- periodx(INT,LINE),
        on(BEGIN,LINE),!,((distg(BEGIN,LINE,X),nearer(BEGIN,X,DPOLIST));
             (ycoord(LOW,U),ycoord(BEGIN,V),!,xvalue(LINE,W),
                  ((U\== V,Q=:=U);(pocket(ALREADY,P),!,ycoord(P,V),minus(V,Q))),
                  declare(W,Q,X))).
```

<and a second clause obtained by interchanging lower-case letters x and y above, and interchanging W and Q as arguments of declare>

Meaning: If BEGIN lies on a line (a "period line") on which INT takes the same value as on the coordinate axis parallel to it, then X is any distinguished point on that line, provided that the part of the line from BEGIN to X contains a point which is no farther than BEGIN or X from the poles in DPOLIST. Alternatively, if the equation of the line is y=constant, then take the x-coordinate of X to be the x-coordinate of LOW if this is not the same as for BEGIN; otherwise take it to be the minus the x-coordinate of the starting-point of the previous line, if any, whose end-point is BEGIN. Similar interpretation on interchange of x and y.

```
tryline(BEGIN,INT,LOW,DPOLIST,[ ],ALREADY,X):-
                         tryfree(BEGIN,INT,LOW,DPOLIST,ALREADY,X).
```

```
tryline(BEGIN,INT,LOW,DPOLIST,[U,..V],_,X):-
                     nearer(BEGIN,U,DPOLIST),!,
                     trydpinf(BEGIN,INT,U,X).
tryline(BEGIN,INT,LOW,DPOLIST,[U,..V],_,X):-
                     trydppole(BEGIN,INT,DPOLIST,U,X).
tryline(BEGIN,INT,LOW,DPOLIST,[U,..V],ALREADY,X):-
                     tryline(BEGIN,INT,LOW,DPOLIST,V,ALREADY,X).
```

Meaning: The distinguished points are tested individually until one participates in the generation of a candidate X. The distinguished points at infinity are given separate treatment by trydpinf. If no distinguished point generates an X, tryfree attempts some geometrical constructions to find X. If this fails, then tryline fails.

```
trydpinf(BEGIN,INT,U,X):-atinf(U),!,
                    (xconstline(BEGIN,U,INT,LINE);yconstline(BEGIN,U,INT,LINE)),
                    makeline(BEGIN,U,Z),!,xn(Z,LINE,X).
xconstline(BEGIN,U,INT,LINE):-xcoord(BEGIN,S),xcoord(U,T),!,
                S==T,periody(INT,LINE).
<definition of yconstline involves interchange of x and y in xconstline>
```

Meaning: If U is a point at infinity, and LINE is a period line, then X is the intersection of LINE and the line from BEGIN to U. Line-elements of contours do not include points at infinity (apart from points on the real axis): that is a job for arc-elements.

```
trydppole(BEGIN,INT,DPOLIST,U,X):-pole(U,INT),!,
                    ((passable(U,INT),!,nearer(BEGIN,U,DPOLIST),X=:=U);
                    (makeline(BEGIN,U,Z),!,(xconstline(BEGIN,U,INT,LINE);
                                             yconstline(BEGIN,U,INT,LINE)),
                                          !xn(Z,LINE,X),!,
                                           between(X,BEGIN,U))).
```

Meaning: If U is a pole of INT, U can be regarded as the end-point of the trial line if the pole is passable (i.e. if it gives rise to no infinities when it is bypassed by an infinitesimal circular arc; the arc is added to the contour at the final stage of processing, later). If U is not passable, then the end-point is the intersection of the line from BEGIN to U and a period line of INT, if any such point exists.

```
tryfree(_,_,_,[],_,_):-fail.
tryfree(BEGIN,_,_,_,ALREADY,X):-tryparallel(BEGIN,ALREADY,X).
tryfree(BEGIN,INT,_,[V,.._],ALREADY,X):-tryangle(BEGIN,INT,V,ALREADY,X).
tryfree(BEGIN,INT,LOW,[_,..L],ALREADY,X):-
                    tryfree(BEGIN,INT,LOW,L,ALREADY,X).
```

Meaning: For each line-element Λ (except the real axis) already placed in the contour, tryparallel constructs a line λ of equal length, distinct from Λ, antiparallel to it, and starting from BEGIN. The end-point of λ is chosen if the integral of INT on Λ is minus the integral of INT on λ. Otherwise, if the previous element of the contour is a line-element, tryangle attempts a geometrical construction which turns the contour back fairly sharply around an enclosable pole of INT.

```
tryangle(BEGIN,INT,V,ALREADY,X):-pocket(ALREADY,A),
                    makeline(A,BEGIN,B),makeline(BEGIN,V,C),
                    angle(B,C,D,),!,D\==0,E is 2*D,
                    inventline(B,BEGIN,E,F,),!,period(INT,LINE,F),
                    xn(F,LINE,X),!,simple(INT,F,LINE).
```

Interpretation: angle determines the angle between two lines; inventline generates a line F from BEGIN making an angle E with the line B; period finds a LINE (not necessarily parallel to a coordinate axis) on which the value of INT is proportional to its value on the positive real axis, X is the intersection of LINE and F, and simple checks (this is purely a heuristic) that the angle between F and LINE has one of a few simple values e.g. 45° and 90°.

Using
META-LEVEL INFERENCE
for Selective Application of Multiple Rewrite Rules
in Algebraic Manipulation

by

Alan Bundy and Bob Welham
Department of Artificial Intelligence
University of Edinburgh

Abstract

In this paper we describe a technique for controlling inference, meta-level inference, and a program for algebraic manipulation, PRESS, which embodies this technique. In PRESS, algebraic expressions are manipulated by a series of methods. The appropriate method is chosen by meta-level inference and itself uses meta-level reasoning to select and apply rewrite rules to the current expression.

The use of meta-level inference is shown to drastically cut down on search, lead to clear and modular programs, aid the proving of properties of the program and enable the automatic learning of both new algebraic facts and new control information.

Acknowledgements

Our thanks are due to Pat Hayes, Ira Goldstein and Soei Tien Tan for some of the original inspiration; Gordon Plotkin, Joel Moses, Alan Robinson, Woody Bledsoe and John Seeley Brown for encouragement and to many other people whose ideas we have unwittingly absorbed over the years.

This work was supported by SRC grants B/SR/22993 and B/RG/94493 and an SRC studentship to Bob Welham.

Keywords

rewrite rules, theorem proving, mathematical reasoning, algebraic manipulation and meta-level reasoning.

1. Introduction

1.1. Meta-Level Inference

This paper describes a technique for controlling inference, meta-level inference, and a program for algebraic manipulation, PRESS, which embodies the technique. In this program inference is conducted at two levels simultaneously: the meta-level and the object-level. The object-level encodes knowledge about the facts of the domain (in this case rules of algebra), while the meta-level encodes control or strategic knowledge (in this case, methods of algebraic manipulation). What are the advantages of this technique?

- The separation of factual and control information enhances the clarity of the program and makes it more modular.

- All the power and flexibility of inference is available for controlling search.

- The meta-level search space is usually much smaller than the object-level space it is controlling and this helps overcome the Combinatorial Explosion.

- The modularity of the program enables the learning of both new factual (object-level) and strategic (meta-level) knowledge.

1.2. Multiple Sets of Rewrite Rules

The object-level knowledge in PRESS consists of rewrite rules organised into several sets. Each set performs a particular job of algebraic manipulation and is associated with a syntactic characterization of the kind of rule it contains. The meta-level knowledge reasons about the job to be performed and the rules available to achieve it, and on this basis selectively applies the rules to the current algebraic expression.

The use of rewrite rules for algebraic manipulation is not new. Single sets of, exhaustively applied, 'simplification' or 'reduction' rules have been used, for instance, in [Martin and Fateman 71, Hearn 67, Bledsoe and Bruell 73]. But this technique is subject to a number of problems [Moses 71]. Our use of:

- Multiple sets of rules

- Selective application of these rules

- and Control by Meta-Level Inference

offers solutions to many of these problems.

What are the problems with the use of single sets of exhaustively applied rewrite rules?

1. Sets of rewrite rules often have an ad hoc flavour because there is no clear basis for including or excluding a particular rule.

2. The restriction of using equations and equivalences one way round means that the system is often incomplete, i.e. some problems are forever out of its reach. [Knuth & Bendix 70] and [Huet 77] give conditions under which a set of rules may be complete and suggest a mechanism for making an incomplete set into a complete one. However, this mechanism is not guaranteed to terminate and can only produce a complete set when the word problem in the formal theory defined by the rules is decidable - a strong limitation!

3. Rules included in the 'simplification' set as essential for some kinds of simplification may be an unwanted embarrassment for others.

4. Automatic learning of new rules is prevented unless some criterion can be found for deciding which way round the rule is to be used.

5. Proof of termination usually involves finding a measure on expressions which decreases every time a rule is applied. This becomes increasingly difficult to specify as the set of rules becomes larger.

How do our techniques solve these problems?

1. The syntactic characterizations provides a non-ad hoc basis for the inclusion of a rule in a set.

2. With multiple sets of rules, a particular axiom may be used in different directions in different sets. With selective application the axiom may even be used in different directions in the same set without any danger of looping.

3. By providing a different set of rules for different kinds of 'simplification'[1] there need be no difficult questions of inclusion.

4. The syntactic characterization of the rules in each set enables us to design a program which can decide in which direction to use a rule. We now have the additional problem of deciding which set to put a rule in, but the syntactic characterization enables this to be decided automatically as well.

5. The syntactic characterizations can also be used to prove termination of the sets, by suggesting a suitable decreasing measure. The fact that application of the rules is selective can also sometimes be used to show that the method terminates, where the exhaustive application of the rules would not.[2]

1.3. The PRESS System

PRESS (PRolog Equation Solving System) is a computer program for solving equations and doing other kinds of algebraic manipulation on transcendental expressions, i.e. expressions involving polynomial, trigonometric, exponential and logarithmic functions. It is based on ideas originally expounded in [Bundy 79] and [Bundy 75].

The program was largely written during 1975 by Bob Welham in the language Prolog (see [Pereira et al 78]). It consists of approximately 250 clauses and occupies 13K, 36 bit, Dec10 words. The Prolog system itself occupies a further 20k words.

The main use of PRESS has been as the algebraic package of the MECHO program (see [Bundy et al 79]). It has been extended from its original purely equation solving role to handle other problems which have arisen in the MECHO project, namely inequality handling and the use of semantic information. Suprisingly, some of the techniques originally developed for equation solving have also found application in inequality handling.

Some equations solved by PRESS can be found in Appendix I. These equations are beyond the scope of existing algebraic manipulation programs, such as MACSYMA and REDUCE. This increase in power is due to the use of the powerful algebraic manipulation methods which are embodied in the multiple rewrite rule sets and the meta-level control information.

---

[1] Indeed by using a more sensitive characterization of the job to be done than 'simplification'

[2] Points 4 and 5 are discussed more fully in the longer version of this paper [Bundy and Welham ng] to appear in the AI Journal

2. Some Algebraic Manipulation Methods

In this section we will explain some of the algebraic manipulation methods used by PRESS. Space allows only an explanation of three of the methods used to solve single equations in one unknown, namely Isolation, Collection and Attraction. [Bundy and Welham 79] contains descriptions of the many other methods available to PRESS. Isolation is a simple, general method for solving equations which only contain one occurrence of the unknown. Collection is designed to reduce the number of occurrences in an expression so that Isolation can apply. Attraction is designed to move occurrences of unknowns closer together so that Collection can apply. We call Isolation a **primary** method because it solves equations directly and Collection and Attraction **secondary** methods because they prepare expressions for other methods. Together they constitute a powerful, general method, called **The Basic Method**, which accounts for much of the success of the system.

We were led to these methods by a study of the behaviour of human mathematicians, during an attempt to understand how they guided their reasoning processes. Equation solving seemed an ideal domain to study this problem because of:

- the large search space of the problem when guidance information is not available,
- and the ease and directness with which human mathematicians can find a solution.

Clearly humans have access to some valuable guidance information and discovering it in this simple case might well shed light on more sophisticated areas of mathematical reasoning. A more detailed account of the discovery of the Basic Method may be found in [Bundy 75].

2.1. Isolation

Isolation is only attempted on equations containing a single occurrence of the unknown, x, e.g. on

```
log(e, x^2 - 1) = 3 [3]
```

but not on

```
e^sin(x) + e^cos(x) = a
```

On such equations it is guaranteed to succeed.

The method consists of 'stripping off' the functions surrounding the single occurrences of x by applying the inverse function to both sides of the equation. This process is repeated until x is isolated on one side of the equation, e.g.

---

[3] The first argument of log is the base

$x^2 - 1 = e^3$

$x^2 = e^3 + 1$

$x = \mathrm{sqrt}(e^3 + 1) \vee x = -\mathrm{sqrt}(e^3 + 1)$

This stripping off is done by applying a system of rewrite rules to the equation, for instance, the rule

u-v = w -> u = w+v [4]

Each of these rules is characterized by the description:

P & f(u1,...ui,...un) = w -> RHS (1)

where RHS is usually of the form ui = fi(u1,...,w,...un), but can also be a disjunction of such formulae, fi is the ith inverse function of f [5] and P is an optional condition.

Isolation is guaranteed to succeed in solving any equation containing only a single occurrence of the unknown, because for each function, f, that PRESS knows about, the ith inverse function is also known about and there are axioms of the form 1 in the Isolation rewrite rule set, that cover every case. This argument can be formalized into a termination proof for Isolation by showing that the depth of the single occurrence of the unknown strictly decreases each time a rule is applied (see [Bundy and Welham ng]).

2.2. Collection

The purpose of Collection is not to solve equations but to reduce the number of occurrences of unknowns in them to one, so that Isolation will apply. For instance, given

log(e,(x+1)*(x-1)) = 3

Collection will apply the axiom

(u+v)*(u-v) = u^2 - v^2 (2)

to obtain

---

[4] We use the convention that the unknown to be solved for in an equation is denoted by x and the variables in a rule are denoted by u, v, w, etc.

[5] An n-ary function has n inverse functions.

log(e,x^2-1^2) = 3

so that Isolation can be applied.

Collection uses a system of rewrite rules like 2 above. All these axioms are characterized by:

P -> LHS=RHS or P -> LHS<->RHS

where some variable, u say, occurs in both LHS and RHS, but fewer times in RHS than in LHS. A consequence of this characterization is that each Collection rule contains more than one occurrence of u and thus the Collection rewrite rule set does not satisfy one of the preconditions of the Huet theorems [Huet 77] about the Knuth-Bendix mechanism.

Most theorem proving programs that use sets of rewrite rules apply them exhaustively, i.e. they attempt to apply each rule to every subexpression in the expression currently being manipulated. Collection uses its rewrite rule set more selectively. The rule is only applied if the variable u matches a term that actually contains the unknown x. This ensures that if Collection is applied it will succeed in reducing the number of occurrences of x.

In addition, the rule is only applied to a least dominating term in x, that is, a term with at least two immediate subterms both of which contain x.

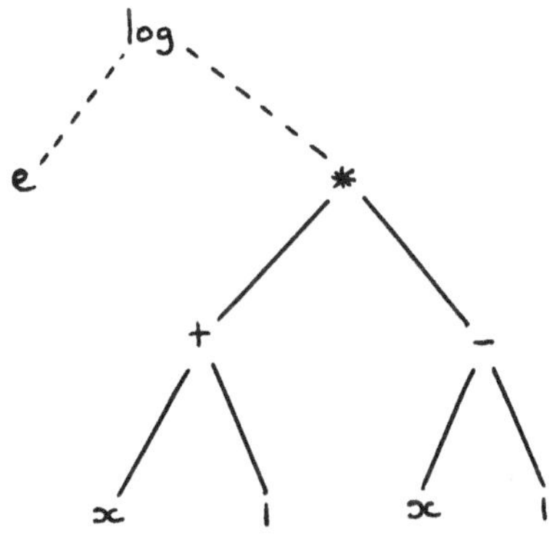

Figure 2-1: A Least Dominating Subterm for x.

(x+1)*(x-1) is a least dominating term, whereas log(e,(x+1)*(x-1)) is not, since only one immediate subterm of it contains x.

It is clear that if a Collection axiom applies to a term which is not least dominating then Isolation can be applied to that axiom to produce a Collection axiom that will apply to a least dominating subterm of the original term. All the

Collection (and Attraction see below) axioms are pre-processed so that they apply only to least dominating terms.

2.3. Attraction

Like Collection, Attraction does not solve equations, but is a secondary method, in this case for Collection. It brings occurrences of unknowns closer together so that they can be collected. For instance, given

log(e,x+1) + log(e,x-1) = 3

Attraction will apply the axiom

log(w,u) + log(w,v) = log(w,u*v) (3)

to obtain

log(e,(x+1)*(x-1)) = 3

so that Collection can be applied.

Attraction also uses a system of rewrite rules of the form

P -> LHS=RHS or P -> LHS<->RHS

but this time both LHS and RHS must contain a marked subset of variables, u, that are 'closer together' in RHS than in LHS, and each variable v in u must occur as often in LHS as it does in RHS. Equation 3 above is an example in which u and v are closer in RHS than in LHS.

The closeness of a set of subterms in an expression is measured by the size[6] of their **least covering tree**. This is the smallest subtree of the expression which includes each member of the set (see figure 2-2). In the case of a doubleton set (as in the example above), this is just the minimum path[7] between the occurrences.

As with Collection, Attraction applies its rewrite rules selectively. For a rule to be applied the following conditions must hold.

- All the marked variables, u, must match terms which contain the unknown, x.

- The application must result in a decrease in the size of the least covering tree of all occurrences of x.

---

[6] Where the size of a tree is the number of arcs it contains.

[7] Definitions of Attraction in previous papers have defined distance between two occurrences only. The least covering tree definition generalizes this.

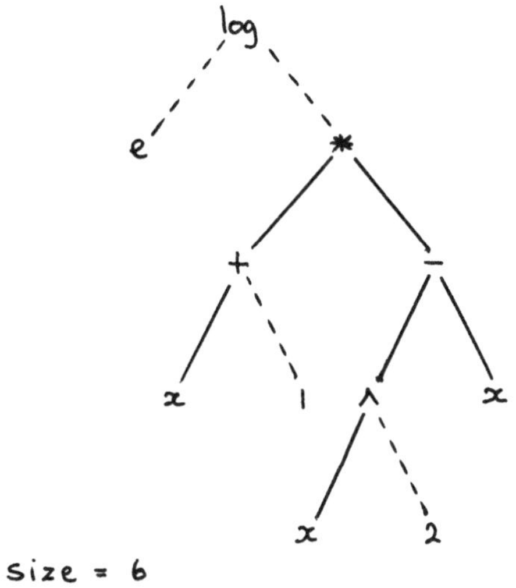

Figure 2-2: The Size of the Least Covering Tree

This last condition is to ensure that there is a global Attraction of the occurrences of x. Without it, the application could, say, draw two occurrences closer, but make these two further from a third, leading to a net increase in the size of the least covering tree. Attraction is also only applied to least dominating terms in x.

These conditions allow an axiom to be used as a rewrite rule in several different ways in the same set, including different orders, provided its variables are marked in different ways. For instance, the axiom:

```
(u^v)^w = u^(v*w)
```

may be used for Attraction left to right if v and w are marked and right to left if u and v are marked. This will not cause looping, as the termination proof in [Bundy and Welham ng] shows.

The same axiom can also be used in different ways by different sets. For instance, the axiom:

```
u*w + v*w = (u+v)*w
```

is used left to right by Collection, marked for w, and left to right by Attraction, marked for u and v.

3. Meta-Level Inference

3.1. Meta and Object-Levels

The object-level in PRESS consists of algebraic axioms describing relationships between real numbers. Given an equation (equations or inequalities) to be solved, these define an object-level search space. This space has a high branching rate, as is usually the case with mathematical theories involving equality.

PRESS consists of a collection of predicate calculus clauses which together constitute a PROLOG program. As well as the procedural meaning attached to these clauses, which defines the behaviour of the PRESS program, they also have a declarative meaning - that is, they can be regarded as axioms in a logical theory. For instance, consider the following clause, extracted from the program and rewritten in standard predicate calculus notation.[8]

```
singleocc(L,X) & freeof(R,X) & isolate(X,L=R,Ans)                        (4)
       -> solve(X,L=R,Ans)
```

The declarative meaning of 4 is:

> If L contains precisely one occurrence of X, R contains none and if the result of isolating X in L=R is Ans then Ans is a solution of L=R with respect to X

The PRESS program uses 4 as a piece of program for trying to solve an equation for an unknown by testing for a single occurrence of the unknown and if successful trying to Isolate the unknown.

Of what formal theory is 4 an axiom? Well, between what kinds of objects does it express relationships? These are not numbers (or matrices or points), but algebraic expressions such as equations, variables and terms. The relationships expressed are syntactic ones, like the number of occurrences of this term in this expression is so much; or the result of applying this axiom to this expression is that expression. Hence axiom 4 represents knowledge about the representation of algebra. This means that the formal theory described by PRESS is the Meta-Theory of Algebra[9].

In the Meta-Theory of Algebra the objects being manipulated are algebraic: variables; constants; functions; formulae; logical connectives etc. So these are all represented as ground-level terms in the meta-theory. In particular, an object-level variable is represented as a meta-level constant and an object-level formula is represented as a meta-level (ground) term.

---

[8] except that upper case letters at the front of a word denote a variable and lower case a constant

[9] The use of the prefix 'meta' throughout this paper is not meant to connote sophistication by association, as some critics have implied. It is merely the correct technical term

3.2. Inferential Control

PRESS consists wholly of meta-theoretic axioms, so as it runs it conducts inference at the meta-level. How does this cause algebraic manipulation to be done, since this requires inference at the object level? The answer is that many of the meta-level predicates express the relation:

Answer is the result of applying Rule to Expression

If this relation is set up as a goal to be satisfied with Rule and Expression bound to particular rules and expressions, then PRESS answers the question by actually applying the rule; and this constitutes a step in the object-level search. Thus, as the meta-level inference continues, it experiments with different object-level steps and if successful finds a proof or solution at the object-level.

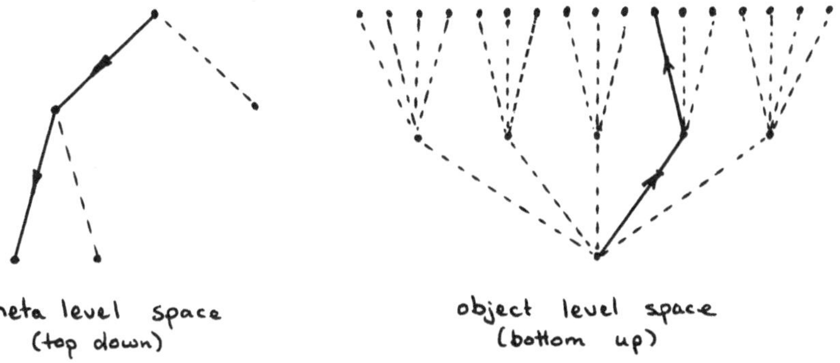

Figure 3-1: Meta-Level Inference controlling Object-Level Inference

By this technique each object-level decision can be based on an arbitary amount of inference at the meta-level. Making a wrong decision and hence getting lost on a fruitless search can become a rare event. For instance, the examples given in Appendix I involved no backup - a correct decision was made at every object-level branch point and the solutions found directly.

But doesn't this technique just trade search at the object-level for an equal amount of search at the meta-level? Apparently not! The meta-level search space is extremely well behaved in terms of its branching rate. Search is involved, but, fortunately, most choices lie between equally successful branches and bad choices rapidly lead to dead ends. This is because choices are usually between different methods of solution and each method uses a terminating rewrite rule set. Thus the choice of a method either results in immediate failure because no rules apply, or rules are applied and change the expression for the better. In no case can a method run away with itself, changing the expression but not terminating.

3.3. Deriving New Control Information

Meta-level inference can be used not only for controlling object-level search, but also for deriving meta-level theorems. This opens up the exciting possibility, which we plan to explore in the future, of using meta-level inference to derive new control information, i.e. define new methods and prove interesting properties of them.[10] What sort of properties are interesting to prove about methods such as Isolation and Collection? We would like to be able to prove automatically that they do the job required of them, that is,

1. that primary methods produce the solutions to equations when the input is of the correct form

2. that secondary methods produce output which is in the correct form for primary methods

3. that secondary methods produce output whose solutions are equivalent to the original input

4. and that, given the right input, all the methods will terminate. (In other words, we will want to prove formal versions of the informal proofs in [Bundy and Welham ng].)

As with the object-level rewrite rules we can give a syntactic characterization of the form of meta-level theorems that would establish this kind of result. For instance, by analogy with 4, property 1 above would have the form:

```
Cond1(Eqn,X)  & Method1(X,Eqn,Ans) -> solve(X,Eqn,Ans)
```

where Method1 is some primary method and Cond1 is the condition under which it succesfully solves equations.

Characterizations of properties 2, 3 and 4 are:

```
Method2(X,Eqn,New) -> Cond1(New,X)

Method2(X,Eqn,New) -> provable(meta,Eqn <-> New)

Cond1(Eqn,X) -> exists Ans Method1(X,Eqn,Ans)
```

where Method2 is a secondary method and provable(meta,Theorem) means Theorem is provable in formal theory meta - the meta-theory of algebra.

Note that these characterizations are at the meta-meta-level, just as the rewrite rule characterizations were at the meta-level. We intend to build meta-meta-level proof methods to guide the derivation of meta-therorems of the form 1-4, just as we have built meta-level equation solving methods to guide object-level search.

We have now described three levels of knowledge: object, meta and meta-meta. Are any further levels required and will this process of meta-level control ever terminate? We have seen above that the sequence of levels may terminate when the control of the topmost one becomes trivial. We have found that the, very simple, PROLOG depth first control regime is quite adequate to control meta-level inference, even when it generates arbitrarily complex control regimes at the object-level.

---

[10] This work is the subject of a new SRC grant, no. GR/B/29252, beginning in October 1980.

Further levels are only needed when the type of inference at the meta-level becomes more sophisticated.
Another way in which the regress might terminate is if the control regime embodied at a particular level were suitable for controlling itself.

Of the meta-theorems of type 1-4 above only a subset would actually be used as meta-level control rules, e.g. type 1. The others (type 2-4) are necessary scaffolding, which justify the use of the type 1 rules, but are not themselves used directly.

4. Related Work

The idea of using Logic to control inference was originally suggested by Hayes in his GOLUX project [Hayes 73] and his ideas were a major influence on the work described here. Where our technique differs from Hayes' is that we have developed a control language specifically geared to the domain (The Meta-Theory of Algebra) rather than a general purpose control language. The GOLUX control language was only used to specify different kinds of Resolution restrictions (Hyperresolution, SL Resolution etc.). This specialization has enabled PRESS to make a much finer match between means and ends and has led to far more directed searches. Indeed PRESS is able to use meta-level inference to select which object-level rule to apply next, whereas GOLUX could only vet the result after the rules had been applied to make sure their application was allowed by the control language.

More recently other researchers have been concerned with meta-level knowledge. For instance, Weyrauch's FOL system [Weyhrauch 79] has proved meta-level theorems and via 'reflexion principles' is able to use them in object-level proofs, but it does not use meta-level inference to guide search. Davis and Buchanan [Davis et al 77] have used meta-knowledge to guide search in the TEIRESIAS control system for MYCIN.

Both the TEIRESIAS and the GOLUX systems were object-level driven, in that an object-level search is monitored at each stage. TEIRESIAS examines the rules before they are applied and GOLUX examines the result afterwards. GOLUX may prune some of the branches; TEIRESIAS both prunes and reorders the rules. The meta-level inference used by TEIRESIAS seems not to be very deep - typically 1 step.

PRESS, on the other hand, is meta-level driven. Quite deep inference at the meta-level will, from time to time, invoke object-level steps. This seems a more suitable organization when only one path through the object-level search space is required, since it enables a lot of thought to go into choosing this path. MYCIN, however, requires all reasonable diagnoses of an illness and needs to be sure that the space has been searched exhaustively (which is easier if the search is object oriented). Thus only a pruning of unreasonable paths is in order. (It is not clear how the reordering of MYCIN rules by TEIRESIAS improves its efficiency.)

The use of sets of rewrite rules for algebraic manipulation and mathematical theorem proving generally is a very common technique (see, for instance, [Martin and Fateman 71, Hearn 67, Bledsoe and Bruell 73, Brown 77]). However, as indicated above, single, heterogeneous sets of rules are usually applied exhaustively as a 'simplification' or 'reduction' technique. There are exceptions: Brown's rules sometimes contain ad hoc control advice within themselves and Boyer and Moore [Boyer and Moore 73] have used several rewrite rule sets. MACSYMA provides several alternative normal forms to the basic 'simplification' procedure, but these must be called explicitly by the user. We know of no work where the author has supplied a characterization of the rule set and hence been able to use this meta-level knowledge to guide search, assimilate new rules or prove termination. Indeed Brown, who specifically attempted to build an 'extensible' rule based system, was reduced to ordering his newly proved rules by hand.

5. Conclusion

In this paper we have described a technique for controlling inference, called **meta-level inference.** This technique has been used to implement an algebraic manipulation program, PRESS. Meta-level inference is used to:

- examine the current expression in the light of the problem to be solved.
- select an appropriate rewrite rule set
- and selectively apply rules from this set to the current expression.

The advantages of this technique are:

- An inferential control regime allows sophisticated control decisions to be made and reduces the amount of search required.
- The use of multiple rewrite rule sets increases the ability of the program by allowing the same axiom to be used in several ways and selectively.
- Meta-level reasoning about the rules enables the assimilation of new theorems and proofs of termination.
- The modular division of the program into object and meta rules aids clarity and makes possible the consideration of the automatic learning of both factual and control information.

I. Results

In this Appendix we give a selection of equations solved by using the methods described in section 2. Examples of other kinds of problems solved by PRESS, e.g. simultaneous equations, inequalities etc, can be found in [Bundy and Welham 79].

<u>Logarithmic Equation</u>

log(e,x+1) + log(e,x-1) = 3

This is the equation used as a working example in [Bundy 75] and in section 2. It requires Attraction followed by Collection and multiple applications of Isolation to produce the answer:

x = (e^3 + 1)^2^-1 v x = -(e^3 + 1)^2^-1

<u>Exponential Equation</u>

2^(x^2)^(x^3) = 2

This equation requires similar processes to the previous ones, but shows that these are applicable to exponential as well as logarithmic functions. The answer found is:

x=1

Trigonometric Equation

((2^(cos(x)^2) * 2^(sin(x)^2))^sin(x))^cos(x) =2^(1/4)

Attraction is applied twice followed by Collection to solve this trigonometric equation, producing the answer:

x = -1^n0*arcsin(2^-1)*2^-1 + n0*90
where n0 is an arbitrary integer.

REFERENCES

[Bledsoe and Bruell 73]
Bledsoe, W.W. and Bruell, P.
A man machine theorem proving system.
In Nilsson, N., editor, Procs of IJCAI3, pages 56-65. Stanford, 1973.

[Boyer and Moore 73]
Boyer, R.S. and Moore J.S.
Proving theorems about LISP functions.
In Nilsson, N., editor, procs. of IJCAI3, pages 486-493. Stanford, August, 1973.
Also available from Edinburgh as DCL memo no. 60.

[Brown 77]
Brown, F.M.
Towards the automation of Set Theory and its Logic.
Research Report 34, Dept. of Artificial Intelligence, Edinburgh., May, 1977.
A shortened version appeared in IJCAI5.

[Bundy and Welham 79]
Bundy, A. and Welham, B.
Using meta-level descriptions for selective application of multiple rewrite rules in algebraic manipulation.
Working Paper 55, Dept. of Artificial Intelligence, Edinburgh., May, 1979.

[Bundy and Welham ng]
Bundy, A. and Welham, B.
Using meta-level inference for selective application of multiple rewrite rules in algebraic manipulation.
Artificial Intelligence , forthcoming.

[Bundy et al 79]
Bundy, A., Byrd, L., Luger, G., Mellish, C., Milne, R. and Palmer, M.
Mecho: A program to solve Mechanics problems.
Working Paper 50, Dept. of Artificial Intelligence, Edinburgh., 1979.

[Bundy 75]
Bundy, A.
Analysing Mathematical Proofs (or reading between the lines).
In Winston, P., editor, Procs of the fourth. IJCAI, Georgia, 1975.
An expanded version is available from Edinburgh as DAI Research Report No. 2.

[Bundy 79]
Bundy, A.
An elementary treatise on equation solving.
Working Paper 51, Dept. of Artificial Intelligence, Edinburgh., 1979.

[Davis et al 77]
Davis, R. & Buchanan, B.G.
Meta-level knowledge: overview and applications.
In Reddy, R., editor, procs of 5th, pages 920-927. IJCAI, 1977.

[Hayes 73]
Hayes, P.
Computation and deduction.
In Proc. of MFCS Symposium. Czech. Academy of Sciences, 1973.

[Hearn 67]
Hearn, A.C.
REDUCE: A user-oriented interactive system for Algebraic simplification.
Academic Press, New York, 1967, pages 79-90.

[Huet 77]
Huet, G.
Confluent reductions: Abstract properties and applications to term rewriting systems.
Rapport de Recherche 250, Laboratoire de Recherche en Informatique et Automatique, IRIA, France, August, 1977.

[Knuth & Bendix 70]
Knuth, D.E. & Bendix, P.B.
Simple word problems in universal algebra.
In Leech, editor, Computational problems in abstract algebra, pages pp 263-297. Pergamon Press, 1970.

[Martin and Fateman 71]
Martin, W.A. and Fateman, R.J.
The MACSYMA system.
In Petrick, S.R., editor, 2nd Symposium on Symbolic Manipulation, pages 59-75. Los Angeles, 1971.

[Moses 71]
Moses, J.
Algebraic simplification, a guide for the perplexed.
In Petrick, S.R., editor, 2nd Symposium on Symbolic Manipulation, pages 282-304. Los Angeles, 1971.

[Pereira et al 78]
Pereira, L.M., Pereira, F.C.N. and Warren, D.H.D.
User's guide to DECsystem-10 PROLOG.
Internal Memo, Dept. of Artificial Intelligence, Edinburgh., 1978.

[Weyhrauch 79]
Weyhrauch, R.W.
Prolegomena to a theory of mechanized formal reasoning.
RWW Informal Note 8., Stanford University, 1979.

Proofs as Descriptions of Computation

C. A. Goad
Department of Computer Science
Stanford University
Stanford, California 94305, USA

0. Abstract

A proof of $\forall x \exists y \varphi(x,y)$ can serve as a description of an algorithm which satisfies the specification given by φ. A variety of techniques from proof theory may be used to execute such a proof - that is, to take the proof and a value for x, and compute a value for y such that $\varphi(x,y)$ holds. Proofs differ from ordinary programs in that they formalize information about algorithms beyond what is needed for the simple execution of the algorithm. This paper concerns (I) the uses of this additional information in the automatic transformation of algorithms, and in particular, in the adaptation of algorithms to special situations, and (2) efficient methods for executing and transforming proofs. A system for manipulating proofs has been implemented. Results of experiments on the specialization of a bin-packing algorithm are described.

1. Introduction

The primary purpose of a proof is to convince - to provide compelling evidence for the truth of a proposition. However, proofs may on occasion have other uses as well.

Suppose that one has a proof that an object with given properties exists. Then the proof can sometimes be used to discover the identity of a particular object with those properties. If restrictions are made on the forms of inference used, then it is possible to guarantee that the proof will (in one sense or another) provide this additional information. For example, a *constructive* proof of $\exists x \varphi(x)$ always "provides" a value v with $\varphi(v)$ in the sense of indicating a method for computing v; the computation may or may not be feasible in practice. However, the restriction to constructivity is too strong. For one thing, a proof of $\exists x \varphi(x)$ may exhibit a value v which satisfies φ, but show that $\varphi(v)$ holds by non-constructive methods. Also, if one restricts the complexity of φ (for example, if φ is a quantifier free formula of first order arithmetic), then any classical proof of $\exists x \varphi(x)$ will provide a realization in the same sense and by the same formal methods as a constructive proof. (By a "realization" of an existential statement $\exists x \varphi(x)$ is meant simply a value which satisfies the predicate φ.)

If an existence proof is given in a formal way - in a way which makes it suitable for mechanical manipulation - then one might hope to mechanize the passage from the proof to the value realizing the existential statement. Work in proof theory has shown that the extraction of realizations from proofs can in fact be mechanized for a variety of formal systems and in a variety of ways. For example; Prawitz's normalization procedure may be used to transform a natural deduction proof of an existential formula into a direct proof of the same formula which will - under rather general conditions - explicitly mention a realization.

Now, if one has a proof of a formula of the form $\forall x \exists y \varphi(x,y)$, the methods from proof theory mentioned just above can evidently be used to compute a function f with $\forall x \varphi(x,f(x))$. To do this, simply apply the general result $\forall x \exists y \varphi(x,y)$ to the input value, and then use normalization (or whatever method one has in hand) to extract a realization. Thus a proof of a formula $\forall x \exists y \varphi(x,y)$ serves the role of a program which computes a function satisfying the "specification" φ.

Given that proofs can be used as programs, what is the interest of this fact for computer science and for practical computing? One answer is as follows.

Existing programming languages are for the most part designed with economy of expression in mind; a program in such a language formalizes exactly the information needed for carrying out the task at hand. A proof, on the other hand, formalizes a great deal of information which is not essential for the simple execution of a computation - such as a description of the task being performed, a verification of the method, and an account of the dependencies between facts involved in the computation. The additional information contained in proofs is useful in the transformation of computing methods - for example in adapting methods to new situations. This should not be surprising, since one expects that the data relevant to the transformation of algorithms will be different and more extensive than the data needed for simple execution.

We shall be concerned with a particular set of transformations on algorithms - called the "pruning transformations". These transformations remove redundant chunks of computation by making use of a kind of dependency information which does not appear in ordinary programs. For the most part, the redundancies removed by pruning are not to be found in proofs generated by people. Thus the pruning transformations will not be of much use when applied to algorithms as originally presented. However, proofs which result from automatic processes tend to include such redundancies.

For example, suppose that one has an algorithm A(x) which is to be used in a situation where it is known in advance that all inputs will have a special form given by the term $t(y_1,\ldots y_n)$. Then A may be automatically adapted to perform efficiently in this special situation by symbolically executing the code for A on the term t, and then applying optimizing transformations to the result. (Ershov[1977] and Sandewall[Beckeman, Haraldsson, Oskarsson, and Sandewall, 1976], among others, have studied this method of specialization as it applies to ordinary programs.) If A is expressed by a proof Π, then the result of symbolically executing Π on the term t will often contain redundancies of the kind removed by pruning even if Π as originally given contained no such redundancies. Thus, the effectiveness of automatic specialization can be increased by adding pruning to the arsenal of optimizations used in the course of specialization.

As they stand, the standard methods of proof theory are not adequate for carrying out the specialization of algorithms in a feasibly efficient way. However, we have devised methods for the execution and pruning of proofs which overcome this problem. This paper concerns these novel methods and their uses in specialization. The methods have been implemented in a proof manipulation system running on the Stanford Artificial Intelligence Laboratory PDP-10 computer. As a preliminary empirical investigation of the usefulness of pruning in the specialization of algorithms, expermiments on the specialization of a bin-packing algorithm have been carried out.

The purpose of this paper is to provide an informal introduction to the ideas invovled in the computational use of proofs. We do not attempt to give a full exposition of our methods. Instead, the central features of pruning and proof execution are illustrated in a simple setting. The reader is referred to the author's PhD thesis [Goad 1980] for further information.

The organization of the paper is as follows. Section 2 presents an example of specialization by pruning applied to a small case analysis proof. The proof manipulation method employed for this example is one of the standard methods of proof theory, namely, Prawitz's normalization procedure. Section 3 concerns the computational use of partially formalized proofs. In section 4, the defects of existing proof theoretic methods are discussed, and our own methods are described in general terms. Section 5 reports results of the experiments with the bin-packing algorithm. We close with a brief discussion of related work in computer science (section 6).

Before concluding this introductory section, we wish to make the following general remarks about the use of proofs as descriptions of computation.

(1) It should be emphasized that the work described in this paper concerns the automatic *manipulation* of existing proofs, and not the automatic construction of new proofs. The bin packing proof used in the experiments was devised "by hand", and was entered by hand into the proof checking component of the proof manipulation system. If one is able to automate, fully or partially, the construction of proofs which describe computational methods, then so much the better. But such matters lie outside the scope of this paper.

(2) It is necessary to keep computational considerations explicitly in mind when constructing proofs which are intended as descriptions of computation. The best proof of a formula $\forall x \exists y \varphi(x,y)$ according to such standard criteria as brevity, elegance or comprehensibility, will often embody a very bad algorithm. Conversely, a proof of $\forall x \exists y \varphi(x,y)$ which formalizes a good algorithm will generally constitute a rather unnatural way of establishing the simple truth of the formula. For the purposes of this paper, proofs are to be regarded as a means of formulating algorithmic ideas. In writing a proof to be used for solving a computational problem, one follows the same procedure as is used in writing an ordinary program. Namely, one first devises a reasonable algorithm, and afterwards formalizes that algorithm (as a proof). If a proof is given in complete detail (which it need not be for computational purposes; see section 3), then it includes a justification for the correctness of the algorithm which it formalizes. As an immediate consequence, formalization of algorithms by proofs provides a means for the mechanical verification of algorithms.

(3) There are other computational applications of proofs considered as descriptions of algorithms, for example: (a) Proofs with unproved "goals" can serve the role of Planner [Hewitt 1971] antecedent or consequent theorems, with the difference that, unlike Planner theorems, they can be proved as well as executed. (b) Run time use of pruning in the execution of proofs has the effect of dependency directed backtracking (London[1978], Doyle[1978]). (c) Pruning can help in the analysis of side effects. However, in this paper, we restrict our attention to the use of proofs in specializing algorithms.

2. An Example

In this section, we give an informal explanation - by means of an example - of the nature and uses of the pruning operations, and of the use of normalization in executing proofs.

The simplest algorithms to which the pruning operations are usefully applicable are pure case analysis algorithms - algorithms which can be expressed by "plain" conditional expressions. In what follows, we present a very small case analysis algorithm which is nonetheless sufficient to illustrate the main points which we wish to make about pruning. These points are: (1) pruning may be used to increase the efficiency of specializations of algorithms, and (2) conventional descriptions of algorithms do not contain the data necessary for the improvements in efficiency realized by pruning. Consider, then, the following algorithm - given as a conditional expression - for computing an upper bound for both the sum and the product of two positive rational numbers x, and y:

$$u(x,y) = \text{if } x \leq 1 \text{ then } y+1 \text{ else } (\text{if } y \leq 1 \text{ then } x+1 \text{ else } 2xy)$$

We will use the bold faced letter **u** to refer both to the algorithm, considered as an abstract method which can be formalized in various ways, and to the above concrete conditional term.

Now, suppose that the value .5 is given for y in advance, and that we wish to optimize **u** given this additional information. The best we can do, if supplied only with the conditional expression as a description of the algorithm, is to symbolically execute the expression on the arguments x, .5. The result is:

$u(x,.5) = \text{if } x \leq 1 \text{ then } 1.5 \text{ else } x+1$

As will be seen below, the formalization of this upper bound algorithm as a constructive existence proof allows u(x,.5) to be automatically simplified, by use of symbolic execution and pruning, to the expression x+1. The fact that x+1 is an upper bound for both x+.5 and .5x does not depend on x being less than one; this dependency information is contained in the proof, and allows the automatic removal of the unnecessary case split according to the size of x. Note that the pruning optimization has the unusual quality that it modifies the function computed by the expression to which it is applied. However, pruning is guaranteed to preserve the validity of an algorithm for the specification embodied in the end-formula of the proof describing the algorithm. Also note that no transformation on conventional computational descriptions can have the same effect as pruning. Conventional descriptions contain information only about the function to be computed, and not about the purpose of the computation, and therefore valid transformations on such descriptions must - unlike pruning - preserve extensional meaning.

In what follows, we first give the constructive existence proof - formalized in a natural deduction proof system - which expresses the upper bound algorithm **u**. We do not assume that the reader is familiar with traditional proof theory, and for this reason an informal explanation of the normalization process as it applies to pure case analysis proofs is given. The reader is referred to Prawitz[1965] for further information on natural deduction systems and on the normalization methods which apply to such systems. Finally, the pruning operations, and their effects on the upper bound algorithm, are described.

A natural deduction proof in the sense of Prawitz represents the deduction of a conclusion from assumptions. We adopt the notational conventions of Prawitz [1965]: proofs are written in tree form, and assumptions are designated by formulas in brackets. For example, the V-elimination rule is written as follows:

$$\vee E\;\frac{A \vee B \qquad \begin{matrix}[A]\\ \Pi_1\\ C\end{matrix} \qquad \begin{matrix}[B]\\ \Pi_2\\ C\end{matrix}}{C}$$

Π_1 represents a proof of the formula C from the assumption A, and Π_2 represents a proof of C from assumption B. The inference rule discharges the assumptions A and B - the meaning of the rule is that if C can be proved from either of the assumptions A,B, and it is known that A∨B holds, then C can be concluded without assumptions A, B.

The following natural deduction proof formalizes the upper bound algorithm u. In the proof and elsewhere $\Psi(x,y,z)$ is used to abreviate the formula $z \geq x+y \wedge z \geq xy$. Leaves of the proof tree which are not surrounded by brackets designate axioms or lemmas. Three unproved lemmas, namely "$x\leq 1 \supset \Psi(x,y,y+1)$", "$y\leq 1 \supset \Psi(x,y,x+1)$", and "$(x>1)\wedge(y>1) \supset \Psi(x,y,2xy)$", appear in the proof. As will be seen, the omission of the proofs of the lemmas does not compromise the computational usefulness of the proof as a whole. (For further comments about the computational use of incompletely given proofs, see section 3 below.) Also, the formulas $x\leq 1 \vee x>1$ and $y\leq 1 \vee y>1$ appear as axioms. We will use the capital letter U to designate the proof.

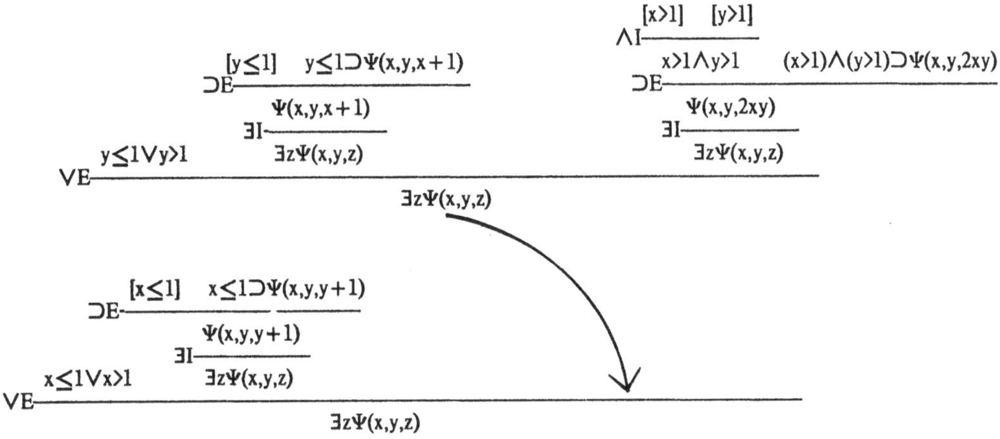

Note that we have neglected to universally quantify the variables x,y so as to arrive at a proof in the standard ∀∃ form mentioned in the introduction. In the current simple context it is more convenient for purposes of exposition to leave the quantification implicit, and to specify that input values to the proof viewed as an algorithm be substituted for the free variables. More precisely, in order to compute an upper bound for the sum and product of two input values v_1 and v_2 by means of normalization, v_1 and v_2 are first substituted for x,y throughout the proof U, and then the proof is normalized.

The normalization procedure is a step-wise proof transformation method which, roughly speaking, has the effect of removing certain kinds of roundabout arguments from proofs. A proof from which all such arguments have been removed is said to be in "normal form". For the current purposes, the crucial property of normal proofs is this: a normal proof of an existential formula exhibits a value which realizes that formula - under the condition that the proof contains no free variables or open assumptions. Thus, whenever concrete values v_1, v_2 are substituted for x and y in the proof U above, the normal form of the resulting proof will supply us with a particular value v_3 for z such that $\Psi(v_1,v_2,v_3)$ holds.

The normalization procedure is given by a collection of reduction rules. In order to normalize a proof, reduction rules are applied repeatedly until no application of a reduction rule is possible. Normalization of simple case analysis proofs such as U corresponds closely to the execution of conditional terms by means of repeated applications of the reduction rules:

C1: (if TRUE then t_1 else t_2) $\Rightarrow$ t_1

C2:. (if FALSE then t_1 else t_2) $\Rightarrow$ t_2

A1: $R(t_1,t_2)$ $\Rightarrow$ TRUE if R is an atomic relation, t_1,t_2 are closed ground terms, and $R(t_1,t_2)$ holds

A2: $R(t_1,t_2)$ $\Rightarrow$ FALSE if R is an atomic relation, t_1,t_2 are closed ground terms, and $\neg R(t_1,t_2)$ holds

We also need reduction rules for simplifying ground terms such as .5+1. However, neither the representation of numbers used, nor the form of the simplification rules for ground terms is relevant to the current discussion.

The following reduction rules for proofs correspond directly to the reduction rules for conditional terms above.

O1:

$$\mathrm{VE}\dfrac{\mathrm{VI}\dfrac{\begin{matrix}\Pi_1\\ A\end{matrix}}{A \vee B} \quad \begin{matrix}[A]\\ \Pi_2\\ C\end{matrix} \quad \begin{matrix}[B]\\ \Pi_3\\ C\end{matrix}}{C} \qquad \Rightarrow \qquad \begin{matrix}\Pi_1\\ [A]\\ \Pi_2\\ C\end{matrix}$$

O2:

$$\mathrm{VE}\dfrac{\mathrm{VI}\dfrac{\begin{matrix}\Pi_1\\ B\end{matrix}}{A \vee B} \quad \begin{matrix}[A]\\ \Pi_2\\ C\end{matrix} \quad \begin{matrix}[B]\\ \Pi_3\\ C\end{matrix}}{C} \qquad \Rightarrow \qquad \begin{matrix}\Pi_1\\ [B]\\ \Pi_3\\ C\end{matrix}$$

P1:

$$\mathrm{Axiom}: R(t_1,t_2) \vee \neg R(t_1,t_2) \qquad \Rightarrow \qquad \mathrm{VI}\dfrac{\mathrm{Axiom}: R(t_1,t_2)}{R(t_1,t_2) \vee \neg R(t_1,t_2)}, \text{ if } R(t_1,t_2) \text{ holds}$$

P2:

$$\mathrm{Axiom}: R(t_1,t_2) \vee \neg R(t_1,t_2) \qquad \Rightarrow \qquad \mathrm{VI}\dfrac{\mathrm{Axiom}: \neg R(t_1,t_2)}{R(t_1,t_2) \vee \neg R(t_1,t_2)}, \text{ if } \neg R(t_1,t_2) \text{ holds}$$

In the above, an expression of the form,

$$\begin{matrix}\Pi_1\\ [F]\\ \Pi_2\end{matrix}$$

denotes the result of substituting Π_1 for open occurences of the assumption F in Π_2. (Note: in the proof U, $t_1 > t_2$ is to be regarded as an abreviation for $\neg(t_1 \leq t_2)$.)

The reduction rules given above may be applied at arbitrary positions within a proof tree. That is to say, wherever an $\vee$-elimination inference appears which matches the template given on the left hand side of rules **O1** or **O2**, that inference may be replaced by the appropriate instantiation of the right hand side of the same rule - and similarly for axioms of the form given by **P1** and **P2**. As indicated above, normalization consists of repeated applications of the rules, in arbitrary order, until no rule is applicable. It has been shown for a variety of proof systems that such a sequence of reductions must always terminate, and that the end result is independent of the order in which the reductions are applied. A result of this kind is normally referred to as a "strong normalization theorem".

Let Π be a proof of a disjunction $A \vee B$ which contains no free variables and no open assumptions. Then the normalization of Π provides a decision as to which of the disjuncts holds. The decision is "coded" by the normal form of Π, which will have one of the two forms

$$\mathrm{VI}\dfrac{\begin{matrix}\Pi\\ A\end{matrix}}{A \vee B} \qquad \text{or} \qquad \mathrm{VI}\dfrac{\begin{matrix}\Pi\\ B\end{matrix}}{A \vee B}$$

Thus the reduction rules **P1** and **P2** together serve as a "primitive subroutine" which decides atomic relations.

As indicated above, and analogously to the case of disjunction, the normalization of a proof (again without open assumptions or free variables) of an existential formula $\exists x\varphi(x)$ provides a value which satisfies φ. The value appears in the normal proof as the argument of the last inference, which will be an existential introduction. That is, the normal form of a proof of $\exists x\varphi(x)$ has the form

$$\exists I\,\frac{\begin{matrix}\Pi\\ \varphi(t)\end{matrix}}{\exists x\varphi(x)}$$

"t" is the value "returned" by the execution of the proof.

Finally, we assume the existence of reductions which simplify ground terms. This is far from a complete set of reduction rules for the full natural deduction proof system, but is adequate for the execution of pure case analysis proofs. The reader can verify, as an example, that the result of normalizing the proof U with inputs 2 for x and .5 for y is the proof

$$\exists I\,\frac{\supset E\,\dfrac{.5<1 \qquad .5\leq 1 \supset \Psi(2,\ .5,\ 3)}{\Psi(2,\ .5,\ 3)}}{\exists z\Psi(2,\ .5,\ z)}$$

The value returned by this proof "3". The normalization parallels the execution of the term u by use of the corresponding reduction rules for conditional terms.

We are now in a position to describe the pruning operations, and the manner in which they may be used in specializing U. The pruning operations for $\vee$-elimination are as follows. (There is also a pruning operation for $\exists$-elimination which, however, will not concern us here.)

$$\vee E\,\frac{\begin{matrix}\Pi_1\\ A\vee B\end{matrix} \qquad \begin{matrix}\Pi_2\\ C\end{matrix} \qquad \begin{matrix}\Pi_3\\ C\end{matrix}}{C} \quad \Rightarrow \quad \begin{matrix}\Pi_2\\ C\end{matrix}$$

if A does not appear as an open assumption in Π_2.

$$\vee E\,\frac{\begin{matrix}\Pi_1\\ A\vee B\end{matrix} \qquad \begin{matrix}\Pi_2\\ C\end{matrix} \qquad \begin{matrix}\Pi_3\\ C\end{matrix}}{C} \quad \Rightarrow \quad \begin{matrix}\Pi_3\\ C\end{matrix}$$

if B does not appear as an open assumption in Π_3.

In order to specialize the algorithm expressed by U to the case where y is fixed at .5, .5 is substituted for y throughout the proof, and the result is normalized. This process yields the following "specialized" proof:

$$\vee E\,\frac{x\leq 1\vee x>1 \qquad \exists I\,\dfrac{\supset E\,\dfrac{[x\leq 1] \qquad x\leq 1\supset\Psi(x,\ .5,\ 1.5)}{\Psi(x,\ .5,\ 1.5)}}{\exists z\Psi(x,\ .5,\ z)} \qquad \exists I\,\dfrac{\supset E\,\dfrac{.5\leq 1 \qquad .5\leq 1\supset\Psi(x,\ .5,\ x+1)}{\Psi(x,\ .5,\ x+1)}}{\exists z\Psi(x,\ .5,\ z)}}{\exists z\Psi(x,\ .5,\ z)}$$

This proof corresponds to the specialized conditional term, "if $x\leq 1$ then 1.5 else x+1". A further optimization is applicable to the specialized proof which is not applicable to the conditional term, namely pruning. The second minor premise of the $\vee$-elimination inference in the specialized proof above does not depend on the assumption x>1. It is this fact about the dependency structure of the computation that the proof U, but not the conditional term u, formalizes, and which allows pruning to take place. The result of applying pruning is:

$$\supset E\frac{.5\leq 1 \quad .5\leq 1\supset\Psi(x,\ .5,\ x+1)}{\exists I\dfrac{\Psi(x,\ .5,\ x+1)}{\exists z\Psi(x,\ .5,\ z)}}$$

This represents the same algorithm as the conditional term "x+1".

Note that, if comparison is a very cheap operation, and adding is very expensive, then it might happen that "x+1" has an average case efficiency which is worse than "if x≤1 then 1.5 else x+1". This illustrates the general point that pruning is not *guaranteed* to increase efficiency. However, pruning often improves the efficiency of an algorithm, and *always* reduces its size. (Size reduction is an important effect of pruning in the experiments on bin-packing; see section 5.)

3. Computing with Incompletely Formalized Proofs

Not every part of a proof of an ∀∃ theorem is used in the computation described by the proof. Proofs of atomic formulas, and more generally of formulas which are built up from negated formulas and atomic formulas by conjunction, implication, and universal quantification, are not needed for computational purposes. We already have an example of this phenomenon in the proof U of the last section, in which several lemmas of this form appear without proof. It is possible to automatically select, from any given proof, that part of the proof tree which is needed in the computational use of the proof.

The methods which are employed in the "non-computational" parts of a proof are not subject to the same restrictions as apply to the computational part. In particular, there is no reason to restrict the use of non-constructive reasoning in the proof of non-computational lemmas. Further, the usual purpose of formalizing proofs of non-computational lemmas is to establish formally the truth of the lemmas. Therefore, if one wishes only to construct a proof which describes a given algorithm without formally verifying the algorithm, then proofs of non-computational lemmas may be omitted.

In the system of programs used for the bin-packing experiments, incomplete proofs - i.e. proofs in which unproved lemmas appear - are accepted by the proof checking part of the system, under the condition that the proofs are "computationally complete". By computational completeness, we mean that enough of the proof is retained to fully describe a computational method.

4. The P-Calculus

Traditional proof theory provides two kinds of methods for the execution of proofs. First there are the methods which operate by transformation of the proofs themselves. Gentzen's cut-elimination procedure [Gentzen 1969] and Prawitz's normalization procedure belong to this class. Second, there are methods which involve extracting "code" of one kind or another from proofs; it is then the code which is executed, and not the proof itself. Examples of methods of the latter kind are the realizability interpretations of Kleene[1945] for arithmetic and Kreisel[1959] for analysis, and the Dialectica interpretation of Godel[1958].

Each of these two approaches is inadequate for the purposes which we have in mind here. The normalization methods are unsatisfactory because of their inefficiency. In all cases of practical interest, a proof will contain much information which is irrelevant both to execution and to pruning. (This is true even if proofs of non-computational lemmas are omitted.) Nonetheless, all of this additional information is subject to extensive manipulation in the course of normalization. On the other hand, the methods which involve extraction of code from proofs retain only the information which is needed for the computation immediately at hand; the additional data needed for the pruning transformations mentioned above is lost. This would not be a problem if we only intended to apply pruning transformations to proofs as they are originally given. However, the use of proofs for the specialization of algorithms requires that the additional data be preserved by symbolic execution.

Our solution to these difficulties involves the use of an extended λ-calculus, which we shall refer to as the p-calculus. The p-calculus is designed to provide expression for just that information contained in natural deduction proofs which is needed for execution and for the pruning operations. P-calculus terms can be extracted from ordinary natural deduction proofs in a straight-forward manner, and executed efficiently by an interpreter of the kind used for λ-calculus based languages such as LISP and SCHEME [Sussman and Steele, 1975].

The p-calculus is arrived at from a standard λ-calculus with pairing by the addition of operators **OE** and **EE** which correspond directly to the $\vee$-elimination and $\exists$-elimination inferences of natural deduction. The **OE** operator may be regarded as a conditional operator which keeps track of the dependency information needed for pruning.

The following inductive clauses are added to the definition of the λ-calculus with pairing in order to arrive at the p-calculus: If x is a variable, and t_1,t_2,t_3 are terms of the p-calculus, then so are

(1) $\mathbf{OE}(x,t_1,t_2,t_3)$, and

(2) $\mathbf{EE}(x,t_1,t_2)$.

The variable x is considered bound in the terms t_2,t_3 of $\mathbf{OE}(x,t_1,t_2,t_3)$, and $\mathbf{EE}(x,t_1,t_2)$, but not in t_1. The variable x represents the open assumptions which are discharged by an $\vee$-elimination or $\exists$-elimination inference, and has the effect of keeping track of the dependency information needed for pruning.

The procedure by which p-calculus terms are extracted from proofs is based on a well known correspondence between the operations by which natural deduction proofs are built up, and the elementary operations - namely, application and abstraction - which are used in constructing descriptions of functions in the λ-calculus. The so-called *calculi of constructions* of Scott[1970], DeBrujin[1970], and Howard[1980] (among others) make this correspondence explicit; in these calculi, the same *formal* operations are used to construct both proofs and descriptions of functions.

The structure of proofs is faithfully preserved in the extraction of p-terms. A p-term is, in effect, a skeleton of a proof in which the inference rules of the proof, but not the formulas to which they are applied, are recorded. Open assumptions in a proof are mapped to free variables in the extracted term, and discharges of assumptions correspond to binding of variables. Further, the treatment of each inference rule except the $\vee$-elimination and $\exists$-elimination rules is the same for our map as it is in the calculus of constructions of Howard[1980]. However, it is this difference in the treatment of $\vee$-elimination and $\exists$-elimination which makes the application of pruning operations to p-terms possible.

Let us designate our extraction map by δ. Then δ, when applied to a proof,

$$\dfrac{\begin{matrix} & [A] & [B] \\ \Pi_1 & \Pi_2 & \Pi_3 \\ A \vee B & C & C \end{matrix}}{C}\ \vee E$$

yields the term $OE(\alpha,\delta(\Pi_1),\delta(\Pi_2),\delta(\Pi_3))$, where the variable α is the variable assigned to represent the assumptions discharged by the $\vee$-elimination inference. The pruning operations for **OE** are as follows:

$\mathbf{OE}(x,t_1,t_2,t_3) \Rightarrow t_2$ if x does not appear free in t_2, and

$\mathbf{OE}(x,t_1,t_2,t_3) \Rightarrow t_3$ if x does not appear free in t_3.

The execution procedure for the p-calculus, like the λ-caluclus and the "calculus" of natural deduction proofs, is given by a set of reductions which are applied repeatedly in order to "execute" (or normalize) a given term. The reductions needed for the p-calculus are just those of the λ-calculus, with the addition of reductions for executing the **OE** and **EE** operators.

The p-calulus, like the λ-calculus, includes terms which cannot be brought into normal form by any finite sequence of reductions. However, those p-calculus terms which are arrived at by extraction from proofs have the strong normalization property: every sequence of reductions which is applied to such a term results after a finite number of steps in the *same* normal form. Pruning operations may be added to the collection of reductions used in normalization, without endangering the termination of reduction sequences. However, this expanded reduction system no longer has the strong normalization property; different normal forms may be arrived at by different choices of the order in which reductions are applied. It should be mentioned that this is an advantage and not a defect of the pruning operations. By dropping the requirement that all reduction orders yield the same result (normal form), one arrives at shorter reduction sequences. This is a corollary of the observation that pruning improves the efficiency of a computation by changing the function which it computes.

The correctness of the extraction method δ is assured by the following theorem.

Theorem: Let Π be a proof containing no open assumptions of a closed formula $\forall x \exists y \varphi(x,y)$. Then $\lambda x.(J1(\delta(\Pi)(x))$ computes a function f with $\forall x \varphi(x,f(x))$.

"J1" in the theorem represents the operator which extracts the first element from a pair. We mean "computes" in the sense of normalization (with or without pruning) in the p-calculus. The above theorem applies to the same class of theories as are treated by the traditional realizability interpretations of proof theory. Among these theories are first order theories of inductive domains - the canonical example being first order arithmetic - and several theories of objects of higher type, such as arithmetic in all finite types [Troelstra 1973], and the theory of species [Tait 1975]. The restriction to computational completeness in the sense of section 3 must be observed.

The same technology used for the efficient interpretation of λ-calculus based languages extends easily to the p-calculus. The implemented interpreter (normalizer) for the p-caluclus used in the bin-packing experiments is modeled directly on the SCHEME interpreter described in [Sussman and Steele, 1975].

5. Specialization of a Bin-Packing Algorithm

The experiments described in this section were performed in order to to obtain a preliminary indication of the frequency with which prunable redundancies occur in the "real" computational world. The experiments concerned the specialization of a first-fit backtracking algorithm for one-dimensional bin packing. This algorithm, when given a list L_1 of block sizes and a list L_2 bin sizes, performs a depth first search for a legal assignment of blocks to bins. The algorithm is referred to as a "first fit" algorithm because, in the course of search, it attempts to place a block in the first bin in which it fits as its initial try. The bin-packing problem is well known to be NP-complete, and this particular algorithm has a worst case running time which is exponential in the size of the input. However, the problem is tractable for small inputs. It is of interest to see how much the algorithm can be sped up in the cases where the inputs are of feasible size.

The bin-packing algorithm was formalized as a natural deduction proof in the first order theory of lists and numbers, and a p-calculus term was extracted from this proof. Several experiments were carried out, each of which involved specializing the algorithm to handle problems of a particular size and structure. For example, a specialized algorithm for packing six blocks given in order of descending size into three bins of equal size was derived from the general bin-packing algorithm by the following steps. (1) The p-calculus term which describes the general algorithm was executed (normalized without pruning) on the symbolic inputs $L_1 = \langle i_1,i_2,i_3,i_4,i_5,i_6 \rangle$, $L_2 = \langle n,n,n \rangle$, where the i_j and n are numeric variables, and where it was assumed further that $i_1 \geq i_2 \geq \ldots \geq i_6$. The resulting p-calculus term had the form of a decision tree. (2) The decision tree was subjected to an optimization involving the elimination of case analyses whose outcome was decided by formulas already assumed on the branch so far taken in the tree. The optimization was carried out by use of the simplex algorithm (all the case analysis predicates in bin-packing have the form of inequalities between sums). The process so far could as easily have been carried out on an ordinary program as on a proof or p-calculus term. However, at stage (3) pruning was applied. The question of central interest was this: what increase in speed and reduction in size would be obtained by the application of pruning?

In practice, it was not feasible to carry out steps 1 and 2 seperately, since the decision tree resulting from step 1 would have been extremely large. Instead, normalization and optimization were applied "in parallel" - the decision tree was optimized in the course of its construction.

Experiments of the kind just described were carried out for all combinations of numbers n_1 of blocks and numbers n_2 of bins with $2 \leq n_2 < n_1 \leq 6$. We will restrict ourselves here to reporting results of one small experiment ($n_1=3$, $n_2=2$), and one typical large experiment ($n_1=6, n_2=3$).

For $n_1=3$, $n_2=2$, the decision tree which results from steps (1) and (2) is as follows - expressed as a conditional rather than a p-calculus term.

```
if i1≤n then
   if i1+i2≤n then
      if i1+i2+i3≤n then <1,1,1> else <1,1,2>
   else
      if i1+i3≤n then <1,2,1>
      else
         if i2+i3≤n then <1,2,2> else <>
else <>
```

The list $\langle a,b,c \rangle$ represents an assignment of blocks i_1,i_2,i_3, to bins numbered a,b,c respectively. In this case there are two bins numbered 1 and 2. $\langle\rangle$ represents the impossibility of any packing.

After pruning, one arrives at:

if $i_1 \leq n$ then
  if $i_2+i_3 \leq n$ then $\langle 1,2,2 \rangle$ else $\langle\rangle$
else $\langle\rangle$

This improved (and optimal) algorithm tries only one packing, namely $\langle 1,2,2 \rangle$. If any packing works, then this one must. This fact is "automatically realized" by the dependency analysis involved in pruning.

For $n_1=6$, $n_2=3$, the decision trees are too large to be given explicitly here. The decision tree which results from steps (1) and (2) has 87 decision nodes and a depth of 14. When pruning is applied, the tree shrinks to 15 decision nodes with a depth of 8. If one measures the running time of a bin-packing algorithm by the number of comparisons which it makes, then the worst case running time of the original algorithm on inputs of the special form currently under consideration is 174. The worst case running time of a decision tree algorithm according to this measure is simply the depth of the tree. Thus the simplex optimization and pruning taken together produce a factor of improvement of nearly 22 in worst case running time (from 174 to 8).

The following conclusions can be drawn from the experiments. (1) The simplex optimization with or without pruning yields a large speed-up of the algorithm. (2) Pruning dramatically decreases the size of the specialized decsion tree algorithm, and produces a moderate improvement in its speed (ie depth). In the largest experiments (where $n_1=6$ and $n_2 \geq 4$), it was not feasible to produce a decision tree algorithm at all without the use of pruning; pruning had to be run in parallel with the simplex optimization and normalization in order to avoid running out of memory space. Thus in this application, the main effect of pruning was to make possible the production of fast specialized algorithms which are of a reasonable size. In devising combinatorial algorithms for handling a finite number of cases, speed is not the only problem, since one can often make use of table look-up to get a very fast but very large algorithm. What is difficult is to produce an algorithm which is *both* small and fast.

6. Related Work in Computer Science

The relationship between the work described here and work in proof theory has been discussed at various points in the preceding sections of the paper. The purpose of this section is to briefly describe some of the work which has been done in computer science on the extraction of information from proofs.

Green [1969] considered the problem of extracting information from resolution proofs. Bishop[1970], and Constable[1971] suggested using constructive proof systems as programming languages. Goto[1979] has implemented Godel's Dialectica intertpretation for intuitionistic first-order arithmetic. Takasu [1978] has discussed computational uses of proofs in the same system by use of Gentzen's [1969] cut-elimination procedure. Miglioli and Ornaghi [1980] describe a method for executing sequent calculus proofs which differs from cut-elimination. In [Manna and Waldinger, 1979], a method for automatic synthesis of programs is described which involves the simultaneous construction of a natural deduction proof of the goal formula and of a program which realizes that formula (in a suitable sense). Bates [1979] develops a constructive "refinement logic", and shows how programs can be extracted from proofs of this logic. A *Prolog* program [Kowalski 1974] is a collection of axioms in Horn clause form. An execution of a Prolog program consists of a search for a proof in a restricted resolution system. The output is a term extracted from the proof. In practice, the output term is constructed during the search for the proof.

It should be emphasized that the aims of the work described above differ fundamentally from the aims of the work presented in this paper. In the former, formal proofs serve as vessels from which computational contents of a standard kind are extracted. In contrast, our concern has been to exploit the *differences* between proofs and conventional descriptions of computation. Specifically, we have shown how new operations on algorithms can be mechanized by making use of the additional information to be found in proofs.

Acknowledgements

This research was supported by the Advanced Research Projects Agency of the Defense Department under contract MDA903-80-C-0102. The author's interest in computational applications of the manipulation of proofs goes back to Professor G. Kreisel's ideas, as expressed for example in the programmatic papers Kreisel[1975] and Kreisel[1977]. The author wishes to thank Kreisel for encouraging him to work on this topic, and for help while the work was in progress.

References

Bates, J.L. *A logic for correct program development,* PhD Thesis, Dept. of Computer Science, Cornell University, August 1979

Beckeman,L., Haraldsson, A., Oskarsson,O., and Sandewall, E.[1976], *A partial evaluator and its use as a programming tool,* Artificial Intelligence Journal 7 pp. 319-357

Bishop, E.[1970], *Mathematics as a numerical language,* Intuitionism and proof theory, Proceedings of the summer conference at Buffalo N.Y., 1968, A. Kino, J. Myhill, R.E. Vesley eds., North Holland, Amsterdam pp 53-71

Constable, R.L.[1971], *Constructive mathematics and automatic program writers,* IFIP Congress 1971

de Brujin, N. G.[1970], *The mathematical language AUTOMATH, its usage, and some of its extensions,* Lecture Notes in Mathematics, vol. 125, Springer Verlag, pp. 29-61

Doyle, J.[1978], *Truth maintenance systems for problem solving,* M.I.T. AI lab technical report AI-TR-419, January 1978

Ershov A.P.[1977], *On the essense of compilation,* IFIP Working Conference on Formal Description of Programming Concepts, Saint Andrews, New Brunswick, vol 1., pp. 1.1-1.28

Gentzen, G.[1969], *The collected papers of Gerhard Gentzen,* (M.E. Szabo ed.), North-Holland, Amsterdam

Goad C.A.[1980], *Computational uses of the manipulation of formal proofs,* PhD Thesis, Stanford University, in preparation

Godel, K.[1958], *Ueber eine bisher noch nicht benutzte Erweiterung des finiten Standpunktes*, Dialectica 12, pp. 280-287

Goto S., [1979], *Program Synthesis from Natural Deduction Proofs*, International Joint Conference on Artificial Intelligence, Tokyo

Green, C.C,[1969], *Application of theorem proving to problem solving*, Proceedings of the International Joint Conference on Artificial Intelligence, Washington DC, pp 219-239

Hewitt, C.[1971], *Procedural embedding of knowledge in planner*, Proceedings of the International Joint Conference on Artificial Intelligence, London

Howard, W.A.[1980], *The formulae-as-types notion of construction*, in *Festschrift on the occasion of H.B. Curry's 80th birthday*, Academic Press, New York, San Francisco, London, to appear

Kleene S.C.[1945], *On the interpretation of intuitionistic number theory*, Journal of Symbolic Logic 10,pp. 109-124

Kowalski, R. [1974], *Predicate logic as a programming language*, Proc. of the IFIP Congress 1974, pp 569-574

Kreisel, G.[1959], *Interpretation of analysis by means of constructive functionals of finite type, in Constructivity in Mathematics*, North-Holland, Amsterdam, pp. 101-128

Kreisel, G.[1975], *Some uses of proof theory for finding computer programs*, Colloque International de Logique, Clermont-Ferrand (July 1975), Colloques Internationaux du Centre National de la Researche Scientifique, No. 249, pp. 123-134

Kreisel, G.[1977], *From foundations to science: justifying and unwinding proofs*, Recueil Des Travaux de L'Institut Mathematique, Belgrade, Nouvelle Serie 1977, Vol. 2, pp. 63-72

London, P.E.[1978], *Dependency networks as a representation for modeling in general problem solvers.* PhD thesis. U Maryland Department of Computer Science Technical Report 698

Manna,Z. and Waldinger, R.[1979], *A deductive approach to program synthesis*, Fourth Workshop on Automatic Deduction, Austin Texas, pp 129-139

Miglioli, P., and Ornaghi, M., *A logically justified computing model*, Fundamenta Informaticae, to appear

Prawitz, D.[1965], *Natural deduction*, Almquist and Wksell, Stockholm

Scott, D.[1970], *Constructive Validity*, Lecture Notes in Mathematics, vol. 125, Springer Verlag, pp. 237-275

Sussman G.J., and Steele, G.L., [1975], *SCHEME, an interpreter for extended lambda calculus*, MIT AI Memo 349

Tait, W.W. [1975], *A realizability interpretation of the theory of species*, Logic Colloquium Proceedings, 1972-1973, A. Dold, B. Eckmann eds., Lecture Notes in Mathematics vol. 453, Springer Verlag, pp. 240-251

Takasu S.[1978], *Proofs and programs*, Proceedings of the Third IBM Symposium on the Mathematical Foundations of Computer Science - Mathematical Logic and Computer Science, Kansai

Troelstra A.S.[1973], *Intuitionistic formal systems, in Metamathematical Investigation of Intuitionistic Arithmetic and Analysis*, A.S. Troelstra, ed., Lecture Notes in Mathematics vol. 344, Springer Verlag, pp. 1-96

PROGRAM SYNTHESIS FROM INCOMPLETE SPECIFICATIONS

Gérard Guiho
Christian Gresse
Laboratoire de Recherche en Informatique
Bât. 490 - Université de Paris-Sud
91405 ORSAY (France) . Tél. 941 66 26

Starting from a recent published paper of W. Bibel we show that it is sometimes useless to specify completely a problem at first. A new method of generalization, useful for output variables, is also exhibited and discussed.

I - INTRODUCTION

Although program synthesis from specifications is a recent development, there are already many approaches.

Some systems like DEDALUS of Manna and Waldinger (7), PSI of Green (3) or that of Bidoit & Al (1) work as a transformation program by using successive rewriting rules. Another system of Manna and Waldinger (8) produces the program to be synthesized from a constructive proof of a theorem.

From these approaches, that of Bibel (2) seems to be original. The problem is specified in predicate logic. Then by application of rules, a deductive process is used to rewrite equivalent logic formulas in order to obtain the logic part of the algorithm to be constructed ; the process of synthesis being guided by a few basic stategies. But no control is done during the synthesis and one has to impose it afterwards to obtain an algorithmic solution, which is not always easy. The problems studied by Bibel are of the form :

$\forall i\ \exists 0\ (IC(i) \rightarrow OC(i,0))$

where i is an input satisfying the condition IC and 0 is the desired output satisfying the condition OC.

Then Bibel chooses a candidate 0' (strategies GUESS and DOMAIN) and examines the two possible alternatives.

- if 0'=0 synthesis is generally finished
- if 0'≠0 (i.e. 0' does not satisfy OC(i, 0')) one tries to merge this new piece of information with the initial one in order to obtain some kind of recursion (Strategies GET.REC and GET.EP).

In fact the Bibel approach is far from an automatic system and some work has to be done to obtain something more than a methodology.

In this paper, we intend to extend the previous results. Our method, which is more procedural than that of Bibel, produces an algorithmic solution directly and focuses on the main point of the synthesis : the interpretation of the condition $0' \neq 0$. A new method of generalization is also exhibited and discussed. The examples synthesized by Bibel belong to two general schemes :

Find e in E such that P(e,E,X)

and

Find E0 in E such that P(E0,E,X)

E can be viewed as a set, E0 as a subset of E, e as an element of E and X as a set of free variables. Further, in the program produced, the abstract data type set will be represented by a more suitable abstract data type (See Guttag (4)).

Let us define E in a recursive way, as Lehman and Smyth do(5).

$E \leftarrow \{\emptyset\}+E x e''$

where $\{\emptyset\}$ is the empty set, + is the dijoint sum and x is the cartesian product. If E is not empty, we can choose an element e' from E, so $E=e' \cup E \backslash e'$ and recursion appears quite obviously.

II - SYNTHESIS OF : *Find e in E such that P(e,E,X)*

(Note that in the whole paper italics do not belong to the final language).

```
Proc :  (e,b)← F(E,X)
            Find e in E such that P(e,E,X)          (1)
endproc.
```

b is a boolean which will be true if F has at least one solution and will be false in the opposite case. The Bibel approach supposes that a problem always has a general solution and in fact that the restriction of the problem to a smaller set always has a solution which can be false relatively to the subproblem when the general solution does not belong to the smaller set.

We don't use this kind of restriction ; therefore we are obliged to go further in the justification of our program transformations.

First, we choose an element e' from E. In an Algol-like language F becomes

```
Proc :  (e,b)←F(E,X)
        If E=∅ then b←false : return  ; endif ;
        choose e' from E ;
        If I(e≠e') then Find e in E\e' such that P(e,E,X)      (2)
                   else (e,b)←(e', true)
        endif ;

endproc
```

I(e≠e') is the interpretation of the condition e'≠e (i.e.e' is not equal to one element e which satisfies P).
In order to find some recursion we try to match the new problem (2) with the initial one (1) . The matching fails because E cannot be matched simultaneously with E in *P(e,E,X) and E\e' (in E\e')*.

So we introduce a new generalized problem (see Wegbreit (9), Kodratoff (6)) :

F(E,X)←G(E,E,X)

G is a problem with specifications :

```
Proc :  (e,b)←G(E1,E2,X)
        Find e in E1 such that P(e,E2,X)          (3)
endproc.
```

In the same way as for the problem F we arrive at :

```
Proc :  (e,b)←G(E1,E2,X)
        If E1=∅ then b←false ; return ; endif ;
        choose e' from E1 ;
        If I(e≠e') then Find e in E1\e' such that P(e,E2,X)    (4)
                  else (e,b)←(e', true)
        endif ;
endproc.
```

Now matching the problem (3) and (4) succeeds and gives the program G1 :

```
Proc :  (e,b)←G(E1,E2,X)
        If E1=∅ then b← false ; return ; endif ;
        choose e' from E1 ;
        If I(e≠e') then (e,b)←G(E1\e',E2,X)
                  else (e,b)←(e', true)
        endif ;
endproc
```

But this program cannot be executed because of the unpleasant clause I(e≠e') where e is unknown. As Bibel does, we can be tempted to transform the program G1 into the following program G2 by a "permutation" between the if condition and the recursive call. (b is essential to a correct evaluation of the condition I(e≠e')).

```
Proc : (e,b)←G(E1,E2,X)
       If E1=∅ then b←false ; return ; endif ;
       choose e' from E1 ;
       (e,b)←G(E1\e',E2,X) ;
       If b∧I(e≠e') then (e,b)←(e,b)
                    else (e,b)←(e', true)
       endif ;
endproc
```

In fact such a transformation requires some strong justifications. Let $\bar{e}$ be an element which satisfies P i.e $P(\bar{e}, E_2, X)$ = true ; the program G1 is equivalent to the program G3.

```
Proc : (e,b)←G(E1,E2,X)
       If E1=∅ then b←false ; return ; endif ;
       choose e' from E1 ;
       (e1,b)←G(E1\e', E2,X) ;
       If b∧I(ē≠e') then (e,b)←(e1,b)
                    else (e,b)←(e', true)
       endif;
endproc.
```

Note here that we don't know if we have $P(e_1, E_2, X)$ = true. In order to eliminate the unknown $\bar{e}$ we are tempted to rewrite $I(\bar{e} \neq e')$ by $I(e_1 \neq e')$. It can be proven that the following conditions are necessary for justifying this transformation.

(a) : $\forall \bar{e}\ (P(\bar{e},E,X) \Rightarrow \forall e'\ (I(\bar{e} \neq e') \equiv \neg P(e',E,X)))$

(b) : $\forall \bar{e}\ (P(\bar{e},E,X) \Rightarrow \forall e'\ (I(e' \neq \bar{e}) \equiv \text{false}))$

Let G4 be the program :

```
Proc : (e,b)←Ḡ(E1,E2,X)
       If E1=∅ then b←false ; return ; endif ;
       choose e' from E1 ;
       (e1,b1)←Ḡ(E1\e',E2,X) ;
       If b1∧I(e1≠e') then (e,b)←(e1,b1)
```

else(e,b)←(e', true)

endif ;

endproc.

If (a) and (b) are true, let us prove the following property

$R : \exists \bar{e} \in E_1 : P(\bar{e},E_2,X) = \text{true} \Rightarrow b = \text{true} \wedge P(e,E_2,X) = \text{true}$

i.e. if the problem has at least one solution $\bar{G}$ ends with a true value for b and with one solution of the problem : e(which is not necessarily equal to $\bar{e}$).

The proof is done by structural induction on $|E_1|$.

α) Assume that $|E_1| = 1$

E_1 has only one element e' : $E_1=\{e'\}$

If $\exists \bar{e} \in E_1 : P(\bar{e},E_2,X) = \text{true}$ so $\bar{e} = e'$

$\bar{G}(E_1 \backslash e',E_2,X)$ is equivalent to $\bar{G}(\emptyset,E_2,X)$ so G4 returns $b_1 = \text{false}$.

$b_1 = \text{false} \Rightarrow b_1 \wedge I(e_1 \neq e') = \text{false} \Rightarrow (e,b) \leftarrow (e', \text{true}) = (\bar{e}, \text{true})$

b = true and e is solution of the problem.

β) Induction hypothesis : R is true for each E'_1 such that $|E'_1| < |E_1|$

If $|E_1| > 1$: e' can be chosen and we have two cases :

First case : $P(e',E_2,X) = \text{true}$ (b) $\Rightarrow$ $I(e_1 \neq e') = \text{false}$

so $b_1 \wedge I(e_1 \neq e') = \text{false} \Rightarrow (e,b)=(e',\text{true})$.

Second case : $P(e',E_2,X) = \text{false}$

$P(e',E_2,X) = \text{false} \Rightarrow e' \neq \bar{e} \Rightarrow \exists \bar{e} \in E_1 \backslash e' : P(\bar{e},E_2,X)=\text{true}$

By induction hypothesis we have $b_1 = \text{true}$ and $P(e_1,E_2,X) = \text{true}$.

Applying (a) gives $b_1 \wedge I(e_1 \neq e') \equiv \neg P(e',E_2,X)$.

We are in the case : $P(e',E_2,X) = \text{false}$ so $b_1 \wedge I(e_1 \neq e') = \text{true}$

G4 returns $(e,b) \leftarrow (e_1,b_1) = (e_1, \text{true})$ with $P(e_1,E_2,W)$ which ends the proof.

$(e,b) \leftarrow \bar{G}\ (E,E,X)$ where $\bar{G}(E_1,E_2,X)$ is synthesized by the program G4.

Notes : It means that when G1 ends with b=true and e such that $P(e,E_2,X) = \text{true}$: G4 ends with a good solution (the same). On the other hand when the problem has no solution, G1 ends with b = false but G4 gives an unpredictible result(with b = true).

Then if we assume that E_1 is not empty and that the problem has at least one solution, $\bar{G}$ can be validly transformed into :

```
Proc :  e← Ḡ(E1,E2,X)
        If E1={e'}  then e←e' ; return ; endif ;
        choose e' from E1 ;
        e←Ḡ(E1\e',E2,X) ;
        If ¬I(e≠e') then e←e' endif ;
endproc
```

It is fundamental to note that in the final program the property P no longer appears. It means that the only relevant information for the problem is given by the condition I(e≠e'). So it seems to be useless to specify completely a problem which satisfies the general scheme.

Find e in E such that P(e,E,X)

Example : If we want to synthesize the well known MAX program, it is clear that a "good" I(e≠e') condition is e' < e (verification of ⓐ and ⓑ properties is left to the reader).

The MAX program becomes :

```
Proc :  e← MAX(E1,E2)
        If E1={e'} then e← e' ; return  ; endif ;
        choose e' in E1 ;
        e←MAX(E1\e',E2) ;
        If e≤ e' then e←e' endif
endproc
```

This program is similar to that obtained by Bibel and can be easily implemented without any recursion (we also note that E_2 is useless and can be deleted from the program).

III - SYNTHESIS OF : *Find Eo in E such that P(Eo,E,X)*

```
Proc : Eo←F(E,X)
         Find Eo in E such that P(Eo,E,X)
Endproc
```

We can obtain Eo by constructing it element by element. By choosing one element the program above becomes.

```
Proc :  Eo←F(E,X)
        If E=∅  then Eo←∅ ; return ; endif ;
        choose e from E ;
        If  I(e∉Eo) then Find Eo in E\e such that P(Eo,E,X)
                    else Find Eo\e in E\e such that P(Eo,E,X)∪e
        endif
endproc
```

I(e∉Eo) is the interpretation of the condition : e ∉Eo.

If we try to match the two problems *Find Eo in E such that P(Eo,E,X)* and *Find Eo in E\e such that P(Eo,E,X)* the matching fails becauses E cannot be matched simultaneously with E and E\e. As with the first example we can introduce a generalized problem G with two variables E_1 and E_2. Thus *Find Eo in* E_1*\e such that P(Eo,*E_2*,X)* is achieved by a recursive call G(E_1\e,E_2,X).

But the matching of *Find Eo\e in* E_1*\e such that P(Eo,*E_2*,X)* with the generalized problem G *Find Eo in* E_1 *such that P(Eo,*E_2*,X)* still fails because Eo cannot be matched with Eo and Eo\e. The generalization we used so far cannot be applied here because it applies only to input variables. Here Eo is an output variable so we use a new generalization underlined in Bibel in the "Minimum spanning tree" example :

Generally, let us assume that in the synthesis of a program we are matching two expressions (1) and (2) which are :

(1) $E_1[x\ ;\ x]$

(2) $E_2[t\ ;\ h(t,u)]$

The matching fails because x has to be matched with two different terms : t and h(t,u).

We generalize by replacing the second occurence of x with h(x,y).

(1) and (2) become (1') $E_1[x\ ;\ h(x,y)]$

(2') $E_2[t\ ;\ h(h(t,u),\ y)]$

If h is associative and admits an identity element ε (2') gives (2").

(2") $E_2[t\ ;\ h(t,h(u,y))]$

Now the matching of (1') and (2") succeeds and gives the following substitutions : $x \to t$; $y \to h(u,y)$.

The initial call is obtained by :

(1)= $E_1[x ; h(x,\varepsilon)]$

For our example expressions (1) and (2) are :

(1) *Find Eo in E_1 such that $P(Eo,E_2,X)$*

(2) *Find Eo\e in E_1\e such that $P(Eo,E_2,X)$*

Eo has to be matched with Eo\e and Eo but (2) corresponds to the case : $e \in Eo$, so Eo can be written Eo\e ∪e.

The set union ∪ is an associative law and admits an identity element ∅ (the empty set). The generalization previously discussed can be applied (h=∪ , x=Eo, t = Eo\e, u = e, y = E_3).

(1') *Find Eo in E_1 such that $P(Eo \cup E_3, E_2,X)$*

(2') *Find Eo\e in E_1\e such that $P((Eo\backslash e \cup e) \cup E_3,E_2,X)$*

Applying the associativity of ∪ gives :

(2'') *Find Eo\e in E_1\e such that $P(Eo\backslash e \cup (e \cup E_3), E_2, X)$*

and the matching of (1') and (2'') succeeds with the substitutions :

$Eo \to Eo\backslash e$; $E_1 \to E_1\backslash e$; $E_2 \to E_2$; $E_3 \to e \cup E_3$; $X \to X$

If we call the generalized problem (1') : $H(E_1,E_2,E_3,X)$ then (2'') is achieved by a recursive call $H(E_1\backslash e, E_2, e \cup E_3, X)$.

In the case $e \notin Eo$ we have to match (1') with the problem (3) :

(3) *Find Eo in E_1\e such that $P(Eo \cup E_3,E_2,X)$*

The matching succeeds with the substitutions :

$Eo \to Eo$; $E_1 \to E_1\backslash e$; $E_2 \to E_2$; $E_3 \to E_3$; $X \to X$

(3) is achieved by the recursive call H $(E_1\backslash e, E_2, E_3, X)$.

In conclusion we have : $Eo \leftarrow H(E_1,E_2,\emptyset,X)$

```
Proc :  Eo←H(E1,E2,E3,X)
        If E1=∅  then Eo← ∅ ; return   endif ;
        choose e in E1 ;
```

```
        If I(e∉Eo) then Eo←H(E1\e, E2,E3,X)
                   else Eo←H(E1\e,E2, e∪E3, X)∪e
        endif
endproc
```

As with the first example the property P no longer appears and the relevant information is given by expliciting the condition $I(e \notin Eo)$. That is particularly significant in the Minimum Spanning tree example of Bibel where the problem is given by :

$$P(E_o,E,c,N) \left\{ \begin{array}{l} \textit{Find Eo in E such that} \\ \forall\, c(G(N,E,c) \rightarrow Eo \subseteq E \wedge T(N,E) \\ \wedge\ \forall\, E'_o(E'_o \subseteq E \wedge T(N,E'_o) \rightarrow C(E_o) \leq c(E'_o))) \end{array} \right.$$

Where $G(N,E,c)$ denote the property that N, E is a graph with a cost function c ; $T(N,E)$ denote the property of N and E forming a tree (N is a set of nodes) and $c(E) = \sum_{e \in E} c(e)$

Then Bibel arrives to the same program H like ours and gives a completely new specification for $I(e \notin Eo)$. Specifying completely the problem at first was in fact useless.

CONCLUSION

In conclusion our method, with a few new ideas, seems to conduct to a system which could be more easily implemented than that of Bibel.

We also pointed that specifying completely the problem at first is sometimes useless. In fact an automatic system could infer the condition $I(e \notin Eo)$ from the specification in some simple problems. In some more difficult problems like the linear pattern matching of Bibel, a major improvement of the algorithm can be done by just looking some relations inside the property P. All that proves that some more work has to be done on the way of specifying problems.

REFERENCES

1 M. BIDOIT, Ch. GRESSE, G. GUIHO : "A system which synthesizes array-manipulating programs from specifications". Proc. 6th IJCAI, p. 63-65, Tokyo 1979.

2 W. BIBEL : "Syntax-directed, semantic - supported program synthesis". proc. 4th Workshop on Automated Deduction, p. 140-147, Austin (Texas) 1979.

3 GREEN. C. Cordell: "A summary of the PSI program synthesis system". Proc. 5th IJCAI. Cambridge 1977. P. 380-381.

4 J. GUTTAG, J. HORNING : "The Algebraic specification of abstract data types". Acta Informatica 10, 1978.

5 D.J. LEHMAN, M.A. SMITH : "Abstract data types : a synthetic approach". Technical Note, Warwick, 1977.

6 Y. KODRATOFF : "Choix d'un programme LISP correspondant à un exemple". Congrès AFCET, Chatenay-Malabry (France), 1978.

7 Z. MANNA, R. WALDINGER : "Synthesis : Dreams→Programs". Technical Note 156. SRI, 1977.

8 Z. MANNA, R. WALDINGER :"A deductive approach to program synthesis". Proc. 6th IJCAI, p. 542-551, Tokyo 1979.

9 B. WEGBREIT : "Goal directed program transformations". IEEE Trans. on Software Eng., vol. 2, N°2, 1976.

A SYSTEM FOR PROVING EQUIVALENCES OF RECURSIVE PROGRAMS

Laurent KOTT

UER de Mathématiques, Université Paris VII

L.A. du CNRS 248 "Informatique Théorique et Programmation"

2, pl. Jussieu F-75221 Paris Cedex 05

Abstract We present a system for proving equivalences of recursive programs based on program transformations, namely the fold/unfold method and a generalisation of this method.

1. Introduction

The last years saw the birth of an original and fruitful system of recursive programs transformations originated by R. Burstall and J. Darlington [5] and based on the so-called fold/unfold method. We have studied [11,12] after the authors the correctness of this system.

We say a transformation, starting from a recursive program P_1 into another one P_2, is correct if and only if P_1 and P_2 are equivalent. So, if we want prove the equivalence of P_1 and P_2, it suffices to exhibit a correct transformation from P_1 (resp. P_2) to P_2 (resp. P_1). We realize that any transformations system (especially the Burstall-Darlington one) is a system for proving equivalence of recursive programs.

Furthermore, proofs are made without any recursion induction principle but by a sequence of identities; each identity corresponds to one rule of the transformation system.

Here, we present two proof methods without any induction principle (see also B. Courcelle and G. Huet [6,10]):

- the first one is based on the Burstall-Darlington system,
- the second is based on a generalisation of the fold/unfold method, called second order replacement method (see [13]).

By combining these two methods we could design a system for proving equivalence of recursive programs.

2. Informal presentation

The best way to give an informal presentation is to present an example. Let us consider the two following recursive programs:

(1) F(m,n) = <u>if</u> m=0 <u>then</u> n <u>else</u> F(m-1,n) + 1

(2) G(m,n) = <u>if</u> m=0 <u>then</u> n <u>else</u> G(m-1,n+1)

Both give a recursive definition of the addition over positive integers. If we want prove the equivalence of (1) and (2) it is quite natural to try to show the equivalence between the terms $F(m-1,n)+1$ and $F(m-1,n+1)$. Let us assume this equivalence holds, then we may replace the subterm of the right handside of (1) - $F(m-1,n)+1$ - by the equivalent term $F(m-1,n+1)$; whence we reach a new program

$$(3)\quad F(m,n) = \underline{\text{if}}\ m=0\ \underline{\text{then}}\ n\ \underline{\text{else}}\ F(m-1,n+1)$$

but (2) and (3) are identical (up to the name of the function symbol) and, from this identity, we deduce the equivalence of (1) and (2).

This is a first method for proving equivalence of recursive programs, called second order replacement method and studied in [12,13].

As mentionned above the proof of this equivalence requires a lemma: the equivalence between the terms $F(m-1,n)+1$ and $F(m-1,n+1)$ or, more generally, between the terms $F(m,n)+1$ and $F(m,n+1)$. For this purpose let us introduce two new function symbols and two recursive programs

$$(4)\quad \begin{cases} H(m,n) = F(m,n) + 1 \\ F(m,n) = \underline{\text{if}}\ m=0\ \underline{\text{then}}\ n\ \underline{\text{else}}\ F(m-1,n) + 1 \end{cases}$$

and

$$(5)\quad \begin{cases} K(m,n) = F(m,n+1) \\ F(m,n) = \underline{\text{if}}\ m=0\ \underline{\text{then}}\ n\ \underline{\text{else}}\ F(m-1,n) + 1 \end{cases}$$

Now we use the fold/unfold method of transformations introduced by R. Burstall and J. Darlington (see also [14]). This example allows us to recall the basic notations of fold/unfold transformations. Starting with the first equation of (4) we perform the following transformations

$$\begin{array}{ll} H(m,n) = F(m,n) + 1 & (\text{ definition }) \\ H(m,n) = (\ \underline{\text{if}}\ m=0\ \underline{\text{then}}\ n\ \underline{\text{else}}\ F(m-1,n) + 1\ ) + 1 & (\text{ unfold }) \\ H(m,n) = \underline{\text{if}}\ m=0\ \underline{\text{then}}\ n+1\ \underline{\text{else}}\ (\ F(m-1,n) + 1\ ) + 1 & (\text{ law }) \\ H(m,n) = \text{if}\ m=0\ \text{then}\ n+1\ \text{else}\ H(m-1,n) + 1 & (\text{ fold }) \end{array}$$

and we obtain the new program

$$(6)\quad \begin{cases} H(m,n) = \underline{\text{if}}\ m=0\ \underline{\text{then}}\ n+1\ \underline{\text{else}}\ H(m-1,n) + 1 \\ F(m,n) = \underline{\text{if}}\ m=0\ \underline{\text{then}}\ n\ \underline{\text{else}}\ F(m-1,n) + 1 \end{cases}$$

Now from the first equation of (5) we perform

$$\begin{array}{ll} K(m,n) = F(m,n+1) & (\text{ definition }) \\ K(m,n) = \underline{\text{if}}\ m=0\ \underline{\text{then}}\ n+1\ \underline{\text{else}}\ F(m-1,n+1) + 1 & (\text{ unfold }) \\ K(m,n) = \underline{\text{if}}\ m=0\ \underline{\text{then}}\ n+1\ \underline{\text{else}}\ K(m-1,n) + 1 & (\text{ fold }) \end{array}$$

and we reach another new program

$$(7)\quad \begin{cases} K(m,n) = \underline{\text{if}}\ m=0\ \underline{\text{then}}\ n+1\ \underline{\text{else}}\ K(m-1,n) + 1 \\ F(m,n) = \underline{\text{if}}\ m=0\ \underline{\text{then}}\ n\ \underline{\text{else}}\ F(m-1,n) + 1 \end{cases}$$

Once again we realize the identity between the programs (6) and (7) up to the names of function symbols. What did we do? By means of the fold/unfold method we transform the programs (4) and (5) in the same third program, say (6). But programs (4) and (6) are equivalent, and programs (5) and (6) are equivalent too; so are the programs (4) and (5).

Finally we have proved the equivalence between H(m,n) and K(m,n) or, in others words between F(m,n)+1 and F(m,n+1). We shall call this method the Mac Carthy method [3,12] because it is an effective way to perform the old recursion induction principle of Mc Carthy. At the end of this informal presentation we should like to say our system is a combination of those two methods.

3. The algebraic theory of recursive program schemes

The theoretical framework used here is the algebraic theory of recursive program schemes initiated by M. Nivat and developped by B. Courcelle, I. Guessarian and by M. Nivat himself [7,9,16]. In the sequel we have three fixed sets:
V the set of variables, A the set of base function symbols, each symbol f of A has an arity a(f), F the set of unknown function symbols equal to $\{F_1,F_2,\ldots,F_N\}$, each F_i has an arity a_i.

We note M(A,V) (resp. M(A ∪ F,V)) the set of finite trees (or terms) well formed on A ∪ V (resp. A ∪ F ∪ V) with respect to arities ($CT_A(V)$ in [1]). Adding to A a distinguished element Ω, with null arity, we define on $M_\Omega(A,V)$ (abbreviation of M(A ∪ {Ω},V)) the syntactic order by

(i) $t \in M_\Omega(A,V)\ \Omega < t$

(ii) $f \in A,\ t_1,\ldots,t_{a(f)},t'_1,\ldots,t'_{a(f)} \in M_\Omega(A,V)$

$$f(t_1,\ldots,t_{a(f)}) < f(t'_1,\ldots,t'_{a(f)}) \leftrightarrow \forall i\ 1 \le i \le a(f)\ t_i < t'_i$$

Recursive Program Schemes (RPS)

On M(A ∪ F,V) a RPS is a system of equations like

$$S \begin{cases} F_i(v_1,\ldots,v_{a_i}) = t_i,\ t_i \in M(A \cup F,\{v_1,\ldots,v_{a_i}\}) \\ 1 \le i \le N \end{cases}$$

S may be considered as a tree grammar over $M_\Omega(A,V)$ if we add the terminal rules

$$\begin{cases} F_i(v_1,\ldots,v_{a_i}) = \Omega \\ 1 \le i \le N \end{cases}$$

We note L(S,t) the language (subset of $M_\Omega(A,V)$) generated by S with the axiom t and $\overset{*}{\underset{S}{\to}}$ the rewriting rule associated to the grammar. A result of M. Nivat states [15] that L(S,t) is a directed subset of $M_\Omega(A,V)$ i.e. for any pair (t,t') of L(S,t) there exists t" of L(S,t) such that t < t" and t' < t".

Interpretations

An interpretation I is an object (D_I, $<_I$, $\{f_I | f \in A\}$) such that

(i) (D_I, $<_I$) is a partial ordered set with a least element ω_I;

(ii) for any f of A, f_I is a monotonous function from $D_I^{a(f)}$ to D_I. Ω_I is

(iii the constant function ω_I;

(iii) any directed subset of D_I has a least upper bound;

(iv) for any f of A, f_I is continuous i.e. preserves the upperbound of directed subsets.

With a given interpretation it is easy to associate (by structural induction) to any t of $M(A,\{v_1,\ldots,v_n\})$ a function, noted t_I, from D_I^n to D_I. Then the function computed by the RPS S with the axiom t (t in $M(A,\{v_1,\ldots,v_n\})$) under the interpretation I is noted $(S,t)_I$ and defined by $(S,t)_I(d_1,\dots,d_n) = \mathrm{Sup}\{t_I(d_1,\dots,d_n) \mid t \in L(S,t)\}$ for any element $(d_1,\dots,d_n)$ in D_I^n.

Class of interpretations

Let be C a class of interpretations, S and S' two RPSs, t and t' two elements of $M(A \cup F,V)$, we define the C-equivalence and the C-inequality by

$$(S,t) =_C (S',t') \leftrightarrow \forall I \in C\ (S,t)_I = (S',t')_I$$
$$(S,t) <_C (S',t') \leftrightarrow \forall I \in C\ (S,t)_I <_I (S',t')_I$$

A class C is said algebraic if and only if the inequality $(S,t) <_C (S',t')$ is valid iff for any s of L(S,t) there exists s' in L(S',t') such that $s <_C s'$.

In this paper we consider only algebraic classes. A very usefull kind of algebraic class is relationnal class. Let be R a subset of the cartesian product $M_\Omega(A,V) \times M_\Omega(A,V)$, we note C_R the class of all the interpretations satisfying R; Whence C_R is the set $\{I \mid \forall (s,s') \in R\ s_I <_I s'_I\}$. As usual we note $\overset{*}{\underset{R}{\dashrightarrow}}$ the rewriting relation over $M_\Omega(A,V)$ generated by R considered as a system of rewriting rules.

If $<_R$ is the reflexive and transitive closure of $< \cup \overset{*}{\underset{R}{\dashrightarrow}}$, a result of I. Guessarian [9] states

$$(S,t) <_{C_R} (S',t') \leftrightarrow \forall s \in L(S,t)\ \exists s' \in L(S',t')\ s <_R s'$$

So, in the sequel, we shall write $<_R$ instead of $<_{C_R}$.

Inequality complexity

Let be S a RPS and t an element of $M(A \cup F,V)$, we note $\{\alpha_S^n t \mid n \in N\}$ the sequence of terms belonging to $M_\Omega(A,V)$ deduced from t by using the parallel outermost computation rule. In the same way we note $\{\sigma_S^n t \mid n \in N\}$ the sequence obtained by using the full substitution. We can rephrase the above result of I. Guessarian in

$$(S,t) <_R (S',t') \leftrightarrow \forall n\ \exists n'\ \alpha_S^n t <_R \alpha_{S'}^{n'} t'$$

or $$(S,t) <_R (S',t') \leftrightarrow \forall n\ \exists n'\ \sigma_S^n t <_R \sigma_{S'}^{n'} t'$$

In [4] we express n' as a function of n, and we write

$$(S,t) <_R (S',t') \leftrightarrow \exists f,\ \text{function from } \mathbb{N} \text{ to } \mathbb{N},\ \alpha_S^n t <_R \alpha_{S'}^{f(n)} t'$$

Then we could say the C-inequality between (S,t) and (S',t') is linear (resp. polynomial, exponential, recursive,...) as soon as the function f is a linear (resp. polynomial, exponential, recursive,...) function from $\mathbb{N}$ to $\mathbb{N}$. In the same way we can define the complexity of a C-equivalence between (S,t) and (S',t') as the maximum of the complexity of $(S,t) <_C (S',t')$ and $(S',t') <_C (S,t)$. We show in [4] there exists equivalence with recursive complexity.

4. The Mac Carthy method

Let us recall the three basic rules of the Burstall-Darlington system for transforming recursive program schemes.

Definition 1 Let be S a RPS, R a set of rewriting rules over $M_\Omega(A,V)$, t and t' two elements of $M(A \cup F,V)$; we get t' from t

(1) by unfolding iff $t \overset{*}{\underset{S}{\dashrightarrow}} t'$

(2) by folding iff $t \overset{*}{\underset{S}{\dashleftarrow}} t'$ (or $t' \overset{*}{\underset{S}{\dashrightarrow}} t$)

(3) by laws iff $t \overset{*}{\underset{R}{\dashrightarrow}} t'$

Definition 2 Let be S and S' two RPSs, R a set of laws; we say that we transform S into S' by a fold/unfold transformation if and only if

(1) S is $\{F_i(v_1,..,v_{a_i}) = t_i,\ 1 \le i \le N\}$
S' is $\{F_i(v_1,..,v_{a_i}) = t'_i,\ 1 \le i \le N\}$

(2) $\forall i\ 1 \le i \le N\ t_i\ (\overset{*}{\underset{S}{\dashrightarrow}} \cup \overset{*}{\underset{R}{\dashrightarrow}} \cup \overset{*}{\underset{S'}{\dashrightarrow}})\ t'_i$

We note FU(R) the set of transformations using the set of laws R and we write $S \underset{FU(R)}{\Longrightarrow} S'$ to say that S is transformed into S' by a transformation of FU(R).

Definition 3 Let be S and S' two RPSs such that $S \underset{FU(R)}{\Longrightarrow} S'$, the transformation is said correct if and only if the equivalence $S =_R S'$ holds.

In [11,12] we have studied the correctness of the Burstall-Darlington system and given several sufficient conditions, easily checkable by the system itself, which the correctness of performed transformations.

Now we can state the Mc Carthy method: let be S_1 and S_2 two RPSs such that

$$S_1 \begin{cases} F_i(v_1,\ldots,v_{a_i}) = t_i^1 \\ 1 \le i \le N \end{cases} \quad \text{and } S_2 \begin{cases} F_i(v_1,\ldots,v_{a_i}) = t_i^2 \\ 1 \le i \le N \end{cases}$$

if there exists a third RPS

$$S_3 \begin{cases} F_i(v_1,\ldots,v_{a_i}) = t_i^3 \\ 1 \le i \le N \end{cases}$$

such that $S_1 \underset{FU(R)}{\Longrightarrow} S_3$ and $S_2 \underset{FU(R)}{\Longrightarrow} S_3$ and, if those transformations are correct, then the equivalence $S_1 <_R S_2$ holds.

Example 1 $V = \{u\}$, $A = \{f,g\}$, where $a(f)=a(g)=1$, $F = \{F_1,F_2\}$, where $a_1=a_2=1$. Let be

$$S_1 \begin{cases} F_1(u) = g(F_1(u)) \\ F_2(u) = g(F_1(u)) \end{cases} \quad \text{and } S_2 \begin{cases} F_1(u) = g(F_1(u)) \\ F_2(u) = f(F_2(u)) \end{cases}$$

and $R = \{<f(g(u)),g(g(f(u)))>,<f(\Omega),g(\Omega)>\}$. If S_3 is the RPS

$$\begin{cases} F_1(u) = g(F_1(u)) \\ F_2(u) = g(g(F_2(u))) \end{cases}$$

we have $S_1 \underset{FU(R)}{\Longrightarrow} S_3$ and $S_2 \underset{FU(R)}{\Longrightarrow} S_3$ because

$$g(F_1(u)) \overset{*}{\underset{S_1}{\dashrightarrow}} g(g(F_1(u))) \overset{*}{\underset{S_1}{\dashrightarrow}} g(g(g(F_1(u)))) \overset{*}{\underset{S_1}{\dashleftarrow}} g(g(F_2(u))) \quad \text{and}$$

$$f(F_1(u)) \overset{*}{\underset{S_2}{\dashrightarrow}} f(g(F_1(u))) \overset{*}{\underset{R}{\dashrightarrow}} g(g(f(F_1(u)))) \overset{*}{\underset{S_2}{\dashleftarrow}} g(g(F_2(u)))$$

The results of [11] ensure us the corrctness of these two transformations, thus we de- the equivalence $S_1 =_R S_2$ and, more precisely, the equivalence $f(F_1(u)) =_R g(F_1(u))$.

This method was stated in [12] and also mentionned in [6] ; it was studied more deeply by D. Begay and L. Kott [3].

Proposition [12] The Mc Carthy method is incomplete; that is to say it is impossible to prove all the equivalences with this method.

The proof is a direct consequence of the fact that fold/unfold method perform only linear equivalences between RPSs.This consideration leads us to introduce a generalisation of the Burstall-Darlington transformations which is more powerfull.

5. Second order replacement method

This method is introduced in [12,13] as a generalisation of the fold/unfold method.

Definition 4 Let be S a RPS and C a class of interpretations, the pair (t,t') of elements of M(A ∪ F,V) is a <S,C>-second order replacement (abbreviated in <S,C>-SOR) if and only if (S,t) and (S,t') are C-equivalent.

Definition 5 Let be S a RPS, C a class of interpretations and a set of <S,C>-SORs $\{(s_i,s'_i) | 1 \le i \le N\}$; assume there is in each t_i, the right handside of the ith equation, an instance of s_i, we call t'_i the term obtained by replacing that instance of s_i the analogous instance of s'_i. Then we reach the RPS $\{F_i(v_1,\ldots,v_{a_i}) = t'_i \ , \ 1 \le i \le N\}$. We shall say that S is transformed into S' by the SOR method and write $S \underset{SOR}{====>} S'$.

Example 2 V = {u}, A = {f,g}, where a(f)=a(g)=1, $F = \{F_1\}$, where $a_1=1$. Let be S $\{F_1(u) = g(F_1(u))\}$ and S' $\{F_1(u) = f(F_1(u))\}$ and R the set of laws $\{<f(g(u)),g(g(f(u)))>,<f(\Omega),g(\Omega)>\}$.

Example 1 implies that ($g(F_1(u)),f(F_1(u))$) is a $<S,C_R>$-SOR such that $S \underset{SOR}{====>} S'$ by this $<S,C_R>$-SOR and the equivalence $S \equiv_R S'$ holds.

Example 2 shows us the SOR method is an actual generalisation of the Burstall-Darlington method; indeed it is impossible, starting from S, to reach S' by syntactic manipulations as folding, unfolding or laws. Furthermore we have the following

Proposition 13 The SOR method is complete for the equivalence of recursive program schemes.

The proof is obvious. But it is only a formal proof which does not any way to find out how and why RPSs are equivalent.

In order to deal with the power of the SOR method and the effective touch of the Mc Carthy method we propose the following system for proving equivalences of RPSs.

6. A system for proving equivalence of RPSs

Now we are ready to describe our system. Let be S and S' two RPSs and R a set of laws; assume we would prove the equivalence $S \equiv_R S'$ then

(1) Find some pairs of terms which transform S into S' if they are $<S,C_R>$-SORs; go to (3)

(2) If they do not exist then S and S' are not equivalent and the system halts

(3) (3.1.) Try to prove these pairs are $<S,C_R>$-SORs by the Mc Carthy method

(3.2.) or go to (1)

(3.3.) if (3.1.) and (3.2.) fail then the system fails.

Unfortunately we have no formal result describing the equivalences of RPSs provable by our system; but examples lead us to conjecture that any equivalence provable by some effective recursion induction principle is provable by our system.

However we believe this system is interesting for two reasons:

- the Burstall-Darlington system exists and the Mc Carthy method is implementing; we hope implement the SOR method (following the ZAP system of M. Feather [8]);

- this system, very flexible, is also a system for developping and transforming recursive programs (see also [2,17]).

7. Bibliography

[1] ADJ "Initial algebra semantics and continuous algebras", J. of ACM, 24, 1, pp. 68-95 (1977)

[2] J. Arsac, Y. Kodratoff "Some methods for transformation of recursive procedures into iterative ones", LITP Report, Université Paris VI (1979)

[3] D. Begay, L. Kott "Preuve de programme sans induction", 5th Coll. de Lille, Rapport de l'Université de Poitiers (1980)

[4] G. Boudol, L. Kott "Recursion induction principle revisited", submitted to TCS (1980)

[5] R. Burstall, J. Darlington "A transformation system for developping recursive programs", J. of ACM, 24, 1, pp. 44-67 (1977)

[6] B. Courcelle "Infinite trees in normal form and recursive equations having a unique solution", L.A. 226 Report 79-06, Université de Bordeaux (1979)

[7] B. Courcelle, M. Nivat "Algebraic families of interpretations", 17th FOCS, Houston (1976)

[8] M. Feather " ZAP program transformation system", Ph. D. Thesis, Edinburgh University (1979)

[9] I. Guessarian "Semantic equivalence of program schemes and its syntactic characterisation", 3rd ICALP, Edinburgh (1976)

[10] G. Huet, B. Lang "Proving and applying program transformations with second order patterns", Acta Inf., 11, pp. 31-55 (1978)

[11] L. Kott "About transformation system: a theoretical study" in "Transformations de Programmes", B. Robinet ed., pp. 232-247 (1978)

[12] L. Kott "Des substitutions dans les systèmes d'équations algébriques sur le magma", Doctoral Dissertation, Université Paris VII (1979)

[13] L. Kott "Second order subtree replacements", submitted to 21st FOCS (1980)

[14] Z. Manna, R. Waldinger "Knowledge and reasoning in program synthesis", Artif. Intel. J., 6, 2, pp. 175-208 (1975)

[15] J. Mac Carthy "A basis for a mathamatical theory of computation", in "Computer Programming and Formal Systems" (1963)

[16] M. Nivat "Interprétation universelle d'un schéma de programme récursif", Informatica, 7, 1, pp. 9-16 (1977)

[17] Z. Manna, R. Waldinger "A deductive approach to program synthesis", ACM ToPLaS, 2, 1, pp. 90-121 (1980)

Variable Elimination and Chaining in a Resolution-based Prover for Inequalities

By

W. W. Bledsoe and Larry M. Hines
The University of Texas, Austin

Abstract. A modified resolution procedure is described which has been designed to prove theorems about general linear inequalities. This prover uses "variable elimination", and a modified form of inequality chaining (in which chaining is allowed only on so called "shielding terms"), and a decision procedure for proving ground inequality theorems. These techniques and others help to avoid the explicit use of certain axioms, such as the transitivity and interpolation axioms for inequalities, in order to reduce the size of the search space and to speed up proofs. Several examples are given along with results from a computer implementation. In particular this program has proved some difficult theorems such as: The sum of two continuous functions is continuous.

1. Introduction

The purpose of this paper is to describe a resolution-based theorem prover which has been designed to prove theorems about linear real inequalities. It is an improvement of the prover described in [1] with a different, more powerful, concept of chaining.

An important motive for building special inequality provers is to avoid the explicit use of axioms such as

TRANSITIVITY: $\forall x \ \forall y \ \forall z \ (x \leq y \wedge y \leq z \rightarrow x \leq z)$
INTERPOLATION: $\forall x \ \forall y \ (x < y \rightarrow \exists z \ (x < z < y))$.

Such axioms tend to lengthen the proof search because they can match with other formulas in so many unproductive ways. Also, the explicit use of the field axioms for the real numbers present similar problems.

To avoid these difficulties special "built-in" procedures have been suggested and used with varying degrees of success. Some of these procedures are

(1) the built-in partial ordering of Slagle and Norton [2];
(2) the ground inequality packages of King [3], Oppen, etal. [4], Shostak [5], Bledsoe, etal. [6] (these tend to be in the Presburger mode);
(3) the methods of Hodes [7];
(4) the Restriction Intervals Method [6].

Even though these provers have met with a degree of success, still further changes are necessary to handle more difficult inequality theorems. Two such changes are the inequality chaining and variable elimination described below.

The class of formulas dealt with here are those from the theory of dense linear order without endpoints. This theory is decidable. But, we also permit arbitrary quantification and uninterpreted function symbols and therefore can encode all of predicate calculus. Our intention, however, was not to provide a general-purpose prover for first-order logic but rather to be able to more easily prove naturally arising inequality theorems.

We also permit the + sign with its usual semantics, but we do not, in this version, allow the + sign to occur within the arguments of an uninterpreted function symbol (e.g., $x + f(x_0)$ is allowed but $f(x + x_0)$ is not). (Such cases could be handled by including an algebraic unifier such as [8].)

The strategy used by this prover is to eliminate variables from literals, so that these ground literals can then be split off and proved by a ground inequality prover. However, in order to eliminate a variable from a clause, it must be eligible in that clause (see Section 2.2), so a first objective is to make variables eligible by removing shielding terms (see Section 2.3). This is done by chaining (only) on shielding terms.

The use of this principle in difficult examples like Example 17 (sum of two continuous functions is continuous), Example 13, and others, apparently makes the difference in whether or not a proof will be obtained in reasonable time.

Except for the ground inequality prover, where equality substitution is used [6], we have avoided the use of the equality symbol by substituting $(a \leq b \wedge b \leq a)$ for each instance of $a = b$.

2. Resolution $\leq$

Resolution$\leq$ is much like ordinary resolution [9], except that in addition to the traditional clauses there is a special clause (only one) called TY which is essentially a conjunction of ground inequality literals, and three different types of resolvents are used. These are

- TY-Resolvents
- Variable-elimination Resolvents
- Chain Resolvents

Ordinary resolvents are not explicitly computed. However, they can be implicitly produced by SELF-CHAIN (see below).

2.1 TY-Resolvents

A TY-Resolvent is obtained by conjoining a ground inequality literal with the special clause TY and checking the result for consistency by calling the routine, GROUND-PROVER. If GROUND-PROVER succeeds, then the resolvent is $\square$; otherwise, it is the augmented TY.

2.2 Variable-elimination Resolvents

A literal, $x \leq a$, is called an RL-literal if x is a variable which does not occur in a, and the variable x is called an RL-variable for that literal. This definition is extended to include the cases $x < a$, $a \leq x$, $a < x$, in a similar way. (As an example of a variable which is not an RL-variable, consider the x in $f(x) \leq c$ or $f(x) < x$.)

If a variable x occurs only as an RL-variable in a clause, it is said to be eligible (and can be eliminated from the clause, as we will see shortly).

We will assume the following interpolation axioms.

$$\exists x\,(x \leq a)$$

$$\exists x\,(a \leq x \leq b) \longleftrightarrow a \leq b$$

$$\exists x\,(a \leq x \leq b \wedge x \leq c) \longleftrightarrow a \leq b \wedge a \leq c$$

etc., where x does not occur in a,b,c. We also assume the appropriate modifications of these axioms, such as

$$\exists x\,(a \leq x \leq b) \longleftrightarrow a < b\ ,$$

when some or all of the $\leq$'s are replaced by $<$'s.

We will (implicitly) use these axioms to eliminate eligible variables in clauses.

Variable Elimination Rule

If x is eligible in a clause C and x occurs in C only in the literals

$$(1) \qquad \begin{array}{ll} a_i \not\leq x & ;\quad i = 1,n \\ x \not\leq b_j & ;\quad j = 1,m \end{array}$$

then C is replaced by its "resolvent" C' which is gotten by removing the literals (1) from C and replacing them by the literals

$$a_i \not\leq b_j\ ; \qquad i = 1,n\ ; \qquad j = 1,m\ .$$

It should be noted that if either n or m is zero, then no literal is added to replace those deleted. The rule is extended appropriately to include the symbol "$<$". It should also be noted that C' would have been obtained by resolving C against the clauses from one of the interpolation axioms.

Example. $C = a \not\leq x \vee x \not< b$
$C' = a \not< b\ .$

Example. $C = a \not\leq x \vee b \not< x \vee \ell\ .$
$C' = \ell\ .$

Example. $C = a \not\leq x \vee x \not\leq b \vee f(x) \leq c\ .$
x is not eligible so it cannot be eliminated.

When an RL-variable is not eligible, as in clause C of this last example, the variable cannot be eliminated. However, it might become eligible in a later resolvent, as, for example, when C is resolved with the clause $(f(x') \not\leq c \vee D)$ where x' does not occur in D.

2.3 Shielding Terms

If L is a literal which is equivalent to $t \leq A$ (or $t < A$, $A \leq t$, $A < t$) and t is of the form $f(t_1, t_2, \ldots, t_n)$, where f is an uninterpreted function symbol, and t contains at least one variable, then we say that t is a shielding term of L. In the examples of Section 2.2, the shielding terms are f(x) and f(x'). A shielding literal is one that contains a shielding term.

The chaining procedure below is designed to remove such shielding terms from clauses.

2.4 Chain Resolvents

Chaining is a procedure which effects a limited application of the transitivity axiom

$$\forall xyz\,(x \leq y \wedge y \leq z \rightarrow x \leq z)$$

so that if b and b' are unifiable, with mgu θ, then $(a < c)\theta$ is inferred from $a < b$ and $b' < c$.

The chaining procedure in this paper is applied only when b is a shielding term.

If C and C' are the two clauses

$$A \leq B \vee P$$

and

$$B' \leq C \vee Q\ ,$$

where B or B' is a shielding term, and B and B' are unifiable with mgu θ, then

$$(A \leq C)\theta \vee P\theta \vee Q\theta$$

is called a chain-resolvent of clauses C and C'. (Similarly for $A < B$, $B' < C$, etc.)

For example, f(y) is a shielding term in the clause

$$C = f(\ell) \leq f(y) \vee y < \ell$$

which can be removed by resolving C against the clause

$$f(s) \leq A \vee b < s$$

to obtain the chain-resolvent

$$R = f(\ell) \leq A \vee y < \ell \vee b < y \ .$$

Although y was not eligible in C, it is in R and, therefore, can be eliminated to obtain

$$f(\ell) \leq A \vee b < \ell \ .$$

Another example is found in Example 17 below, where clauses

$$C_1 = f(z_y) + g(z_y) < f(x_0) + g(x_0) + \varepsilon_0 \vee y \leq 0 \ ,$$
$$C_2 = f(x_0) + \varepsilon \leq f(s)$$

are resolved to obtain the chain-resolvent

$$f(x_0) + \varepsilon + g(z_y) < f(x_0) + g(x_0) + \varepsilon_0 \vee y \leq 0 \ .$$

(Actually, in order to use the definition of chain-resolvent as given above, we had to first rewrite C_1 in its equivalent forms $f(z_y) < -g(z_y) + f(x_0) + g(x_0) + \varepsilon_0 \vee y \leq 0$.)

We also allow "self-chaining" resolvents (see SELF-CHAIN) whereby the resolvent $(P\theta)$ is obtained from the clause $(a < a' \vee P)$ if θ is the mgu of a and a'.

2.5 Processing Resolvents

Of the three types of resolvents only one, chain resolvents, are constructed during the regular resolution cycle; the other two, TY-resolvents and variable-elimination resolvents, are produced by processing other resolvents and preprocessing the initial clauses.

A "splitting" procedure is also used (see below) which insures that, among other things, ground literals occur only in unit clauses. This greatly enhances the usefulness of the TY clause.

When a new resolvent, R, is formed:

1. If R is $\square$ the proof is successfully terminated.

2. If R is a unit inequality ground clause it is "resolved" with TY.

3. If R has an eligible variable, that variable is eliminated by the methods of Section 2.2.

4. Otherwise R is simplified and added to the set of clauses (with new standardized apart variables). Ordinary subsumption and tautology removal are used.

5. If R can be split into two or more independent (no variables in common) subclauses $R_1, \ldots, R_n$, the process is continued for each of $S \cup \{R_i\}$, for $i = 1, 2, \ldots, n$.

2.6 Sequencing

At the beginning of a proof, the theorem to be proved is converted to clausal form. All unit ground inequality clauses are "conjoined" together to form the special clause, TY. These unit clauses are also retained as separate clauses. TY is checked for consistency by a call to the function GROUND-PROVER. If it is inconsistent, the proof is successfully terminated.

Also, at the beginning, any variable x that is eligible in a clause is eliminated in that clause by the procedure of Section 2.2. And splitting is performed where possible. (It should be noted at this point that ground clauses can be split completely, and that this causes an excessive amount of splitting when the set S contains only ground clauses. However, this prover is designed to handle difficult non-ground theorems where very little splitting takes place. See the examples of Section 4.)

The procedure is to compute resolvents by a method similar to linear resolution.

A top clause C is chosen from S which is not ground. There is such a C, since otherwise S is a set of unit ground clauses and would have already been processed by GROUND-PROVER. Furthermore, at least one shielding term occurs in a literal L of C.

Loosely speaking, such an L is selected from C and the chaining algorithm is applied to the shielding terms of L. As variables become eligible, they are eliminated. If ground literals arise they are split off and resolved against TY. Also, each new resolvent R is simplified by REDUCE and duplicate literals are merged, and the process is repeated with R as the new top clause.

3. The Principal Parts

(RESOLUTION≤ Th)

This is the top-level function, where Th is the theorem to be proved. It returns T or NIL.

1. Convert Th to clausal form, getting the set S of clauses.

2. Call INITIAL-RL

   This eliminates any eligible variables from the original clauses. Then each clause is REDUCED and then ordered, with RL-literals last. If $\square$ is obtained, the calculation is successfully terminated. The result is a set S of clauses. Subsumption and tautology removal are also used.

3. Call (SPLIT S)

   If L is a literal of a non-unit clause C of S, and L has no variable in common with $C \sim \{L\}$, then call both

$$(\text{SPLIT } (S \sim \{C\} \cup \{\{L\}\}))$$

   and

$$(SPLIT\ (S \sim \{C\} \cup \{C \sim \{L\}\}))\ .$$

4. Call INITIAL-TY
   This constructs the special clause TY (not a member of S). If TY is □ the calculation is successfully terminated.

5. Return (REMOVE-SHIELDING-TERMS S).

Once the splitting of S (if any) is completed in step 3 above, and the TY is initialized, the program trys to produce "chain resolvents" which are processed to produce, in some cases, TY-resolvents and variable elimination resolvents.

The algorithms presented here in the routines REMOVE-SHIELDING-TERMS, CHAIN, and SELF-CHAIN, are given to show one way that these new resolvents can be used in an actual prover. This implementation resembles linear resolution with ordering of literals. Completeness is not claimed.

(REMOVE-SHIELDING-TERMS S)

This is called by RESOLUTION≤ and by PROCESS-RESOLVENT. S has at least one non-ground clause. (Otherwise, GROUND-PROVER would have already handled such a case.)

1. While S ≠ NIL do

1.1 Select a non-ground clause, C from S (a new one not yet chosen). C has at least one shielding literal. (Otherwise C is either ground or any variable occurring in it would be eligible and hence would have been eliminated.) Put S = S ~ {C}. Put C' = C.

1.2 While C' ≠ NIL do

1.21 Select a new shielding literal L from C'. Put C' = C' ~ {L}. Let STR be the set of shielding terms of L. Let LE be the predicate of L (i.e., "<" or "≤"). Let LS be the "left side" of L. (i.e., LS is such that (LE LS 0) is equivalent to L).

1.22 While STR ≠ NIL do

1.221 Select a new shielding term TR from STR. Put STR = STR ~ {TR}.

1.222 Call (CHAIN TR LS LE L C S). If it returns T, return T.

END;

END;

END;

2. Return NIL.

(CHAIN TR LS LE L C S)

This is called by REMOVE-SHIELDING-TERMS. The literals in each clause have already been put into a "left side" form (e.g., $A \leq B$ is transformed to $A - B \leq 0$).

1. Set SS = S ~ {C}.

2. While SS ≠ NIL do

2.1 Let CC be the next clause in SS. Put SS = SS ~ {CC}.

2.2 While CC ≠ NIL do

2.21 Let LL be the next literal in CC. Put CC = CC ~ {LL}. Let TRS be the set of terms in LL which are not variables or numbers. Let LEE be the predicate of LL. Let LSS be the "left side" of LL (i.e., LSS is such that (LEE LSS 0) is equivalent to LL).

2.22 While TRS ≠ NIL do

2.221 Let TRR be the next term in TRS. Put TRS = TRS ~ {TRR}.

2.222 Put θ = (UNIFY TR (- TRR)).
If θ ≠ NIL then

2.2221 Let LE be "≤" if both L and LL have "≤" as their predicates. Else let LE be "<". Let New-L be (LE (+ LSθ LSSθ) 0). Let R be {New-L} ∪ (Rest of C)θ ∪ (Rest of CC)θ. Call (PROCESS-RESOLVENT R). If it returns T, then return T.

END;

END;

END;

3. Return (SELF-CHAIN C L TR).

(SELF-CHAIN C L TR)

Called by CHAIN.

1. Let TRS be the set of terms in L which are not variables or numbers. Put TRS = TRS ~ {TR}.

2. While TRS ≠ NIL do
   2.1 Let TRR be the next term in TRS. Put TRS = TRS ~ {TRR}.
   2.2 Put θ = (UNIFY TR (- TRR)).
   2.3 If θ ≠ NIL and (PROCESS-RESOLVENT Cθ) ≠ NIL then return T. END;

3. Return NIL.

For example, if $C = (f(x)-f(x_0) \leq 0 \vee x \leq a)$, $L = (f(x)-f(x_0) \leq 0)$, $TR = f(x)$, then $R = (f(x_0)-f(x_0) \leq 0 \vee x_0 \leq a)$ which is simplified to $x_0 \leq a$.

(PROCESS-RESOLVENT R)

This is called by the routines CHAIN and SELF-CHAIN, when a new resolvent R has just been produced.

- Put R = (REDUCE R).
- If R = $\square$, return T.
- If R is a ground inequality unit, call (GROUND-PROVER TY R).
- Put R = (ELIMINATE-VARIABLES R).
- If R = $\square$, return T.
- If R is a tautology, return NIL.
- If R can be split on L (i.e., C ~ {L} is not empty and L and C ~ {L} have no variable in common).

  (PROCESS-RESOLVENT {L})

  and

  (PROCESS-RESOLVENT (C ~ {L})).
- (SUBSUME R S).

  Returns NIL if R is subsumed by S, and removes from S clauses subsumed by R.
- Put R = (SORT R).

  Sort the literals of R so that RL-literals are last.
- Replace C by R in S (but leave C in S).
- Return (REMOVE-SHIELDING-TERMS S).

(REDUCE R)

This is a procedure which rewrites certain formulas as others [10]. For example, each of the formulas $(0 < 1)$, $(A+5 \leq A+6)$ is rewritten as T, whereas each of $(2 \leq 1)$, $(f(x)+1 \leq f(x))$, is rewritten as $\square$, and $(2 \leq 1) \vee (A \leq B))$ is rewritten as $((A \leq B))$.

An algebraic simplifier is used in various parts of the program.

THE SPECIAL CLAUSE TY

TY is a conjunction of ground inequality literals which may be altered, as the proof proceeds, by conjoining onto it additional ground inequality units. The initial value of TY is gotten by a call to INITIAL-TY which combines all the ground inequality unit clauses of S into one conjunction. If TY is or becomes contradictory, then the proof is successfully terminated. A function (GROUND-PROVER TY L) is called to determine whether TY is indeed contradictory. If L is not NIL, it is first conjoined onto TY before the determination is made.

GROUND-PROVER is called by INITIAL-TY and called as (GROUND-PROVER TY R) by PROCESS-RESOLVENT in the case when the resolvent R is a ground inequality unit clause. In that case TY is augmented, and this new value of TY is retained in the remainder of the proof. If GROUND-PROVER does not return $\square$, it might infer from TY a set E of equality units (as, for example, would be the case if R was the unit $(\leq A\ B)$ and TY already had the conjunct $(\leq B\ A)$). In this case, these equality units are applied to TY and all of S by a special equals substituting routine.

Any ground inequality package such as those described in [4,5,6] can be used to handle the functions of GROUND-PROVER. Our implementation has used the one described in [6, pp. 7-8].

(ELIMINATE-VARIABLES C)

This is called by INITIAL-RL and PROCESS-RESOLVENT. If the clause C has variables which are eligible in C (see Variable-Elimination Resolvents, Section 2.2), then they are removed from C using the methods of Section 2.2, the resultant clause is returned, and C is discarded.

4. Examples

Here we list some examples which have been proved by our LISP program. In most cases, the prover follows closely the presentation given here, with few non-useful resolvents being generated.

The first few examples are trivial and are listed only to illustrate the methods.

Example 1. $(a < b \rightarrow a \leq b)$

1. $a < b$
2. $b < a$

} original set of clauses

TY: $[a < b,\ b < a]$ added by preprocessing.

Since TY is inconsistent, $\Box$ is obtained during preprocessing and the proof is complete.

Example 2. $(\forall x\ (f(x) \leq c) \rightarrow f(a) \leq c \wedge f(b) \leq c)$

1. $f(x) \leq c$
2. $c < f(a) \vee c < f(b)$.

Preprocessing splits clause 2, getting the two cases 2.1 and 2.2.

Example 2.1.

1. $f(x) \leq c$
2. $c < f(a)$

---

3. $\Box$ 1, 2, a/x

Example 2.2.

1. $f(x) \leq c$
2. $c < f(b)$

---

3. $\Box$ 1, 2, b/x

Example 3. $a \leq b \rightarrow \exists x\ (x \leq a)$

1. $a \leq b$
2. $a < x$

---

3. $\Box$ 2, variable elimination (note that x is eligible in clause 2 and that clause 1 is not used).

Example 4. $(a \leq b \rightarrow \exists x\,(a \leq x \leq b))$

1. $a \leq b$
2. $x < a \vee b < x$

---

3. $b < a$ — 2, variable elimination (x is eligible in clause 2)
4. $\square$ — 1, 3

In this example we omitted writing the special clause TY since it was the single clause 1. The actual procedure is as follows:

TY:	$[a \leq b]$	Preprocessing
3.	$b < a$	2, variable elimination
TY:	$[a \leq b,\ b < a]$, $\square$	process-resolvent, 3.

Example 5. $(\forall x\ \forall y\,(f(x) \leq f(y) \rightarrow x \leq y) \wedge f(a) \leq f(b) \rightarrow a \leq b)$

1. $f(y) < f(x) \vee x \leq y$
2. $f(a) \leq f(b)$
3. $b < a$

TY:	$[b < a,\ f(a) \leq f(b)]$	
4.	$f(a) < f(x) \vee x \leq b$	1, 2 b/y (removing f(y))
5.	$a \leq b$	4, Self-chain
TY:	$[b < a,\ f(a) \leq f(b),\ a \leq b]$, $\square$	process-resolvent, 5.

Notice that we did not resolve upon the literal $x \leq y$ of clause 1 (because it is an RL-literal), but did chain on the shielding term $f(y)$.

Example 6. $(f(\ell) < 0 \wedge 0 \leq f(b) \wedge \ell < c \wedge b \leq \ell$
$\rightarrow \exists y\,[\forall z\,(z \leq b \wedge f(z) \leq 0 \rightarrow z \leq y) \wedge y < \ell])$

1. $f(\ell) \leq 0$
2. $0 \leq f(b)$
3. $\ell < c$
4. $b \leq \ell$
5. $z_y \leq b \vee \ell \leq y$
6. $f(z_y) \leq 0 \vee \ell \leq y$
7. $y < z_y \vee \ell \leq y$

TY:	$[f(\ell) < 0,\ 0 \leq f(b),\ \ell < c,\ b \leq \ell]$	
8.	$y < b \vee \ell \leq y$	7, 5 (removing z_y)
9.	$\ell \leq b$	8, variable elimination
10.	$\square$	9, TY.

Here z_y is a skolem expression, i.e., a skolem function applied to the variable y. We will use this method of representing skolem expressions throughout this paper. The reader can determine which symbols are variables by the quantification in the statement of the theorem.

Note that when clause 9 is added to TY, the program first infers that $b = \ell$ and then uses that to reach the contradiction, $0 \leq f(b) < 0$.

Example 7. $a \le 2 \le b \to \exists x\, (0 \le x \le 5 \wedge a \le x)$

1. $a \le 2$	TY:	$[a \le 2,\ 2 \le b]$	
2. $2 \le b$	4.	$0 \not\le 5 \vee a \not\le 5$	3, variable elimination
3. $0 \not\le x \vee x \not\le 5 \vee a \not\le x$†	5.	$a \not\le 5$	4, REDUCE
	6.	□	5, TY.

Example 8. $(\forall y\, (y < \ell \to \exists z\, (y < z \le b)) \wedge a \le \ell \to a \le b)$

1. $z_y \le b \vee \ell \le y$	TY:	$[a \le \ell,\ b < a]$	
2. $y < z_y \vee \ell \le y$	5.	$y < b \vee \ell \le y$	1, 2 (removing z_y)
3. $a \le \ell$	6.	$\ell \le b$	5, variable elimination
4. $b < a$	7.	□	6, TY.

Example 9. $(\forall \varepsilon\, (0 < \varepsilon \to A \le B + \varepsilon) \to A \le B)$

1. $A \le B + \varepsilon \vee \varepsilon \le 0$ (ε is a variable)	TY:	$[B < A]$	
2. $B < A$	3.	$0 \le B - A$	1, eliminate variable ε
	4.	□	3, TY.

Example 10. $\exists \varepsilon\, [(0 < \varepsilon \to A \le B_\varepsilon + \varepsilon) \wedge B_\varepsilon \le C \to A \le C]$

1. $A \le B_\varepsilon + \varepsilon \vee \varepsilon \le 0$	TY:	$[C < A]$	
2. $B_\varepsilon \le C$	4.	$A \le C + \varepsilon \vee \varepsilon \le 0$	1, 2 (removing B_ε)
3. $C < A$	5.	$A \le C$	4, variable elimination
	6.	□	5, TY.

Example 11. $\exists \varepsilon\, [(0 < \varepsilon \to A_\varepsilon \le B_\varepsilon + \varepsilon) \wedge B_\varepsilon \le C \to A_\varepsilon \le C]$ (Not a Theorem)

1. $A_\varepsilon \le B_\varepsilon + \varepsilon \vee \varepsilon \le 0$
2. $B_\varepsilon \le C$
3. $C < A_\varepsilon$

---

4. $A_\varepsilon \le C + \varepsilon$	1, 2 (removing B_ε)	
5. $C < C + \varepsilon \vee \varepsilon \le 0$	4, 3 (removing A_ε)	(Tautology)
4. $C < B_\varepsilon + \varepsilon \vee \varepsilon \le 0$	1, 3 (removing A_ε)	
5. $C < C + \varepsilon \vee \varepsilon \le 0$	4, 2 (removing B_ε)	(Tautology)

FAILURE.

---

†In this presentation we sometimes use the notation $x \not\le y$ instead of its equivalent $y < x$, and $x \not< y$ instead of $y \le x$, but the program always converts such expressions so that no negations (of inequalities) are used.

Example 12. $[\forall \varepsilon\ (\varepsilon > 0 \rightarrow f(\ell) \le f(z_\varepsilon) + \varepsilon \wedge f(z_\varepsilon) \le f(t_0)) \rightarrow f(\ell) \le f(t_0)]$

1. $f(\ell) \le f(z_\varepsilon) + \varepsilon \vee \varepsilon \le 0$
2. $f(z_\varepsilon) \le f(t_0) \vee \varepsilon \le 0$
3. $f(t_0) < f(\ell)$
4. $f(\ell) \le f(t_0) + \varepsilon \vee \varepsilon \le 0$ — 1, 2 (removing $f(z_\varepsilon)$)
5. $f(\ell) \le f(t_0)$ — 4, variable elimination
6. $\square$ — 5, TY.

TY: $[f(t_0) < f(\ell)]$

In Example 13 and later examples, we often indicate the literal being resolved upon by outlining it with a rectangle, $\boxed{L}$. This is usually the first literal of the top clause after it has been processed and sorted.

Example 13. $[a \le \ell \le b \wedge a \le t_0 < \ell$
$\wedge \forall \varepsilon\ (\varepsilon > 0 \rightarrow \exists r\ (r < \ell \wedge \forall s\ (r \le s \le \ell \rightarrow f(\ell) \le f(s) + \varepsilon)))$
$\wedge \forall y\ (a \le y < \ell \rightarrow \exists z\ (y < z \le \ell \wedge \forall t\ (a \le t < z \rightarrow f(z) \le f(t))))$
$\rightarrow f(\ell) \le f(t_0)]$

1. $a \le \ell$
2. $\ell \le b$
3. $a \le t_0$
4. $t_0 < \ell$
5. $r_\varepsilon < \ell \vee \varepsilon \le 0$
6. $f(\ell) \le f(s) + \varepsilon \vee \varepsilon \le 0 \vee s < r_\varepsilon \vee \ell < s$
7. $y < z_y \vee y < a \vee \ell \le y$
8. $z_y \le \ell \vee y < a \vee \ell \le y$
9. $\boxed{f(z_y) \le f(t)} \vee y < a \vee \ell \le y \vee t < a \vee z_y \le t$
10. $f(\ell) \not\le f(t_0)$

TY: $[a \le \ell,\ \ell \le b,\ a \le t_0,\ t_0 < \ell,\ f(t_0) < f(\ell)]$

11. $f(\ell) \le f(t) + \varepsilon \vee \boxed{z_y \le t} \vee t < a \vee \ell \le y$
$\vee y < a \vee z_y < r_\varepsilon \vee \ell < z_y \vee \varepsilon \le 0$ — 9, 6 (removing $f(z_y)$, z_y/s)
12. $y < t \vee f(\ell) \le f(t) + \varepsilon \vee t < a \vee \ell \le y$
$\vee y < a \vee \boxed{z_y < r_\varepsilon} \vee \ell < z_y \vee \varepsilon \le 0$ — 11, 7 (removing z_y from $(z_y \le t)$)
13. $y < r_\varepsilon \vee f(\ell) \le f(t) + \varepsilon \vee y < t \vee t < a \vee \ell \le y$
$\vee y < a \vee \boxed{\ell < z_y} \vee \varepsilon \le 0$ — 12, 7 (removing z_y from $(z_y < r_\varepsilon)$)
14. $y < r_\varepsilon \vee y < t \vee f(\ell) \le f(t) + \varepsilon$
$\vee t < a \vee \ell \le y \vee y < a \vee \varepsilon \le 0$ — 13, 8 (removing z_y from $(\ell < z_y)$)
15. $\ell \le r_\varepsilon \vee f(\ell) \le f(t) + \varepsilon \vee \ell \le t \vee \boxed{\ell \le a}$
$\vee t < a \vee \varepsilon \le 0$ — 14, variable elimination y

Clause 15 splits on $(\ell \le a)$ into clauses 16.1 and 16.2.

16.1. $\ell \leq a$ 17. $\square$ 16.1, TY

16.2. $\boxed{\ell \leq r_\varepsilon} \vee f(\ell) \leq f(t) + \varepsilon \vee \ell \leq t \vee t < a \vee \varepsilon \leq 0$

17. $f(\ell) \leq f(t) + \varepsilon \vee \ell \leq t \vee t < a \vee \varepsilon \leq 0$ 16.2, 5 (removing r_ε)

18. $\boxed{f(\ell) \leq f(t)} \vee \ell \leq t \vee t < a$ 17, variable elimination ε

19. $\ell \leq t_0 \vee t_0 < a$ 18, 10 (removing $f(t)$)

Clause 19 splits on $(\ell \leq t_0)$ into clause 20.1 and 20.2.

20.1. $\ell \leq t_0$ 20.2 $t_0 < a$

21. $\square$ 20.1, TY   21. $\square$ 20.2, TY.

Examples 14 and 15 arise in the search for counterexamples in a proof.

Example 14. $\exists a \exists b \forall x \exists u ([(x < b \rightarrow u \leq a) \wedge x \leq u] \wedge (u \leq a \vee x \neq b))$

1. $x_0 < b \vee x_0 \not\leq u \vee u \not\leq a$
2.1. $x_0 < b \vee x_0 \not\leq u \vee x_0 \leq b$
2.2. $x_0 < b \vee x_0 \not\leq u \vee b \leq x_0$
3. $u \not\leq a \vee x_0 \not\leq u \vee u \not\leq a$
4.1. $u \not< a \vee x_0 \not\leq u \vee x_0 \leq b$
4.2. $u \not< a \vee x_0 \not\leq u \vee b \leq x_0$

(1–4.2:) a, b, u are variables, x_0 stands for the expression x_{ab}.

Clauses 4.1 and 4.2 are subsumed by clause 3, and clause 2.2 is a tautology.

5. $x_0 < b \vee x_0 \not\leq a$ 1, eliminate u
6.1. $x_0 \leq b$ 2.1, eliminate u
7. $x_0 \not\leq a$ 3, eliminate u
8. $a < b$ 7, 6.1 (removing x_0)
9. $\square$ 8, eliminate a (or b)

Example 15. $\exists a \exists b \exists \ell \exists w \forall s \exists u$

$$(a \leq b \wedge [u \leq 0 \vee s \neq a] \wedge [0 \leq u \vee s \neq b] \wedge [s \leq w \vee 0 < u] \wedge w \leq \ell)$$

The proof of this example, which is like that of Example 14 but longer, shows the power of variable elimination in reducing a messy, but not hard, problem to manageable size.

Example 16. $a''(z_0) < b''(z_0) \wedge \forall u\, (a(u) < b(u) \wedge a'(u) < z_0 < b'(u))$

$$\rightarrow \exists x \exists y \exists z\, (a'(y) < z < b'(y) \wedge (a(x) < y < b(x) \wedge a''(z) < x < b''(z)))$$

1. $\boxed{a'(y) \not< z} \vee z \not< b'(y) \vee a(x) \not< y \vee y \not< b(x) \vee a''(z) \not< x \vee x \not< b''(z)$

2. $a''(z_0) < b''(z_0)$ 4. $a'(u) < z_0$

3. $a(u) < b(u)$ 5. $z_0 < b'(u)$.

x, y, z, and u are variables. Notice that no variable is eligible in clause 1.

6. $\boxed{z \not< b'(y)} \vee a(x) \not< y \vee y \not< b(x) \vee z < z_0$ $\vee a''(z) \not< x \vee x \not< b''(z)$ — 1, 4, y/u, (removing a(y))

7. $a(x) \not< y \vee y \not< b(x) \vee a''(z) \not< x \vee x \not< b''(z)$ $\vee z_0 < z \vee z < z_0$. — 6, 5 (removing b'(y))

Now y is eligible in 7.

8. $\boxed{a(x) \not< b(x)} \vee a''(z) \not< x \vee x \not< b''(z)$ $\vee z_0 < z \vee z < z_0$ — 7, eliminate y

9. $a''(z) \not< x \vee x \not< b''(z) \vee z_0 < z \vee z < z_0$ — 8, 3, x/u (removing b(x))

Now x is eligible in 9.

10. $a''(z) \not< b''(z) \vee z_0 < z \vee z < z_0$ — 9, eliminate x

11. $\square$ — 10, 2.

Example 17. (Sum of two continuous functions is continuous.)

If
$$\lim_{x \to x_0} f(x) = f(x_0) \quad \text{and} \quad \lim_{x \to x_0} g(x) = g(x_0) ,$$
then
$$\lim_{x \to x_0} [f(x) + g(x)] = f(x_0) + g(x_0) ,$$
or
$$\begin{aligned} & \forall \varepsilon (\varepsilon > 0 \to \exists \delta (\delta > 0 \wedge \forall y (|x_0 - y| < \delta \to |f(x_0) - f(y)| \leq \varepsilon))) \\ \wedge & \forall \varepsilon (\varepsilon > 0 \to \exists \delta (\delta > 0 \wedge \forall y (|x_0 - y| < \delta \to |g(x_0) - g(y)| \leq \varepsilon))) \\ \wedge & \quad \varepsilon_0 > 0 \to \exists \delta (\delta > 0 \wedge \forall x (|x_0 - x| < \delta \to |(f(x) + g(x)) - (f(x_0) + g(x_0))| \leq \varepsilon_0)). \end{aligned}$$

In the following we have used $(-E \leq A \wedge A \leq E)$ instead of $(|A| \leq E)$, in order to avoid the use of the absolute value sign.

1. $f(x_\delta) + g(x_\delta) + \varepsilon_0 < f(x_0) + g(x_0) \vee f(x_0) + g(x_0) + \varepsilon_0 < f(x_\delta) + g(x_\delta) \vee \delta \leq 0$
2. $0 < \varepsilon_0$
3. $0 < \delta_\varepsilon \vee \varepsilon \leq 0$
4. $0 < \delta'_{\varepsilon'} \vee \varepsilon' \leq 0$
5. $f(x_0) \leq f(y) + \varepsilon \vee \delta_\varepsilon + x_0 < y \vee \delta_\varepsilon + y < x_0 \vee \varepsilon \leq 0$
6. $f(y) \leq f(x_0) + \varepsilon \vee \delta_\varepsilon + x_0 < y \vee \delta_\varepsilon + y < x_0 \vee \varepsilon \leq 0$
7. $g(x_0) \leq g(y) + \varepsilon' \vee \delta'_{\varepsilon'} + x_0 < y \vee \delta'_{\varepsilon'} + y < x_0 \vee \varepsilon' \leq 0$
8. $g(y) \leq g(x_0) + \varepsilon' \vee \delta'_{\varepsilon'} + x_0 < y \vee \delta'_{\varepsilon'} + y < x_0 \vee \varepsilon' \leq 0$
9. $x_0 \leq x_\delta + \delta \vee \delta \leq 0$
10. $x_\delta \leq x_0 + \delta \vee \delta \leq 0$

---

11. $g(x_\delta) + \varepsilon_0 < g(x_0) + \varepsilon \vee f(x_0) + g(x_0) + \varepsilon_0 < f(x_\delta) + g(x_\delta) \vee \delta \leq 0$ $\underbrace{\vee \delta_\varepsilon + x_0 < x_\delta \vee \delta_\varepsilon + x_\delta < x_0 \vee \varepsilon \leq 0}_{\alpha_1}$

1, 5 x_δ/y (removing $f(x_\delta)$ from $(f(x_\delta) + g(x_\delta) + \varepsilon_0 < f(x_0) + g(x_0))$)

12. $\varepsilon_0 < \varepsilon+\varepsilon' \vee f(x_0) + g(x_0) + \varepsilon_0 < f(x_\delta) + g(x_\delta) \vee \delta \leq 0 \vee \alpha_1$
$\underbrace{\vee\ \delta'_{\varepsilon'} + x_0 < x_\delta \vee \delta'_{\varepsilon'} + x_\delta < x_0 \vee \varepsilon' \leq 0}_{\alpha_2}$

11, 7 x_δ/y (removing $g(x_\delta)$ from $(g(x_\delta) + \varepsilon_0 < g(x_0) + \varepsilon)$)

13. $g(x_0) + \varepsilon_0 < g(x_\delta) + \varepsilon \vee \varepsilon_0 < \varepsilon + \varepsilon' \vee \delta \leq 0 \vee \alpha_1 \vee \alpha_2$

12, 6 x_δ/y (removing $f(x_\delta)$ from $(f(x_0) + g(x_0) + \varepsilon_0 < f(x_\delta) + g(x_\delta))$)

14. $\varepsilon_0 < \varepsilon + \varepsilon' \vee \delta \leq 0 \vee \delta_\varepsilon + x_0 < x_\delta \vee \delta_\varepsilon + x_\delta < x_0 \vee \varepsilon \leq 0 \vee \alpha_2$

13, 8 x_δ/y (removing $g(x_\delta)$ from $(g(x_0) + \varepsilon_0 < g(x_\delta) + \varepsilon)$)

15. $\varepsilon_0 < \varepsilon + \varepsilon' \vee \delta \leq 0 \vee \delta_\varepsilon < \delta \vee \delta_\varepsilon + x_\delta < x_0 \vee \varepsilon \leq 0 \vee \alpha_2$

14, 10 (removing x_δ from $(\delta_\varepsilon + x_0 < x_\delta)$)

16. $\varepsilon_0 < \varepsilon + \varepsilon' \vee \delta \leq 0 \vee \delta_\varepsilon < \delta \vee \varepsilon \leq 0 \vee \delta'_{\varepsilon'} < \delta \vee \delta'_{\varepsilon'} + x_\delta < x_0 \vee \varepsilon' \leq 0$

15, 9 (removing x_δ from $(\delta_\varepsilon + x_\delta < x_0)$)

17. $\varepsilon_0 < \varepsilon + \varepsilon' \vee \delta \leq 0 \vee \delta_\varepsilon < \delta \vee \varepsilon \leq 0 \vee \delta'_{\varepsilon'} < \delta \vee \delta'_{\varepsilon'} + x_\delta < x_0 \vee \varepsilon' \leq 0$

16, 10 (removing x_δ from $(\delta'_{\varepsilon'} + x_0 < x_\delta)$)

18. $\varepsilon_0 < \varepsilon + \varepsilon' \vee \delta \leq 0 \vee \delta_\varepsilon < \delta \vee \varepsilon \leq 0 \vee \delta'_{\varepsilon'} < \delta \vee \varepsilon' \leq 0$

17, 9 (removing x_δ from $(\delta'_{\varepsilon'} + x_\delta < x_0)$)

19. $\varepsilon_0 < \varepsilon + \varepsilon' \vee \delta_\varepsilon \leq 0 \vee \delta'_{\varepsilon'} \leq 0 \vee \varepsilon' \leq 0 \vee \varepsilon \leq 0$

18, variable elimination

20. $\varepsilon_0 < \varepsilon + \varepsilon' \vee 0 < 0 \vee \delta'_{\varepsilon'} \leq 0 \vee \varepsilon' \leq 0 \vee \varepsilon \leq 0$
19, 3 (removing δ_ε)

21. $\varepsilon_0 \leq \varepsilon' \vee \delta'_{\varepsilon'} \leq 0 \vee \varepsilon' \leq 0$ — 20, variable elimination

22. $\varepsilon_0 \leq \varepsilon' \vee \varepsilon' \leq 0$ — 21, 4 (removing $\delta'_{\varepsilon'}$)

23. $\varepsilon_0 \leq 0$ — 22, variable elimination

24. □ — 23, TY.

5. Comments

The Special Clause TY

TY is used to collect together all ground inequality literals. (Because of splitting, ground literals can only occur in unit clauses.) One could get the same effect by not using TY at all but instead collecting together all ground inequality literals each time a new ground inequality is produced as a resolvent and checking for a contradiction. We prefer the TY arrangement because it lets us use the sup-inf procedures of [5] to speedily process ground inequalities. A similar speed advantage can be obtained by the use of the ground inequality package of [4].

In any of these methods, a set of ground equality units might be inferred by TY, and these are applied to TY and S by an equality substitution mechanism.

If one did not use splitting, then a method could be devised whereby special TY-literals (a conjunction of ground inequality literals) would occur in clauses.

Chaining

The chaining we employ is, of course, similar to that used by Slagle and Norton [2], except that we do not chain across variables. In fact, we chain only across shielding terms, (see Sections 2.3, 2.4), thus greatly reducing the search space.

Completeness

This system is not complete. However we believe it will become complete is we add paramodulation [11] to handle equality substitution, and another inference rule whereby the clause $(a = b \vee P \vee Q)\theta$ is inferred from $(a \leq b \vee P)$ and $(b' \leq a' \vee Q)$, if θ is the mgu of $\{b,b'\}$. (This rule is like Slagle and Norton's modification of Rule 3 [2].)

No Added Axioms

Each of the proofs related here was done completely automatically, without human intervention. Also, no additional axioms or lemmas were added or needed, just the statement of the theorem in each case.

Example 17, on the sum of continuous function, was proved automatically several years ago using a special limit heuristic. Also Wos, Overbeek, Lusk and Winker, have proved a simplified version of it with their hyper-resolution program at Argonne Laboratory. But, it appears that this is the first time that a general purpose prover, without special heuristics, has proved this theorem or other inequality theorems of like difficulty.

Limitations

We were surprised by the fact that no non-used clauses (except tautologies which were immediately discarded), were generated in the proof of Example 17, or any of the other proofs in this paper.

However, it would be misleading to claim too much since the family of theorems dealt with includes all of first order logic, it is not surprising that this prover (any prover) will inevitably find difficulty in proving a wide class of theorems. Indeed, many non-used clauses are generated in some recent proofs of other (harder) theorems. So there is much to be done.

Resolution vs Natural Deduction

Resolution is particularly suited for the "variable elimination" method because each clause has its own unique variable. Some other advantages of Resolution are that no substitution needs to be returned from the proof of a subgoal, no backtracking is needed, the clausal data type is uniform and simple, and completeness results are easier to obtain. It remains to be seen whether these advantages offset disadvantages that have been articulated elsewhere, but it seems a safe bet that a well-tailored resolution system will be best for inequality theorems of limited difficulty where human interaction is not required, and it is hoped that such a limited capacity prover can be coded on a mini-computer to work in parallel with and support a larger system.

Acknowledgment

This work was supported by NSF Grant MCS77-20701. We wish to thank Mike Ballantyne and Peter Bruell for their ideas and help.

References

1. W.W. Bledsoe. A Resolution-based Prover for General Inequalities. University of Texas, Math Department Memo ATP-52, July 1979.

2. J.R. Slagle and L. Norton. Experiments with an Automatic Theorem Prover Having Partial Ordering Rules. CACM 16(1973), 683-688.

3. J.C. King. A Program Verifier. Ph.D. Thesis, Carnegie-Mellon University, 1969.

4. Greg Nelson and Derek Oppen. A Simplifier Based on Efficient Decision Algorithms. Proc. 5th ACM Symp. on Principles of Programming Languages, 1978.

5. Robert Shostak. A Practical Decision Procedure for Arithmetic with Function Symbols. JACM, April 1979.

6. W.W. Bledsoe, Peter Bruell and Robert Shostak. A Prover for General Inequalities. University of Texas, Math Department Memo ATP-40A, February 1979. Also IJCAI-79, Tokyo, Japan, August 1979.

7. Louis Hodes. Solving Programs by Formula Manipulation in Logic and Linear Inequality. Proc. IJCAI-71, London, 1971, pages 553-559.

8. M.A. Stickel. A Complete Unification for Associative-Commutative Functions, IJCAI-75, Tbilisi, USSR, 1975, pages 71-76.

9. J.A. Robinson. A Machine-oriented Logic Based on the Resolution Principle. JACM 12(1965), 23-41.

10. W.W. Bledsoe. Splitting and Reduction Heuristics in Automatic Theorem Proving. AEJ 2(1971), 55-77.

11. L. Wos and G.A. Robinson. Paramodulation and Set of Support. Proc. Symp. Automatic Demonstration, Versailles, France. Springer-Verlag, New York, 1968, 276-310.

DECISION PROCEDURES FOR SOME FRAGMENTS OF SET THEORY

A. Ferro, E. G. Omodeo, and J. T. Schwartz
Courant Institute of Mathematical Sciences, New York University

1. INTRODUCTION.

The present note describes several decision techniques applicable to elementary set-theoretic statements of forms likely to occur within extended proof-verifications formulated in any language which allows use of set theoretic constructs. (A proof verifier for such a language is currently being developed at our university). The algorithms to be described can determine the validity of arbitrary collections of statements involving the propositional connectives, the elementary set-boolean operations intersection, union, difference, etc., the set comparisons $a = b$ and $a \subseteq b$, the membership relation $a \in b$, the singleton-former $\{a\}$, and the set cardinality operator $\#a$, together with integer addition, subtraction, and comparison. Two algorithms are presented; the first, which has been implemented in the programming language SETL with some useful optimizations, handles only the restricted case in which neither the singleton-former nor the cardinality operator appears, but is more efficient than the second algorithm. In the last section it is shown that the algorithms presented can be extended to handle statements involving map operators such as domain, range, and the inverse operator f^{-1} together with the special map predicates singlevalued(f) and one-one(f). See Pratt [7] for a statement emphasizing the importance of built-in knowledge of important special cases to the design of broad-spectrum proof verifiers.

2. SYNTAX AND SEMANTICS OF MULTILEVEL SYLLOGISTIC.

The symbols of the language of multilevel syllogistic are:

(1) The parentheses) and (;

(2) a denumerable sequence of *variables* $x_1, x_2, \ldots$;

(3) the *constant* $\emptyset$;

(4) the binary *operators* $\cup$, $\cap$, $-$;

(5) the binary *relators* $=$, $\subseteq$, $\in$;

(6) the usual logical *connectives* $\neg$, &, $\vee$, $\rightarrow$, $\leftrightarrow$.

In what follows, * will indicate any of the binary operators.

The set T of *terms* is the smallest superset of $\{\emptyset, x_1, x_2, \ldots\}$ such that if t_1 and t_2 are in T then $(t_1 * t_2)$ is also in T.

The *atomic formulae* are the expressions of the form

$$(t_1 = t_2),\ (t_1 \subseteq t_2),\ (t_1 \in t_2)\ ,$$

where t_1 and t_2 are terms.

Negated and unnegated atomic formulae are called *literals*.

The set of *formulae* is the smallest set containing all the atomic formulae and closed with respect to the logical connectives. $t_1 \neq t_2$ and $t_1 \notin t_2$ abbreviate $\neg(t_1 = t_2)$ and $\neg(t_1 \in t_2)$ respectively.

Example of a formula:

$$(x \cup z \notin (\emptyset - (z-x)))\ \&\ (z \in y \cup x) \rightarrow (x \notin \emptyset) \vee (x \cap z \neq y) \vee (y \subseteq \emptyset).$$

An *interpretation* M is an assignment of a set to each variable. Once these sets are assigned, constant and operator signs of our language are interpreted in the usual set-theoretic sense, namely $\emptyset$, $\cup$, $\cap$, $-$, $\subseteq$, $\in$, designate the empty set, union, intersection, set difference, inclusion, and membership respectively.

With these rules every interpretation assigns a unique set value to each term and a unique truth value to each formula. M is said to be a *model* for each formula which is true in M. If every (some) interpretation is a model for the formula p, then p is said to be *valid* (*satisfiable*).

If $p \leftrightarrow q$ is valid then p and q are said to be (logically) *equivalent*.

If $p \rightarrow q$ is valid then q is said to be a (logical) *consequence* of p.

M is called a *singleton model* if every set variable is interpreted as a subset of $\{\emptyset\}$.

3. A DECISION PROCEDURE FOR MULTILEVEL SYLLOGISTIC.

Our problem is to check a formula p for validity. It can easily be shown that this is equivalent to checking satisfiability of a conjunction q of literals having one of the following forms:

(*) $x = y * z$

($\neq$) $x \neq y$

($\in$) $x \in y,\quad x \notin y$

where x,y,z are variables.

Let q^* be the conjunction of all the literals in q having the form (*) and let $V_\in$ be the set of all the variables x in q which appear as left-hand sides of literals of the form ($\in$). For each x in $V_\in$ consider the conjunction L_x of all the literals in q of the form ($\in$) whose left-hand side is x. Let L'_x be obtained from L_x by replacing $x \in y$ and $x \notin y$ by $y \neq \emptyset$ and $y = \emptyset$ respectively. The following algorithm (see [3]) decides the satisfiability of q:

STEP 1. For every pair of distinct variables x,y in q, determine if the equality $x = y$ is a logical consequence of q^*, by testing q^* & $x \neq y$ for the existence of a singleton model. Put x and y into the same equivalence class whenever q^* entails $x = y$. Choose a representative variable z from each class and replace all the other variables in the class by z in q.

STEP 2. If any of the literals in q has the form $x \neq x$ then return 'UNSATISFIABLE'.

STEP 3. If $V_\in$ is empty then return 'SATISFIABLE'.

STEP 4. If, for each x in $V_\in$, we can find a singleton model M_x for q^* & L'_x in such a way that the relationship $x < y$ defined by $x < y$ iff $M_x\ y \neq \emptyset$ extends to a total ordering on all the variables of q, then return 'SATISFIABLE'; otherwise return 'UNSATISFIABLE'.

Example: Let p be $(x \in y \cup z)$ & $(y \in w)$ & $(z \in w) \rightarrow w \notin x$. Then $\neg p$ is $(x \in y \cup z)$ & $(y \in w)$ & $(z \in w)$ & $(w \in x)$ and q is

$$u = y \cup z \ \& \ x \in u \ \& \ y \in w \ \& \ z \in w \ \& \ w \in x \ .$$

The only singleton models for q^* & L'_x are

Case (1): $u \neq \emptyset$, $y \neq \emptyset$, z arbitrary; Case (2): $u \neq \emptyset$, $z \neq \emptyset$, y arbitrary.

In case (1) $x < u$, $x < y$ and in case (2), $x < u$, $x < z$. In both cases $w < x$, $y < w$ and $z < w$. Thus in the first case we have a cycle $x < y < w < x$ whereas in the second case we have a cycle $x < z < w < x$. Hence we can conclude that q is unsatisfiable, i.e. that p is valid.

Remark 1. It is important to note that the set of statements q^* & L'_x appearing in Step 4 has a singleton model iff the result of substituting

$$\begin{array}{lll} x \leftrightarrow y & \text{for} & x = y \\ x \vee y & & x \cup y \\ x \,\&\, y & & x \cap y \\ x \,\&\, \neg y & & x - y \\ \neg x & & x = \emptyset \end{array}$$

in $q^* \,\&\, L'_x$ has a propositional model (the same assertion holds for the set of statements $q^* \,\&\, x \neq y$ mentioned in step 1). Thus if a singleton model for $q^* \,\&\, L'_x$ (or $q^* \,\&\, x \neq y$) exists it can be found by any standard decision procedure for propositional calculus [2].

Remark 2. The convexity (in Oppen's sense, see [6]) of the theory we consider is a consequence of the following easily proved assertion:

If q is satisfiable and $x_1 = y_1 \vee x_2 = y_2 \vee \ldots \vee x_n = y_n$ is a logical consequence of q then for some $i = 1,2,\ldots,n$, $x_i = y_i$ is a logical consequence of q^*.

Indeed, if q is satisfiable whereas $q \,\&\, x_1 \neq y_1 \,\&\, \ldots \,\&\, x_n \neq y_n$ is unsatisfiable, then since this unsatisfiability will be necessarily detected at Step 2, it must be caused by some literal $x_i \neq y_i$ for which $x_i = y_i$ is a logical consequence of q^*.

4. AN EXTENDED MULTILEVEL SYLLOGISTIC.

In this section we describe an algorithm for deciding formulae of an extended multilevel syllogistic which includes the singleton-former and cardinality operator #x together with arithmetic addition, subtraction, and comparison in addition to the operators considered previously. In this extended language set terms are built as described in section 2, but we have an additional set construction, namely

{t} is a set term (designating a singleton set) whenever t is a set term. Integer terms are built up using the binary operators + and - from a new infinity of integer variables, the constants 0,1, and terms of the form #t, where t is a set term. We also allow two new kinds of atomic formulae $j \leq k$ and $j = k$, where j and k are integer terms.

The validity problem for the new language reduces to that of checking satisfiability of a conjunction of literals of the following forms:

(*) $\qquad x = y * z$

($\neq$) $\qquad x \neq y$

({ }) $\qquad y = \{x\}$

(#) $j = \#x$

(A) $0 \leq j,\ j = k-\ell,\ j = 1,\ j = 0$

where x,y,z are set variables and j,k,ℓ are integer variables.

The following theorem (see [3]) defines our decision algorithm in rough outline:

THEOREM 1. Let q be a conjunction of literals having one of the above forms. Let V be the collection of all the set variables in q and let P be the collection of all nonempty subsets S of V such that the interpretation

$$Mx = \begin{cases} \emptyset & \text{if } x \notin S \\ \{\emptyset\} & \text{otherwise} \end{cases}$$

is a singleton model for q^*. Associate an integer variable νS with each S in P and put $\Delta(a,b) = \sum_{\substack{a\in S \\ b\notin S}} \nu S + \sum_{\substack{a\notin S \\ b\in S}} \nu S$.

Then q is satisfiable if and only if there is a partial ordering < on P such that the following system of arithmetic conditions has a solution in integers such that

if $(y = \{x\})$ is in q, $x \in S$, $y \in S'$, and $S \nless S'$ then either $\nu S = 0$ or $\nu S' = 0$:

(1) take all the conditions of the form (A) in q ;

(2) add the conditions $\Delta(x,y) > 0$, for all $(x \neq y)$ in q ;

(3) add the conditions $\Delta(x,\emptyset) = j$, for all $(j = \#x)$ in q;

(4a) add the conditions $\Delta(y,\emptyset) = 1$, for all $(y = \{x\})$ in q ;

(4b) add the conditions $\Delta(x,x') = 0 \leftrightarrow \Delta(y,y') = 0$, for all $(y = \{x\})$, $(y' = \{x'\})$ in q ;

(5) finally add the conditions $\nu S \geq 0$, for all S in P.

It follows from the decidability of Presburger arithmetic [8] that the satisfiability of these conditions is mechanically verifiable. In view of the decidability of the additive theory of cardinals [13], this remains true if we allow arbitrary cardinal numbers to appear in place of integers. For application of the decision method of this section it is worth noting that finite set formers and ordered tuples can be expressed using the singleton-former, as is shown by the following rather standard recursive definitions:

(a) $\{\ \} = \emptyset$

$\{e_1,e_2,\ldots,e_n\} = \{e_1\} \cup \{e_2,e_3,\ldots,e_n\}$;

(b) $\langle e\rangle = e$

$\langle e_1,e_2,\ldots,e_n\rangle = \{\{e_1\}, \{e_1, \langle e_2,e_3,\ldots,e_n\rangle\}\}$.

5. A FURTHER EXTENSION OF MULTILEVEL SYLLOGISTIC.

We can also extend multilevel syllogistic by introducing maps as a new sort of variable, together with domain, range, and restricted range operators which take maps into sets. For this extension, we introduce a new infinity of map variables f,g,..., and two new map operators D and R. Set terms can be built as before but we have the additional set terms Df, Rf, and f[t] where f is any map variable and t is a set term (here Df and Rf respectively denote the domain and the range of f; f[t] denotes the range of the restriction of f to t).

The validity problem for the language thus defined reduces to that of checking satisfiability of a conjunction q of literals of the following forms:

(*) $x = y * z$

(≠) $x \neq y$

(∈) $x \in y, \quad x \notin y$

([ ]) $x_{f_y} = f[y]$

(D,R) $x_{Df} = Df, \quad x_{Rf} = Rf$,

where $x, y, z, x_{f_y}, x_{Df}, x_{Rf}$ are set variables.

Without loss of generality we can suppose that f occurs in q iff literals involving f of each of the three types ([ ]), and (D,R) occur in q. For every f in q let $y_1, y_2, \ldots, y_k$ be all the variables appearing on the right-hand side of literals of the form x = f[y] in q, and let y^+ and y^- stand for $y \cap x_{Df}$ and $x_{Df} - y$ respectively. Also for each such f introduce new set variables $s_{i_1 i_2 \ldots i_k}$, where each i_j takes one of the two values +,−, representing $f[y_1^{i_1} \cap y_2^{i_2} \cap \ldots \cap y_k^{i_k}]$, and add the following formulae to q;

(1) $$y_1^{i_1} \cap y_2^{i_2} \cap \ldots \cap y_k^{i_k} = \emptyset \leftrightarrow s_{i_1 i_2 \ldots i_k} = \emptyset ;$$

(2) $$x_{f_{y_j}} = \bigcup_{i_j = +} s_{i_1 i_2 \ldots i_k} ;$$

(3) $$x_{Rf} = \bigcup_{i_1 i_2 \ldots i_k} s_{i_1 i_2 \ldots i_k} ;$$

Delete all the literals of the forms ([ ]) and (D,R). Let $\bar{q}$ be the resulting formula.

THEOREM 2. q is satisfiable iff $\bar{q}$ is satisfiable.

Proof: If q has a model M, then M can be extended to a model of $\bar{q}$ by putting

$$Ms_{i_1\ldots i_k} = (Mf)\ [My_1^{i_1} \cap \ldots \cap My_k^{i_k}]\ .$$

Conversely, let M be a model for $\bar{q}$ and let f be any map variable in q. Let Mf be any map whose domain and range are Mx_{Df} and Mx_{Rf} respectively, and such that

$$(Mf)\ [My_1^{i_1} \cap \ldots \cap My_k^{i_k}] = Ms_{i_1 i_2 \ldots i_k} \quad \text{for each } i_1,\ldots,i_k.$$

The existence of such a map Mf follows from conditions (1) and (3) by observing that Mx_{Df} is the disjoint union of all sets $My_1^{i_1} \cap \ldots \cap My_k^{i_k}$. This choice of Mf plainly satisfies all literals of type (D,R) in q. Moreover, from condition (2) and by noting that My_j^+ is the disjoint union of all sets $My_1^{i_1} \cap \ldots \cap My_k^{i_k}$ with $i_j = +$, it follows that literals of type ([ ]) in q are correctly modeled. This completes the proof of the theorem. Q.E.D.

The decidability of this extension follows from this theorem, using the decidability of multilevel syllogistic.

Remarks. (a) We can extend our language by adding the inverse operator f^{-1} on maps and also by allowing literals of the form $y = f^{-1}[x]$ to appear in q. To do this, we have only to note the following equivalence:

$$y = f^{-1}[x] \leftrightarrow y \subseteq Df \ \&\ f[y] = x \cap Rf \ \&\ f[Df - y] \cap x = \emptyset\ .$$

(b) We can add the predicate singlevalued( ) also allowing literals of type

singlevalued(f), ¬singlevalued(f)

to appear in q. If for some f in q both literals singlevalued(f) and ¬singlevalued(f) are in q, then q is clearly unsatisfiable. If at most one of those literals is in q for each map variable f, and if for each literal ¬singlevalued(f) we add the formula

$$x_{Df} \neq \emptyset$$

to $\bar{q}$, then Theorem 2 still holds. Indeed, let M be a model for $\bar{q}$ and let $x_1, x_2, \ldots, x_h$ be all variables in $\bar{q}$. Then, without loss of generality we can assume that all nonempty disjoint parts of the Venn diagram determined by the sets

$$Mx_1, Mx_2, \ldots, Mx_h ,$$

are countably infinite (just add arbitrary new individuals to these disjoint parts). Next, take Mf, as in the proof of Theorem 2, to be either singlevalued or not singlevalued according to the corresponding literal in q (arbitrarily if there is no such literal).

(c) The predicate one-one( ) can be treated in much the same way. For each literal ¬one-one(f) add $x_{Df} \neq \emptyset$, and for each literal one-one(f) add the set of formulae

$$(4) \quad s_{i_1 \ldots i_k} \cap s_{j_1 \ldots j_k} = \emptyset \quad \text{for every } (i_1, \ldots, i_k) \neq (j_1, \ldots, j_k) .$$

(d) Finally we observe that the multilevel syllogistic described in Section 2, and extended to include all the constructs described in Sections 4 and 5 remains decidable because Theorem 2 can be extended to this case. If the predicate singlevalued( ) is also introduced then each literal ¬singlevalued(f) is replaced by the set of formulae

$$z \in x_{Df} \ \& \ w = \{z\} \ \& \ w' = f[w] \ \& \ \#w' > 1,$$

where z, w and w' are new variables, whereas if singlevalued(f) is in q then condition (1) must be replaced by

$$(1') \qquad \#(y_1^{i_1} \cap y_2^{i_2} \cap \ldots \cap y_k^{i_k}) \geq \#s_{i_1 i_2 \ldots i_k} \ \& \ (s_{i_1 i_2 \ldots i_k} = \emptyset \rightarrow y_1^{i_1} \cap y_2^{i_2} \cap \ldots \cap y_k^{i_k} = \emptyset) .$$

Similarly, ¬one-one(f) is replaced by the set of formulae

$$\neg \text{singlevalued}(f) \ \vee \ (z \subseteq x_{Df} \ \& \ \#z > 1 \ \& \ w = f[z] \ \& \ \#w = 1) ,$$

where w, z are new variables, whereas if one-one(f) is in q then we introduce condition (4) of Remark (c) and replace (1) by:

$$(1'') \qquad \#(y_1^{i_1} \cap y_2^{i_2} \cap \ldots \cap y_k^{i_k}) = \#s_{i_1 i_2 \ldots i_k} .$$

REFERENCES

[1] Behmann, H., Beiträge zur Algebra der Logik insbesondere zum Entscheidungsproblem, Math. Annalen 86 (1922) 163-220.

[2] Davis, M., Putnam, H., A computing procedure for quantification theory, J. ACM 7 (1960) 201-215.

[3] Ferro, A., Omodeo, E. G., Schwartz, J. T., Decision procedures for elementary sublanguages of set theory. I. Multilevel syllogistic and some extensions, Comm. Pure Appl. Math. (to appear).

[4] Łukasiewicz, J., Aristotle's Syllogistic, Clarendon Press, 1952.

[5] Nelson, C. G., Oppen, D. C., A simplifier based on efficient decision algorithms, Fifth Ann. Symp. on Principles of Programming Languages (1978) 141-150.

[6] Oppen, D. C., Complexity of combinations of quantifier-free procedures, Workshop on Automatic Deduction, Austin, Texas (1979).

[7] Pratt, V. R., On specifying verifiers, (to appear).

[8] Presburger, M., Über die Vollständigkeit eines gewissen Systems der Arithmetik ganzer Zahlen, in welchem die Addition als einzige Operation hervortritt, Comptes-rendus du premier Congrès des mathématiciens des Pays Slaves, Warsaw (1929), 92-101.

[9] Quine, W.V.O. Methods of Logic, Henry Holt & Co., New York, 1950.

[10] Schwartz, J. T., Instantiation and decision procedures for certain classes of quantified set-theoretic formulae. Inst. for Computer Applic. in Science and Engr., NASA Langley Research Center, Hampton, Virginia, Report Number 7810 (1978).

[11] Schwartz, J. T., A survey of program proof technology, Courant Institute, Comp. Sci. Dept., Report No. 1, September 1978.

[12] Shepherdson, J. C., On the interpretation of Aristotle's syllogistic, J. Symb. Log. 21 (1956) 137-147.

[13] Tarski, A., Ordinal Algebras, North-Holland, Amsterdam, 1956.

[14] Ville, F., Décidabilité des formules existentielles en théorie des ensembles, C. R. Acad. Sc. Paris, t. 272, Série A (1971), 513-516.

ACKNOWLEDGEMENTS

This work was supported in part by NSF Grant Number MCS 79-04377 and by U. S. Department of Energy Contract Number DE-AC02-76ER03077/V. The authors acknowledge support by C.N.R. of Italy.

Simplifying Interpreted Formulas

D. W. Loveland
Department of Computer Science
Duke University
Durham, North Carolina 27706

R. E. Shostak
Computer Science Laboratory
SRI International
Menlo Park, California 94025
(Supported in part by AFOSR Contract F49620-79-C-0099)

Abstract

A method is presented for converting a decision procedure for unquantified formulas in an arbitrary first-order theory to a simplifier for such formulas. Given a quantifier-free d.n.f. formula, the method produces a simplest (according to a given criterion) d.n.f. equivalent from among all formulas with atoms in the original formula. The method is predicated on techniques for minimizing purely boolean expressions in the presence of "don't-care' conditions. The don't-cares are used to capture the semantics of the interpreted literals in the formula to be simplified.

Two procedures are described: a primitive version of the method that advances the fundamental idea, and a more refined version intended for practical use. Complexity issues are discussed, as is a nontrivial example illustrating the utility of the method.

1. Introduction

The problem of simplifying logical expressions was first addressed in the early 1950s in the form of boolean minimization. The motivation at that time was to reduce, as much as possible, the number of components needed to realize a given switching circuit. Minimization techniques were developed to operate according to a variety of criteria, including the fewest literals in a sum-of-products or product-of-sums expression, the fewest terms, or the fewest terms and occurrences of literals.

The problem of simplifying logical expressions has resurfaced in the last few years in connection with program verification, synthesis, and allied concerns in artificial intelligence. In these applications, the expressions to be simplified are no longer merely propositional; they may contain interpreted predicates or function symbols. Even the problem of defining useful simplicity criteria for such formulas can be tricky, since the usual syntactic measures are sometimes misleading. For example, the formula $y \geq x \vee 5y \leq x+10$ (where x and y are understood to range over positive integers) is much more concise than the equivalent $(x=1 \wedge y=1) \vee (x=1 \wedge y=2) \vee (x=2 \wedge y=2)$, even though the latter is likely to be more useful in many theorem-proving situations.

Ideally, one would like a general-purpose method for simplifying formulas in arbitrary nonlogical theories with respect to arbitrary simplification measures. Though such a method is clearly too much to hope for, the approach described herein is a step in the direction of this goal. Our method may be viewed as a practical way of converting a decision procedure for unquantified formulas in an arbitrary first-order theory to a simplifier for such formulas. Given a quantifier-free formula in d.n.f., it produces a simplest (according to any given reasonable criterion) d.n.f. equivalent from among all formulas whose atoms occur in the original formula. By "reasonable" criterion, we mean one according to which the deletion of a literal from a term or of a term from a disjunction always produces a simpler formula.

Before describing the approach, we might point out that simplification can often be accomplished merely by eliminating unsatisfiable disjuncts in a disjunctive normal form. (Note, in particular, that this technique necessarily reduces all unsatisfiable formulas to "false.") The elimination of such disjuncts is not, however, sufficient to produce a simplest form for nonvalid formulas. The difficulty is illustrated by the following formula from the theory of Presburger arithmetic with function symbols:

$$F \equiv (y \neq z) \vee (x \leq y \wedge x+y \leq 0) \vee (x \leq 1 \wedge f(z) \neq f(y)+1)$$

While none of the disjuncts of F is unsatisfiable, F does have a much simpler equivalent, namely

$$y \neq z \vee x \leq 1 \quad .$$

Isolated consideration of the terms in the d.n.f. expression is thus insufficient.

Our method is presented in four parts. Section 2 describes the <u>standard procedure</u>, a primitive version that advances the fundamental idea. A much more efficient version, called the <u>modified procedure</u>, is given in Section 3. Section 4 gives a brief analysis of the computational complexity of the two versions, and the last section summarizes a nontrivial example that illustrates the utility of the modified method.

2. The Standard Procedure

The procedure given in this section takes as input a quantifier-free d.n.f(c.n.f.) formula in a first-order theory and returns an equivalent d.n.f (c.n.f.) expression with the property that no other such expression with atoms from the original formula is simpler with respect to a given reasonable (in the sense given earlier) measure of simplicity. The procedure works with any first-order theory for which the satisfiability of quantifier-free conjunctions of literals can be tested.

One can view the method as a nonlogical counterpart of the systematic minimization techniques developed for purely proposition formulas. In fact, the technique makes use of the method of prime implicants first described by Quine and McCluskey [3,4].

Our treatment assumes that a d.n.f. expression is to be found. One can obtain c.n.f. expressions using a dual method.

We begin with a brief review of Quine's method of prime implicants for purely propositional expressions. A more detailed account is given in [1].

Defn. A term is a conjunction of literals.

Defn. A term t_1 subsumes a term t_2 if each literal of t_2 is also a literal of t_1.

Defn. An implicant of a formula F is a term that implies F.

Defn. A prime implicant of a formula F is a term that implies F and subsumes no shorter term that implies F.

The fundamental interest of prime implicants is that any simplest d.n.f. equivalent G for a propositional formula F must be a disjunction of prime implicants of F. To see this, suppose that some term t of G is not a prime implicant of F. Because t implies F but is not a prime implicant, t must subsume a shorter term t' that also implies F. The expression obtained from G by replacing t with t' is still equivalent to F, contradicting the assumption that G is simplest.

Several methods can be used to determine the set of prime implicants of a formula F. One such, called the method of iterated consensus [5,6] begins with the set of terms in a d.n.f. form of F. The nontautological resolvents of terms in the set are repeatedly formed and added to the set. At the same time, subsuming terms are deleted. When no new terms can be added that do not subsume existing terms, the set of prime implicants has been obtained.

Consider, for example, the formula F given by

$$F \equiv \bar{p}rs \vee \bar{p}qr\bar{s} \vee pqrs \quad .$$

Resolving $\bar{p}rs$ and $\bar{p}qr\bar{s}$ gives rise to $\bar{p}qr$. Because $\bar{p}qr\bar{s}$ subsumes $\bar{p}qr$, the former can be deleted. Next, by resolving $\bar{p}rs$ with pqrs, one obtains qrs; pqrs can thus be deleted. Because no more terms can be added or deleted, the remaining terms, $\bar{p}rs$, $\bar{p}qr$, and qrs, are the prime implicants of F.

Once the prime implicants of a formula have been found, a simplest d.n.f. expression can be obtained by determining a simplest subset of prime implicants whose disjunction is implied by the formula. Note that simplest disjunctions need not be unique; frequently several different combinations of prime implicants give rise to simplest equivalents. To discover these combinations, it is useful to classify the prime implicants into three catagories:

- *Core* implicants are those that must appear in any such combination. If a given implicant does not imply the disjunction of all other implicants, it must be a member of the core.
- *Absolutely eliminable* implicants are those that imply the disjunction of the core implicants, and so can be ignored.
- *Eliminable* implicants are those that are neither core nor absolutely eliminable.

The various simplest equivalents differ only in their selection of eliminable implicants.

The most straightforward method of finding these equivalents involves constructing a table T whose rows are labeled by prime implicants and whose columns are labeled by the terms in the perfectly developed d.n.f. (in the perfectly developed d.n.f., each letter atom occurs (either signed or unsigned) in each term of the formula to be simplified.) A '1' is placed at T(t,u) if the prime implicant t is subsumed by term u, and a '0' otherwise. The core implicants are easily identified as those subsumed by at least one term that subsumes no other implicant; absolutely eliminable implicants are those subsumed only by terms that subsume at least one core implicant. All rows labeled by core and absolutely eliminable implicants are then cancelled (deleted from the table), as well as all columns labeled by terms that subsume core implicants. The subsets of remaining implicants sufficient to cover the remaining columns are then enumerated exhaustively and a simplest one is selected.

Our procedure for simplifying interpreted expressions depends on an elaboration of the method just described that can handle so-called "don't-care" conditions. In the application of minimization techniques to digital design it is sometimes useful to exploit situations in which certain assignments to the variables of an expression to be simplified are not actually realized. For such assignments, the value of the simplified expression can be arbitrary. As one might expect, greater simplification can often be obtained if one relaxes the requirement that the simplified expression be equivalent to the original, so as to necessitate equivalence only for assignments other than the don't-cares.

The treatment of don't-care conditions requires two slight modifications of the basic method. First, for purposes of generating prime implicants, the d.n.f. form of the formula to be simplified is augmented by disjoining to it a term for each don't-care condition. If, for example, p=T, q=F, r=T is a don't-care input, the term $p\bar{q}r$ is added. Second, the terms in the perfectly developed d.n.f. that imply don't-care conditions are omitted from the prime-implicant matrix.

Suppose it is wished, for example, to simplify the formula $F \equiv p \vee qr$ with respect to don't-care conditions {p=F, q=T, r=F} and {p=T, q=F, r=T}. We first find the prime implicants of the augmented formula $p \vee qr \vee \bar{p}q\bar{r} \vee p\bar{q}r$. Using the method of iterated consensus, $p\bar{q}r$ can be eliminated immediately because it is subsumes p. Resolving p against $\bar{p}q\bar{r}$, $q\bar{r}$ is obtained. Since $\bar{p}q\bar{r}$ subsumes $q\bar{r}$, $\bar{p}q\bar{r}$ can now be eliminated. Resolving $q\bar{r}$ against qr yields q, which permits the elimination of both $q\bar{r}$ and qr. We are therefore left with the prime implicants p and q. The prime implicant table will contain rows for p and q and columns for all the terms in the perfectly-developed d.n.f. for F (namely, pqr, $pq\bar{r}$, $p\bar{q}r$, $p\bar{q}\bar{r}$, $\bar{p}qr$) other than the don't-care term $p\bar{q}r$. It is easy to verify that both p and q are core implicants (q is subsumed by $\bar{p}qr$ and p by the remaining terms), hence the simplified form is just $p \vee q$.

Our application of this method to the problem of simplifying interpreted expressions is predicated on the use of don't-care conditions to encode the semantics of the terms appearing in the expressions. The basic idea is to treat the interpreted formula to be simplified as if it were purely propositional (i.e., as if interpreted terms were actually uninterpreted), except that all unsatisfiable (with respect to the interpreted semantics) conjunctions of literals with atoms occurring in the formula are treated as don't-cares.

The procedure is easily understood in the context of a small example. Suppose, then, that the formula F to be simplified is just

$$x \leq y \vee (z>0 \wedge x+2z-y>3) \quad ,$$

where all variables range over nonnegative integers.

If we let p, q, r denote the atoms $x \leq y$, $z>0$, and $x+2z-y>3$, respectively, F can be written $p \vee qr$.

Now consider the eight possible assignments of truthvalues to p, q, r: pqr, $pq\bar{r}$, $p\bar{q}r$,..., $\bar{p}\bar{q}\bar{r}$. If each term were submitted to a refutation procedure for quantifier-free Presburger arithmetic, it would be found that all assignments other than $\bar{p}q\bar{r}$ and $p\bar{q}r$ are satisfiable. The question of simplifying F thus becomes that of finding the simplest propositional equivalent of p v qr subject to the don't-care conditions $\bar{p}q\bar{r}$ and $p\bar{q}r$. Having solved this problem in the propositional example above, we may conclude that p v q, i.e., $x \leq y$ v $z>0$, is a simplest equivalent. (Note, incidentally, that since p and q are core implicants, the fact that p v q is simplest does not depend on the simplicity measure.)

The standard method may be summarized as follows:

1. Let A be the set of atoms occurring in the formula F to be simplified, and let T be the set of terms representing the $2^{|A|}$ truth assignments to A. Using a refutation procedure for the theory in question, determine the unsatisfiable subset U of T.

2. Using the method of prime implicants, find a simplest (with respect to the desired reasonable measure) formula that is truth-functionally equivalent to F modulo the don't-care set U.

A proof that this procedure does in fact produce a simplest semantic equivalent for F among all formulas with atoms in A is given in a more detailed version of this paper[2].

3. The Modified Procedure

Because the problem solved embeds the satisfiability question for propositional formulas, any version of the procedure requires (at least) exponential time in input formula length in the worst case (based on present-day knowledge). This section details refinements of the standard procedure, however, that improve performance greatly in many situations. The standard procedure may nevertheless be preferable when there is a substantial number of multiple occurrences of atoms of F.

Our measure of effort will be taken as the number of calls to the refutation procedure. That this is the best measure is arguable since some refutation procedures can be so quick as to have the boolean manipulation dominate the cost. However, our methods are independent of the refutation procedure used and most such procedures require a significant interval of time per call (which may be only a second, but is nevertheless significant when hundreds of calls are made). Moreover, except for the iterated consensus (resolution)section, total effort is proportional to the number of calls.

The greatest potential for performance gain follows from the requirement of the standard procedure that all conjunctions to be processed must be evaluated by the refutation procedure before serious boolean processing begins. Although we improve the "worst-case" situation somewhat (worst-case with respect to the various chances that simplification may occur), we greatly improve the cost of processing a typical formula, especially when no simplification does occur. We are left at least with the situation that high cost is associated with definite gain.

For purposes of explanation, it is convenient to consider the standard procedure as consisting of two phases: in Phase 1, the unsatisfiable truth assignments are determined and the prime implicants of the formula augmented by don't-care terms are generated; in Phase 2, the prime-implicant table is created and a simplest set of implicants implied by the original formula is chosen. The improved procedure refines both of these phases.

The main improvement to Phase 1 turns upon the observation that it is unnecessary to test all truth assignments for satisfiability. In particular, the assignments that subsume terms of the original formula need not be tested, since these assignments would be discarded in the iterated consensus procedure anyway. In our earlier example, for instance, five of the eight assignments (namely pqr, $pq\bar{r}$, $p\bar{q}r$, $p\bar{q}\bar{r}$, and $\bar{p}qr$) subsume either p or qr, leaving only three ($\bar{p}q\bar{r}$, $\bar{p}\bar{q}r$, $\bar{p}\bar{q}\bar{r}$) to be submitted to the refutation procedure.

Described in [2] is another refinement of Phase 1 that further lowers the required number of calls to the procedure, but at the cost of possibly missing significant simplifications.

The improved Phase 2 procedure is equivalent to the standard one, but is substantially more efficient in most cases. It appears not to have been considered for boolean minimization because "don't-care" conditions are traditionally given rather than computed.

The procedure is defined using an auxilliary predicate P(X,Y), where X and Y are sets of terms. Letting $Y=\{t_1,t_2...t_k\}$, P(X,Y) is computed by enumerating all terms of the form

$$C \wedge L_1 \wedge L_2 ... \wedge L_k \quad ,$$

where C is the conjunction of all terms in X and each L_i is the complement of some literal in t_i. The enumerated terms are tested one by one for satisfiability. P returns "true" if one is found to be satisfiable, and returns "false" otherwise. The key property of P is that P(X,Y)=false iff $C \Rightarrow t_1 \vee t_2 \vee ... \vee t_k$.

If for example, X={a,bc} and Y={cde,gh}, the terms $abc\bar{c}\bar{g}$, $abc\bar{c}\bar{h}$, $abc\bar{d}\bar{g}$, $abc\bar{d}\bar{h}$, $abc\bar{e}\bar{g}$, $abc\bar{e}\bar{h}$ are enumerated. Note that the first two of these are syntactically unsatisfiable, and so do not require calls to the refutation procedure. If it were found, for instance, that $abc\bar{d}\bar{g}$ is satisfiable, the evaluation could terminate after this one call, returning "true."

The improved Phase 2 procedure is as follows. Let I be the set of prime implicants computed by Phase 1, and let I' be obtained by deleting from I all of its unsatisfiable members. (Computing I' from I thus requires applying the refutation procedure to each member.) A modified prime-implicant table T_m is now constructed whose rows are labeled with members of I' and whose columns are labeled by sets of terms. The columns are created dynamically in the following way:

1. Initialize the table by creating a column for each term in I', with the singleton set of that term as label.

2. Fill in each new column as follows. If P(X,I'-X) evaluates to false, where X is the set labeling the column to be filled in, enter '*' in each row position (indicating a cancelled column). Otherwise, for the row labeled by implicant u, enter '1' if $u \in X$ and '0' if $u \notin X$.

3. For each two cancelled columns with labels X_1, X_2, create a new column, if one does not already exist, labeled by $X_1 \cup X_2$.

4. Repeat Steps (2) and (3) until no new columns can be added.

5. Select prime implicants to define a simplest equivalent to F as in the standard procedure- i.e., choose a simplest set S of prime implicants such that for every uncancelled column X, there exists an $s \in S$ such that $T_m(s,X)=1$.

We illustrate the modified procedure with the earlier example:

$$F \equiv a \vee bc \vee de$$

where

a: $y \neq z$
b: $x \leq y$
c: $x+y \leq 0$
d: $x < 1$
e: $fz \neq fy+1$

Phase 1.

The truth assignments not subsuming terms in F are $\bar{a}b\bar{c}d\bar{e}$, $\bar{a}b\bar{c}\bar{d}e$, $\bar{a}b\bar{c}\bar{d}\bar{e}$, $\bar{a}\bar{b}c\bar{d}e$, $\bar{a}\bar{b}cd\bar{e}$, $\bar{a}\bar{b}c\bar{d}\bar{e}$, $\bar{a}\bar{b}\bar{c}d\bar{e}$, $\bar{a}\bar{b}\bar{c}\bar{d}e$, and $\bar{a}\bar{b}\bar{c}\bar{d}\bar{e}$. Of these nine, all but $\bar{a}b\bar{c}\bar{d}e$, $\bar{a}\bar{b}c\bar{d}e$, and $\bar{a}\bar{b}\bar{c}\bar{d}e$ are found by a refutation procedure to be unsatisfiable. The iterated-concensus process is applied to F augmented by the six "don't-cares" to obtain the set I = {a, bc, d, $\bar{e}$} of prime implicants.

Phase 2.

Each member of I is tested and found satisfiable, so I' = I. The modified table T_m is initialized with rows and columns labeled by members of I'. Steps (2) and (3) of the Phase 2 procedure are now applied repeatedly to form the table shown below.

	{a}	{bc}	{d}	{$\bar{e}$}	{bc,$\bar{e}$}
a	1	*	0	*	*
bc	0	*	0	*	*
d	0	*	1	*	*
$\bar{e}$	0	*	0	*	*

Justification for the table is summarized below:

1. Initialize, creating columns labeled {a}, {bc}, {d}, {$\bar{e}$}

2. Fill in columns:

Column{a}:
P({a}, {bc, d, $\bar{e}$}):
conjunction tested: $a\bar{b}\bar{d}e$ satisfiable

= true
Fill in standard way

Column{bc}
P({bc}, {a, d, $\bar{e}$})
conjunction tested: $bc\bar{a}\bar{d}e$ unsatisfiable
= false
cancel column

Column{d}:
P({d}, {a, bc, $\bar{e}$}):
conjunction tested: $d\bar{a}\bar{b}e$ satisfiable
= true
Fill in standard way

Column {$\bar{e}$}:
P({$\bar{e}$}, {a, bc, d})
conjunctions tested: $\bar{e}\bar{a}\bar{b}\bar{d}$ unsatisfiable
= false $\bar{e}\bar{a}\bar{c}\bar{d}$ unsatisfiable
cancel column

3. Create new column labeled {bc,$\bar{e}$}

2. (Repeated). Fill in new columns:

Column {bc, $\bar{e}$}
P({bc, $\bar{e}$}, {a, d})
conjunction tested: $bc\bar{e}\bar{a}\bar{d}$ unsatisfiable
= false
cancel column

3. (Repeated). No new columns

5. Core implicants a,d cover all uncancelled columns.

The simplified form is thus a v d, i.e., F $\equiv$ y≠z v x=1.

(Note that here only 19 calls to the refutation procedure were required, as against 32 for the standard procedure.)

A proof of correctness of the modified procedure is given in [2].

4. Complexity Issues

While it is difficult to obtain quantitative measures of the improvement afforded by the modified procedure, some calculations can be made under certain simplifying assumptions. Our analysis will consider that the formula F to be simplified has n terms, each with m literals, and that no atoms in F have multiple occurrences.

For the standard procedure, exactly 2^{mn} calls to the refutation procedure are made in Phase 1 and, of course, none are made in Phase 2.

For the modified procedure, calls are made in both phases. In Phase 1, a call is made for each truth assignment (to the mn atoms of F) that does not subsume a term of F. Each truth assignment may be viewed as a choice, for each term, of one of 2^m assignments to the atoms of that term. Because all but one of these 2^m assignments are permissible, a total of $(2^m-1)^n$ calls is made in Phase 1.

The number of calls made in Phase 2 depends on the set I of prime implicants discovered in the first phase. To obtain a rough idea of Phase 2 behavior, let us assume I contains p implicants, each with q literals, and that $p \leq n$, $q \leq m$. (We have found this assumption to be valid in practice.)

Phase 2 first requires that each of the p implicants be tested for satisfiability. The remainder of Phase 2 may require zero calls (if all prime implicants are unsatisfiable). Assuming that p prime implicants are satisfiable, we may need as few as p more calls (if all tested conjunctions are satisfiable) or as many as $(q+1)^p-1$ more calls (if all tested conjunctions are unsatisfiable). The lower bound holds because for each singleton set X, P(X,I'-X) will return "true" after one call and Step 3 provides no new columns beyond the p initial columns. Thus, a total of p calls is made. The upper bound holds because each conjunction tested contains for each of the p prime implicants either the prime implicant itself or the complement of one of the q literals of the implicant. In the one unrealizable case, no prime implicant occurs in the conjunction. Using $p \leq n$, $q \leq m$, we have a worst-case bound of $(m+1)^n-1$, and a best-case bound of 2n.

It is worth noting that the total worst-case cost for the modified procedure is almost always less than that for the standard procedure ($(2^m-1)^n+n+(m+1)^n-1$ versus 2^{mn}) for reasonable m and n. However, the primary value of the modified procedure is that often m is small enough (typically about 1.5) so that Phase 1 cost is moderate. Moreover, a general mix of candidate formulas includes many that are not simplifiable and with the cost of Phase 2 close to 2n.

5. An Example

We conclude with a summary of a less trivial example. The example illustrates that quite striking reductions can be obtained in innocent-looking formulas.

Consider

$$F \equiv y>\max(2,z) \ \vee\ y>1+z \ \vee\ (y\neq 0 \wedge y\leq -1) \ \vee\ (y\neq 0 \wedge y\neq z) \ \vee\ y=0 \ \vee\ (z\neq 1 \wedge y\neq 1) \ \} \quad .$$

Phase 1: Use of modified procedure requires 3 calls, and results in prime-implicant set:

$$\{y>\max(2,z),\ y>1+z,\ y=0,\ y\leq -1,\ y\neq z,\ z\neq 1,\ y\neq 1\} \quad .$$

Phase 2: Modified procedure requires 63 calls.

Result: $F \equiv z\neq 1 \vee y\neq 1$.

The standard procedure requires 128 calls.

To balance this example, we consider two formulas with similar structure to F, but where little simplification occurs. The letters A, B, ... represent semantically unrelated atoms.

$$F_1 \equiv A \vee B \vee (\bar{C} \wedge D) \vee (\bar{C} \wedge E) \vee C \vee (G \wedge H)$$

$$\text{(which simplifies to } F_1 \equiv A \vee B \vee D \vee E \vee C \vee (G \wedge H)) \quad ,$$

$$F_2 \equiv A \vee B \vee (C \wedge D) \vee (W \wedge E) \vee Z \vee (G \wedge H) \quad .$$

F_1 produces 3 Phase 1 calls and 12 Phase 2 calls. F_2 produces 27 phase 1 calls and 12 Phase 2 calls. The standard procedure requires 128 and 512 calls , respectively.

Truly low-cost maximal simplification using refutation decision procedures is unlikely. However, we believe this paper shows that, given the speed of the best existing refutation procedures, simplification of expressions that occur in practice is currently feasible.

6. References

1. Bartee, T. C., Lebow, J. L., Reed, I. S., Theory and Design of Digital Machines, McGraw-Hill, New York (1962).

2. Loveland, D. W., Shostak, R. E., "Simplifying Interpreted Expressions," to appear as SRI and IBM technical reports.

3. McCluskey, E. J., "Minimization of Boolean Functions," Bell System Tech. Journal, Vol. 35, pp. 1417-1444 (Nov. 1956).

4. Quine, W. V., "The Problem of Simplifying Truth Functions," Am. Math. Monthly, Vol. 59, pp. 521-531 (Oct. 1952).

5. Quine, W. V., "On Cores and Prime Implicants of Truth Functions," Am. Math. Monthly, Vol. 66, pp. 755-760 (Nov. 1959).

6. Samson, E. W., and Mills, B. E., "Circuit Minimization: Algebra and Algorithm for New Boolean Canonical Expressions," AFCRC-TR-56-110, Cambridge, Massachusetts (1954).

SPECIFICATION AND VERIFICATION
OF REAL-TIME, DISTRIBUTED SYSTEMS
USING THE THEORY OF CONSTRAINTS

Frederick C. Furtek

The Charles Stark Draper Laboratory, Inc.
Cambridge, Massachusetts 02139

ABSTRACT: A technique for the specification and verification of real-time, distributed systems is proposed. It provides a unified representation for both internal design and externally observable behavior and an automated procedure for deriving the external behavior associated with a design. The approach is applicable to both hardware and software, and can handle systems in which timing, concurrency, indeterminacy, and ongoing behavior are important considerations.

Although the approach contains elements of switching theory and automata theory, it diverges from other models with the definition of a *constraint* as an incompatibility among successive states. The proposed verification technique centers around an algorithm (incorporating the resolution principle and coded in LISP) that takes as input an acceptor for a set of design constraints and generates as output an acceptor for the set of *prime constraints*. A description of external behavior is obtained from this acceptor by pruning out all constraints involving hidden variables. It is anticipated that from this description it will be possible to verify properties relating to consistency, equivalence, deadlock, computer security, and fault tolerance.

1. INTRODUCTION

A formal design process begins with the specification of requirements for a system's externally observable behavior. An internal design is then specified, followed by a verification to show that the externally observable behavior of the design meets the requirements. The principal methods of verification are testing, simulation, and formal analysis. If a verification fails, then either the requirements or the design, or both, have to be changed, and the process repeated. It is seen that a formal design process requires: (1) a language to specify design, (2) a language to specify external behavior, and (3) a procedure to derive external behavior from design.

For real-time, distributed systems, testing and simulation are currently the primary methods of verification. Because of the complexity of these systems, however, it is impossible to do either exhaustive testing or exhaustive simulation. Moreover, in the case of testing, the time and expense needed to create the test environment imposes an additional burden. The success of formal techniques in verifying real-time, distributed systems has been limited by the special demands of these systems. The issues involved are likely to be independent of the hardware/software distinction, and to include:

Timing
Concurrency
Indeterminacy
Ongoing Behavior

Timing encompasses both the relative ordering of events and the elapsed time between events. Concurrency denotes the absence of ordering among events. Indeterminacy refers to the situation in which the outcome of a choice is left unspecified. Ongoing behavior is contrasted with initial/terminal behavior, where only the initial and terminal states of a system are of interest.

Most of the effort in specification and verification has focused on providing semantics for programming languages and proving properties about programs [2,3,9,18,22, 29,30]. Because these approaches are software-oriented, they do not provide the hardware/software neutrality that is highly desirable in real-time, distributed situations. Moreover, they are limited in their ability to handle timing, concurrency, indeterminacy, and ongoing behavior. (See Manna and Waldinger [18].)

To address these issues, two broad classes of models have been proposed. In one class – typified by Petri nets [11,23], occurrence graphs [10], partial orders [8], path expressions [1,31], and related formalisms [16] – behavior is viewed as a set of orderings (either total or partial) on event occurrences. Timing is concerned with the relative order of events – what comes before what – and not with the elapsed time between events. Concurrency is expressed by allowing two occurrences to appear in either order within a total order, or by leaving two occurrences unordered within a partial order. The major shortcoming of these efforts is that they do not provide for verification in the sense described above.

In contrast to the event-oriented models are the state-oriented formalisms of temporal logic [24,25], dynamic logic [26,27], loosely coupled systems [14,17,33], and constraints [4,5,6,20,21]. Here one is concerned with establishing properties pertaining to the allowed state-sequences of a system. Temporal logic and dynamic logic deal specifically with proving properties about the execution sequences generated by a program, while loosely coupled systems address the problem of coordinating access to a distributed set of resources by a set of concurrent processes. Although the original interest in constraints was motivated by applications to computer security, the same principles provide the foundation for the specification and verification approach described below.

While the work described here has many of the same aims as temporal logic and dynamic logic, it operates at a more primitive level – dealing not with processes, formulae, and modalities, but rather with the logical dependencies among variables. The level of description is similar to that of the loosely-coupled-system model. In fact, with a change in terminology, the constraint model becomes a generalization of loosely coupled systems. In both cases, a system is characterized in terms of excluded behavior – that is, in terms of states and state sequences that are disallowed. But while loosely coupled systems permit restrictions on just states and (certain) state transitions, the constraint model permits restrictions on state sequences of arbitrary length. This extension provides a general method for describing both internal design and externally observable behavior, and it allows resolution to be used as a rule of inference. Resolution is a key element in an algorithm that automatically generates a representation of external behavior from a specification of design.

Outlined below is a specification and verification approach based on the theory of constraints. After formulating a model in terms of conditions, variables, states, and simulations, the design of a resource monitor is considered. It is shown that both the monitor's design and the requirements on its external behavior can be characterized by prime constraints representing excluded behavior. A verification is carried out using an algorithm coded in LISP and incorporating the resolution principle. An acceptor for a set of design constraints is supplied to the algorithm, which returns an acceptor for the set of external prime constraints. The suitability of this technique for real-time, distributed systems is then discussed, and suggestions for future work are presented.

2. THE MODEL

There are four basic concepts in our model: conditions, variables, states, and simulations. A condition may be interpreted as the assignment of a 'value' to a variable. (A binary variable may be viewed as a predicate, the two associated conditions indicating whether the predicate is true or false.) Although there are several ways to formalize the relationship between conditions and variables, the following approach captures the essential property of that relationship: that a condition belong to a unique variable.

> POSTULATE: Associated with a system is a finite set of conditions and a partition on this set, the blocks of which are called variables. Each variable has at least two conditions.

> DEFINITION: A state is a set of conditions containing exactly one condition from each variable.

The behavior of a system is represented by the set of allowed state-sequences, which are called simulations.

POSTULATE: Associated with a system is a set of finite state-sequences called simulations. Simulations satisfy two properties:

1. The set of simulations is a regular language.*
2. Every subsequence** of a simulation is also a simulation.

The regularity assumption restricts our model to finite systems. Note, however, that it is not necessary that different instances of the same state be equivalent with respect to future behavior. The state sequence preceding a particular state can determine, in part, which state sequences may follow that state. The second assumption reflects the fact that if a particular state sequence is allowed, then every subsequence must also be allowed.

To illustrate the model and to provide motivation for later concepts, a simple resource-sharing problem is considered. The variables in the example are intentionally chosen to be binary, permitting a simpler formulation of the subsequent ideas.

3. A RESOURCE-SHARING EXAMPLE

Consider the problem of designing a monitor to mediate access to a single resource by two users. The inputs to the monitor are in the form of two binary variables, {W1, NW1} and {W2,NW2}, where the conditions have these interpretations:

W1 – User 1 wants the resource.

NW1 – User 1 does not want the resource.

W2 – User 2 wants the resource.

NW2 – User 2 does not want the resource.

The output of the monitor is represented by a single binary variable, {G1,G2}, where the two conditions determine who gets the resource:

G1 – Resource granted to User 1.

G2 – Resource granted to User 2.

The resource is to be granted upon request, subject to the restriction that it be taken away from a user only after he has had it for at least two time units (cycles). An

---

* A regular language is any set of strings (sequences) accepted by a finite automaton. See Hopcroft and Ullman [12], p. 27.

** α is a subsequence of the sequence β if and only if there exist sequences γ and δ such that $\beta = \gamma\alpha\delta$.

additional requirement is that the resource not be taken away from one user unless the other user wants it. Although somewhat awkward to state in natural language, these requirements find a simple formalization, as is seen in Section 5.

In order to implement a monitor that meets these requirements, a variable internal to the monitor, {T,NT}, is introduced. The two conditions indicate whether the resource can be transferred from one user to the other:

T – Resource can be transferred.

NT – Resource cannot be transferred.

A system's design is specified by a set of implications, each implication being of the form,

$$C1 \cdot C2 \cdot \ldots \cdot Cm \longrightarrow Cn$$

where the Ci are conditions. This expression means that a state containing C1, C2, ..., and Cm can be immediately followed only by a state containing Cn. The implications describing the monitor's design are listed in Table 1. A simulation is any state sequence that is consistent with each of these implications. Timing is introduced by associating each state of a simulation with a unit of time. The implications thus determine how conditions in one time-frame influence conditions in the next time-frame.

To show that the implications in Table 1 support the desired behavior, we turn to the concept of a constraint.

Table 1: Implications Representing Monitor Design

$G1 \cdot NW2 \longrightarrow G1$	(1)
$G2 \cdot NW1 \longrightarrow G2$	(2)
$G1 \cdot NT \longrightarrow G1$	(3)
$G2 \cdot NT \longrightarrow G2$	(4)
$W1 \cdot G2 \cdot T \longrightarrow G1$	(5)
$W2 \cdot G1 \cdot T \longrightarrow G2$	(6)
$W1 \cdot G2 \cdot T \longrightarrow NT$	(7)
$W2 \cdot G1 \cdot T \longrightarrow NT$	(8)
$NT \longrightarrow T$	(9)
$G1 \cdot NW2 \longrightarrow T$	(10)
$G2 \cdot NW1 \longrightarrow T$	(11)

4. CONSTRAINTS

Although our model permits variables of differing sizes, the notion of a constraint is especially easily to formulate when all of the variables are binary. Accordingly, in what follows it is assumed that each variable has exactly two conditions. The general case is treated in [6] and [7].

DEFINITION: A <u>partial state</u> is a set of conditions containing at most one condition from each variable.

DEFINITION: A sequence of sets α is <u>embedded</u> in a sequence of sets β if and only if there exists a subsequence γ of β such that,*

$$|\gamma| = |\alpha|$$
$$\alpha(i) \subseteq \gamma(i) \qquad \text{for } 1 \leq i \leq |\alpha|$$

DEFINITION: A <u>constraint</u> is a finite sequence of partial states that is not embedded in any simulation. A constraint that has no other constraints embedded within it is said to be <u>prime</u>.

A constraint represents, in effect, an incompatibility among successive states of a simulation – a pattern of behavior that cannot occur. A prime constraint is an 'essential' incompatibility since it cannot be reduced without yielding a nonconstraint.

Consider the monitor specification of the preceding section. The sequence of partial states,

$$<\{W1,G2,T\},\ \{G2\}>$$

is an example of a prime constraint. That it is a constraint follows from Implication 5, which says that if User 1 wants the resource (W1), User 2 has the resource (G2), and the resource is transferable (T), then in the next state the resource must be granted to User 1 (G1). A state containing W1, G2, and T, therefore, cannot be followed by a state containing G2. Or equivalently, a simulation may contain the state sequence <S1,S2> only if,

$$NW1 \in S1 \ \vee\ G1 \in S1 \ \vee\ NT \in S1 \ \vee\ G1 \in S2$$

---

* For a sequence α, $|\alpha|$ denotes the length of α and $\alpha(i)$ denotes the i'th component of α.

To see that the aforementioned constraint is prime, we observe that removing a condition or removing a partial state from either the left side or the right side yields a nonconstraint. Deleting the G2 from the first partial state, for example, yields a sequence that is embedded in the simulation,

<{W1,W2,G1,T}, {W1,W2,G2,NT}>.

An example of a constraint that is not prime is,

<{G1,W2}, {W1,G2,T}, {G2,T}, {NW2}>.

Embedded in the middle of this sequence is the prime constraint given previously.

One of the key features of prime constraints is that they provide a characterization for the set of simulations. (Proofs of theorems are given in [7].)

> THEOREM: A state sequence is a simulation if and only if it has no prime constraints embedded within it.

But the simulations of the monitor have already been defined by a set of implications. Converting an impliocation into a constraint is straightforward. The antecedent provides the conditions for the first partial state and the 'complement' of the consequent provides the condition for the second partial state of a 'two-place' constraint. When applied to the implications in Table 1, this transformation produces the (prime) constraints in Table 2.

Table 2: Design Constraints for Monitor

<{G1,NW2}, {G2}>	(1)
<{G2,NW1}, {G1}>	(2)
<{G1,NT}, {G2}>	(3)
<{G2,NT}, {G1}	(4)
<{G2,W1,T}, {G2}>	(5)
<{G1,W2,T}, {G1}>	(6)
<{G2,W1,T}, {T}>	(7)
<{G1,W2,T}, {T}>	(8)
<{NT}, {NT}>	(9)
<{G1,NW2}, {NT}>	(10)
<{G2,NW1}, {NT}>	(11)

5. EXTERNAL BEHAVIOR

In the monitor example, the variables {W1,NW1}, {W2,NW2}, and {G1,G2} are externally observable, while the variable {T,NT} is internal to the monitor and not visible to the outside. It is assumed that such a distinction can always be made.

POSTULATE: The variables are partitioned into internal and external variables.

The notion of externally observable behavior follows naturally.

DEFINITION: An external state-sequence is a sequence of partial states in which each partial state contains exactly one condition from each external variable and no other conditions. An external simulation is an external state-sequence that is embedded in a (total) simulation.

The external simulations thus represent the externally observable component of total system behavior.

We have already seen that (total) simulations are characterized by the set of prime constraints. A similar result holds for external simulations.

DEFINITION: An external constraint is a constraint containing only conditions from external variables.

<{G1,NW2}, {G2}> is an example of an external constraint, but <{G1,NT}, {G2}> is not because NT belongs to an internal variable.

THEOREM: An external state sequence is an external simulation if and only if it has no external prime constraints embedded within it.

What this result says is that in order to determine the external behavior of a system, we need consider only those prime constraints that are restricted to the external variables. All other constraints – i.e., those containing conditions from hidden variables – may be disregarded.

Since requirements are concerned with external behavior, they can be expressed in terms of external constraints. The external constraints representing the requirements on the monitor are given in Table 3. The first two constraints say that if one user wants the resource and the other user has had it for two time units, then the resource is transferred. The third and fourth constraints prohibit the granting of the resource for only one time unit. The last two constraints prohibit the resource from being transferred to a user that does not want it. The task now is to show that these requirement constraints follow from the design constraints.

Table 3: Requirement Constraints for Monitor
<{G1}, {G1,W2}, {G1}>
<{G2}, {G2,W1}, {G2}>
<{G1}, {G2}, {G1}>
<{G2}, {G1}, {G2}>
<{G1,NW2}, {G2}>
<{G2,NW1}, {G1}>

6. RESOLUTION

Although the constraints describing the monitor's design are all prime, they represent only a fraction of all the prime constraints. There are many additional prime constraints – some of length two and others of greater length – and they are all deducible from the constraints in Table 2. To generate these other constraints, an extension of the resolution principle [13,15,19,28,32] is used. The attractiveness of resolution as a rule of inference for constraints stems from two properties: (1) The (successful) resolution of two constraints yields another constraint. (2) Resolution alone is sufficient to generate every prime constraint following from a set of constraints.

Consider the two constraints

<{G1,NW2}, {NT}>

<{G1,W2,T}, {G1}>

from Table 2. The NT in the first constraint is to be resolved with the T in the second constraint. (The NW2 can also be resolved with the W2.) We note, however, that the NT appears in the second partial state while the T appears in the first partial state. To bring the two into alignment and to produce constraints of equal length, a { } is appended to the right side of the first constraint and to the left side of the second constraint, giving the two new constraints:

<{G1,NW2}, {NT}, { }>

<{ }, {G1,W2,T}, {G1}>.

Paralleling conventional resolution, the two resolved conditions are deleted, and then the componentwise union of the resulting sequences is taken, giving

<{G1,NW2}, {G1,W2}, {G1}>.

The new sequence is checked to see that only partial states result from the component-wise union. If any component of the new sequence contains both conditions from a single variable, then the resolution fails. In the example, we see that each of the three components is a partial state, and so the resolution is successful.

7. PCGRAPH

If a finite set of constraints gave rise to only a finite number of prime constraints, then the entire set of prime constraints could be generated by a finite number of resolutions. This, however, is not the case.

Consider just the single prime constraint

<{G1,NW2}, {G2}>.

By duplicating this constraint and then resolving the G1 in one copy with the G2 in the other, the new prime constraint

<{G1,NW2}, {NW2}, {G2}>

is generated. The G2 in this new constraint, in turn, can be resolved with the G1 in the original constraint producing,

<{G1,NW2}, {NW2}, {NW2}, {G2}>.

This process can be repeated ad infinitum, generating an infinite series of (prime) constraints. (What this series tells us is that the resource will remain with one user as long as the other user does not want it.)

To deal with this situation, we rely upon a result that follows from the regularity assumption for simulations.

THEOREM: The set of prime constraints is a regular language.

This result means that the set of prime constraints can be represented by a finite-state acceptor. PCGRAPH, an algorithm incorporating resolution and coded in LISP, generates such a graph. It takes as input an acceptor for a set of constraints and generates as output an acceptor for the set of prime constraints that follow from the input constraints. (A detailed description of the algorithm and supporting mathematics may be found in [7].) When the constraint <{G1,NW2}, {G2}> is fed to PCGRAPH, the acceptor shown in Figure 1 is generated. The starting nodes are those nodes without entering arcs, and the accepting nodes are those nodes without emerging arcs.

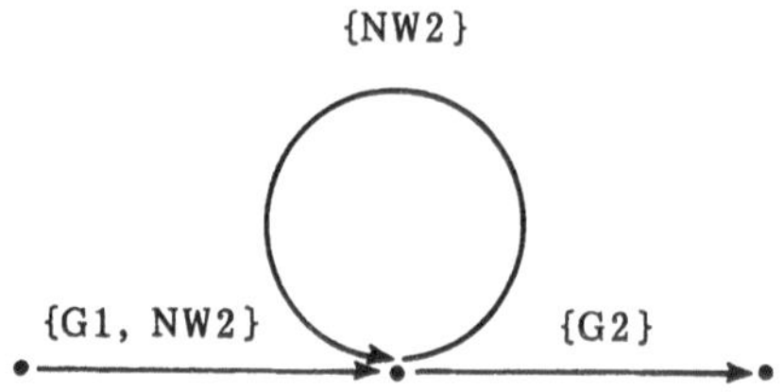

Figure 1: Prime-Constraint Acceptor

Once a (complete) prime-constraint acceptor has been generated, an acceptor for just the external prime constraints can be obtained by pruning out those arcs that mention internal variables and those arcs that no longer appear in a path from a starting node to an accepting node. Thus, to verify the monitor design, PCGRAPH is applied to the design constraints in Table 2 and the resulting graph pruned to eliminate all mention of {T,NT}. The final graph is examined to see that each of the requirement constraints in Table 3 is either present or represented by a simpler prime constraint.

The only disadvantage of this approach is that the graphs generated tend to be quite large. An alternative approach exploits the fact that, in general, only a subset of the design constraints is needed to prove an individual requirement constraint. Thus, instead of supplying PCGRAPH with all the design constraints simultaneously, several runs are made through PCGRAPH, each with a different subset of design constraints. Of course, one still has to determine an appropriate subset for each requirement, but insight into the problem usually suggests one. For example, insight suggests that the subset of constraints in Table 4 might provide a basis for the requirement.

$$<\{G1\},\ \{G2\},\ \{G1\}>.$$

When PCGRAPH – followed by pruning – is applied to this subset, the acceptor shown in Figure 2 is generated. Since it contains the desired constraint, the requirement is verified. Similar verifications can be carried out for each of the remaining requirements. (The last two requirement constraints follow trivially from the first two design constraints.)

Table 4: Subset of Design Constraints
<{G1,NW2}, {G2}>
<{G1,NT}, {G2}>
<{G2,NT}, {G1}>
<{G1,W2,T}, {T}>

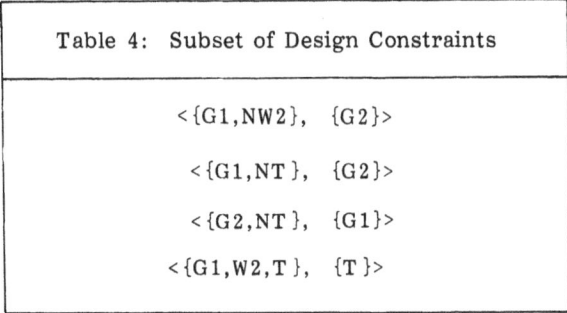

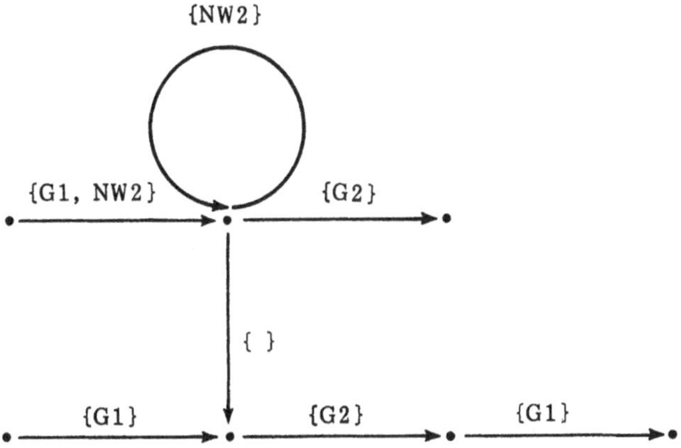

Figure 2: Prime-Constraint Acceptor

8. CONCLUSIONS

The preceding sections have shown how constraints can be used to specify both the design of a system and the requirements on external behavior, and verify that the design meets the requirements. While the approach has broad applicability, it is especially suited to real-time, distributed systems. The issues that must be addressed in these systems have been identified as: timing, concurrency, indeterminacy, and ongoing behavior.

As was shown in the monitor example, timing relationships can be introduced by interpreting the states of a simulation as time slices, each one associated with a unit of time – much like the frames of a motion-picture film. With this interpretation, timing constraints become indistinguishable from logical constraints, and the same techniques become applicable to both.

Although concurrent activity is not readily apparent in a set of state sequences, it becomes apparent in the set of prime constraints because they represent only essential dependencies. The presence of conditions from two distinct variables in a prime constraint of length n indicates an interaction between the two variables spanning a sequence of n states. Prime constraints thus provide a measure of how loosely coupled two variables are. In the extreme case where a system consists of two completely independent components, the set of prime constraints will be partitionable into two subsets, each pertaining to the variables of only one component.

In most system-modeling techniques, indeterminacy must be explicitly mentioned, but with constraints indeterminacy is implicitly assumed. Unless one of two alternate choices is excluded by a constraint, then either may occur. Adding constraints thus reduces indeterminacy. While this approach may place an additional burden on the system specifier, it provides considerable flexibility.

Because constraints deal with arbitrary state sequences rather than pairs of initial and terminal states, ongoing behavior is readily expressible.

9. FUTURE WORK

Because a modest set of constraints can give rise to an extremely complex acceptor for the set of external constraints, the possibility of generating a 'basis' acceptor for the external constraints should be explored. Such a basis acceptor would be equivalent to the complete external acceptor since any external prime constraint could be generated, if needed, from the basis.

Another approach to reducing complexity would be to make PCGRAPH goal-directed. If the object of a verification is to prove the existence of a constraint, or set of constraints, then criteria may be established for resolutions contributing towards that goal. For example, if the maximum length of the constraints to be proved is n, then those resolutions yielding constraints of length greater than n may be omitted. PCGRAPH can, of course, be made interactive with the user deciding which resolutions to perform.

With the ability of constraints to represent external behavior, there already exists the foundation for a hierarchical methodology. The theory for dealing with mappings between different levels of description needs to be developed, with special attention given to representing different granularities of time.

Although constraints excel in representing the timing and synchronization aspects of behavior, they are not as well suited to representing the functional aspects of behavior. The feasibility of integrating the constraint model with an applicative model, such as the lambda calculus, should be explored.

REFERENCES

1. Campbell, R.H. and A.N. Habermann, "The Specification of Process Synchronization by Path Expressions", Lecture Notes in Computer Science, Vol. 16, Springer-Verlag, 1974, pp. 90-102.

2. Dijkstra, E.W., A Discipline of Programming, Prentice-Hall, Englewood Cliffs, N.J., 1976.

3. Floyd, R.W., "Assigning Meanings to Programs", Mathematical Aspects of Computer Science, Vol. 19, Schwartz, J.T., ed., American Mathematical Society, Providence, R.I., 1967, pp. 19-32.

4. Furtek, F.C., "Constraints and Compromise", Foundations of Secure Computation, (DeMillo, R.A., D.P. Dobkin, A.K. Jones, R.J. Lipton, ed.), Academic Press, 1978, pp. 189-204.

5. Furtek, F.C., A Validation Technique for Computer Security Based on the Theory of Constraints, ESD-TR-78-182, Electronic Systems Division, AFSC, Hanscom AFB, MA, December 1978.

6. Furtek, F.C., "Doing Without Values", Ninth International Symposium on Multiple-Valued Logic, Bath, England, 1979, pp. 114-120.

7. Furtek, F.C., "The Theory of Constraints", In Preparation.

8. Greif, I., "A Language for Formal Problem Specification", Commun. Assoc. Comput. Mach., Vol. 20, No. 12, December 1977, pp. 931-935.

9. Hoare, C.A.R., "An Axiomatic Basis of Computer Programming", Commun. Assoc. Comput. Mach., Vol. 12, October 1969, pp. 576-580.

10. Holt, A.W., et al., Final Report of the Information System Theory Project, RADC-TR-68-305, Rome Air Development Center, Griffis AFB, NY, September 1968.

11. Holt, A.W. and F.C. Commoner, Events and Conditions, Applied Data Research, Inc., New York, 1970.

12. Hopcroft, J.E. and J.D. Ullman, Formal Languages and Their Relation to Automata, Addison-Wesley, Reading, MA, 1969.

13. Knuth, D.E., Letter in SIGACT News, Vol. 9, No. 1, Jan.-March 1977, pp. 8-9.

14. Lautenbach, K. and H. Wedde, "Generating Control Mechanisms by Restrictions", Lecture Notes in Computer Science, Vol. 45, Springer-Verlag, 1976, pp. 416-422.

15. Luckham, D., "The Resolution Principle in Theorem-Proving", Machine Intelligence 1, (Collins, N.L. and D. Michie, ed.), American Elsevier, N.Y., 1967, pp. 47-61.

16. Lynch, N.A. and M.J. Fischer, On Describing the Behavior and Implementation of Distributed Systems, Technical Report GIT-ICS-79/03, School of Information and Computer Science, Georgia Institute of Technology, May 1979.

17. Maggiolo-Schettini, A., H. Wedde, and J. Winkowski, "Modeling a Solution for a Control Problem in Distributed Systems by Restrictions", Lecture Notes in Computer Science, Vol. 70, Springer-Verlag, 1979, pp. 226-248.

18. Manna, Z. and R. Waldinger, "The Logic of Computer Programming", IEEE Transactions on Software Engineering, Vol. SE-4, No. 3, May 1978, pp. 199-229.

19. McCluskey, E.J., "Minimization of Boolean Functions", Bell System Tech. J., Vol. 35, No. 5, November 1956, pp. 1417-1444.

20. Millen, J.K., "Constraints and Multilevel Security", Foundations of Secure Computation, (R.A. DeMillo, D.P. Dobkin, A.K. Jones, R.J. Lipton, ed.), Academic Press, 1978, pp. 205-222.

21. Millen, J.K., Causal System Security, ESD-TR-78-171, Electronic Systems Division, AFSC, Hanscom AFB, MA, October 1978.

22. Parnas, D.L., "A Technique for Module Specification with Examples", Commun. Assoc. Comput. Mach., Vol. 15, No. 5, May 1972, pp. 330-336.

23. Petri, C.A., Communication with Automata, Supplement 1 to RADC-TR-65-377, Vol. 1, Rome Air Development Center, Griffiss AFB, New York, 1966. [Originally published in German: Kommunikation mit Automaten, Schriften des Rheinisch-Westfalischen Institutes fur Instrumentelle Mathematik an der Universitat Bonn, Hft. 2, Bonn, 1962.]

24. Pnueli, A., "The Temporal Logic of Programs", 19th Ann. Symp. on Found. of Comput. Sci., Providence, R.I., November 1977, pp. 46-57.

25. Pnueli, A., "The Temporal Semantics of Programs", Lecture Notes in Computer Science, Vol. 70, Springer-Verlag, 1979, pp. 1-20.

26. Pratt, V.R., "Semantical Considerations on Floyd-Hoare Logic", Proc. 17th Ann. IEEE Symp. Found. of Comp. Sci., 1976, pp. 109-121.

27. Pratt, V.R., "Process Logic: Preliminary Report", Conf. Record of 6th Ann. Symp. on Prin. of Prog. Lang., San Antonio, Texas, January 1979, pp. 93-99.

28. Robinson, J.A., "A Machine-Oriented Logic Based on the Resolution Principle", J. Assoc. Comput. Mach., Vol. 12, 1965, pp. 23-41.

29. Robinson, L. and K.N. Levitt, "Proof Techniques for Hierarchically Structured Programs", Commun. Assoc. Comput. Mach., Vol. 20, No. 4, April 1977, pp. 271-283.

30. Scott, D., "Mathematical Concepts in Programming Language Semantics", SJCC 72, AFIPS Press, 1972, pp. 225-234.

31. Shields, M.W., "Adequate Path Expressions", Lecture Notes in Computer Science, Vol. 70, Springer-Verlag, 1979, pp. 249-265.

32. Tison, P., "Generalization of Consensus Theory and Application to the Minimization of Boolean Functions", IEEE Trans. Computers, Vol. EC-16, No. 4, August 1967, pp. 446-456.

33. Wedde, H. and J. Winkowski, "Determining Processes by Violations", Lecture Notes in Computer Science, Vol. 53, Springer-Verlag, 1977, pp. 549-559.

Reasoning by Plausible Inference *

Leonard Friedman
Jet Propulsion Laboratory, C.I.T.
Pasadena, CA 91103, USA

ABSTRACT

The PI (Plausible Inference) system has been implemented with the ability to reason in both directions, unlike previous AI systems for inexact reasoning. This is combined with truth-maintenance, dependency-directed backtracking, and time-varying contexts to permit modelling dynamic situations. Four rules of inference are employed in PI, Modus Ponens, Modus Tollens, Confirmation, and Denial. Each of these causes credibility as well as truth value to be propagated in a semantic network of assertions. Applications are foreseen in 'guessing' the answers to problems in formal deduction so that search trees may be pruned, and in the flexible selection of strategies for planning, as well as diagnostic or trouble-shooting models involving less-than-certain assertions.

* This paper represents the results of one phase of research carried out at the Jet Propulsion Laboratory, California Institute of Technology, under contract NAS7-100, sponsored by the National Aeronautics and Space Administration.

'A person has a background, a machine has not. Indeed, you can build a machine to draw demonstrative conclusions for you, but I think you can never build a machine that will draw plausible inferences.'

George Polya, in 'Patterns of Plausible Inference', p.116
Princeton University Press, Princeton, 1954

1. The ERIS System and Plausible Inference

The need for augmenting the power of Propositional Calculus and First-Order Predicate Calculus has long been evident to the designers of Automatic Deduction systems. In particular, it would be most desirable to represent degrees-of-belief other than certainty, and to obtain the performance of inference rules known to exist in common-sense reasoning, but absent in formal logic. Although formal logic can represent many domain-specific situations, it has been difficult to introduce domain knowledge into the search mechanisms involved in deduction. Features augmenting the power of deduction and planning systems that have recently been introduced include truth-maintenance, and dependency-directed backtracking. (Fikes '75, Stallman and Sussman '77, Doyle '78, London '78). The ERIS system, under development at JPL for several years, is attempting to integrate all of these capabilities into a single unified knowledge-engineering system. The problem domain is Mission Operations, the process of controlling a spacecraft from Earth. This involves, among other activities, generating and verifying computer command sequences which are relayed to the on-board spacecraft computer so that it may run new scientific experiments. Scheduling of experiments is also of prime importance for events like planet encounter. These are problems that have resisted being automated, and are candidates for ERIS solutions.

An ERIS deduction system has been implemented for propositional calculus that maintains consistency or discovers either contradiction or loss of support. See proceedings of this conference (A. Thompson, '80). A planning system that is built from an early version of the deduction system is running. Similarly, PI (for Plausible Inference), a system for common-sense reasoning and deduction, has been implemented starting with an early version of the deduction system. PI permits continuous degrees of belief and, unlike previous plausibility systems, it reasons in both directions. PI will make it possible to use a general method for introducing domain-specific knowledge into formal logic searches. The underlying theory of PI, a brief comparison of its performance with other systems and the direction of future improvements will be given here. Efforts are under way to integrate the systems, and theoretical investigations with regard to extending the deduction techniques used in ERIS to First-Order Predicate Calculus indicate that this is feasible, and it will be undertaken soon. When that has been accomplished, Planning and Plausible Inference will be extended to First-Order logic.

The basic concepts of Plausible Inference were introduced by George Polya (Polya, '54). He abstracted several inference rules not present in formal logic from the methods actually employed by people when reasoning (even mathematicians!). I call these the rule of confirmation and the rule of denial. Together with Modus Ponens and Modus Tollens they define all the possibilities for propagation of truth value in implications. However, Polya's rules characterized beliefs by terms such as 'more-likely' and 'less-likely'. Such non-quantitative descriptions are difficult to use in a computer formalism. Shortliffe and Buchanan used the rule of confirmation in MYCIN (Shortliffe and Buchanan, '75 ; hereafter referred to as S and B). To make the rule calculable they introduced a metric for belief propagation which they defined in terms of functions of conditional probabilities. Although the metric was defined in this way, S and B never used probabilities while operating the system. Instead, they relied on the inputs of human 'experts' who supplied numbers that were not probabilities, but measures of the belief in truth or falsity of assertions. Bellman and Zadeh had earlier proposed a similar measure (Bellman and Zadeh, '70). We have adopted S and B's metric and system of adding and combining beliefs, and will review it here in our notatation, shorn of any reference to probability. The interested reader may refer to their original paper or (Shortliffe '76) for a complete description of their rationale in developing the metric.

2. Credibility Propagation

Each assertion in the ERIS knowledge base has an associated property called 'Truth Value' (TV), with a value set of true, false, or unknown, {T,F,U}. In Plausible Inference, we introduce an additional property called 'credibility' which represents the degree-of-confidence in the truth or falsity of the assertion. Either truth value or credibility will be referred to as a belief. Let C(A) be the credibility of any assertion A. The value of C(A) is a number that ranges from -1 to

+1 with 0 representing the unknown, +1 representing complete certainty and -1 complete disbelief. In formal logic C(A) is restricted to the three values {-1,0,+1}.

PI operates so as to maintain consistency between the notions of truth of an assertion, and the degree of belief in the truth of that assertion. If credibility becomes positive, the system regards the assertion as true with the credibility just established. If credibility becomes negative, the assertion is regarded as false with that credibility. The convention applies to truth values as well when they change. The rules are:

For any assertion A,

$$[TV(A)=T] \equiv [C(A) > 0],$$
$$[TV(A)=F] \equiv [C(A) < 0],$$

This equivalence is called the C-TV Consistency rule.

There are two distinctively different ways for propagating credibility in Plausible Inference. The first, which applies only to the antecedent and consequent of an implication, is directed propagation. Antecedents transmit all or a fraction of their credibility (belief or disbelief) to specific consequents when they change truth value. Consequents do the same to antecedents. The amount depends on their own credibility reduced by a fixed factor, $\Delta 1$, defined below. $\Delta 1$ is a directed quantity, somewhat in the sense that a vector is, while credibility is like a scalar. In addition, for antecedents and consequents, evidence is weighed for or against an assertion when it is both supported and denied by other assertions. This is done by adding the increments to its credibility in a non-linear way, then the decrements and finally taking the difference of all credibilities propagated to it in this directed fashion.

The second way for propagating credibility is non-directed and applies to expressions using the logical connectives, AND, OR, and IMPLIES. The collective credibility of such expressions is given by the maximum or minimum credibility of one of their constituent assertions. (For NOT, the credibility increments and decrements are exchanged so that belief becomes disbelief and vice versa.) This difference in the way credibility is propagated models our intuitive notion that the believability of the members of an implication is linked; whereas, the member assertions of a conjunction or disjunction are treated as independently believable factors. This applies even in the case that a disjunction is an axiom that must be true. If all but one of its member assertions are believed to be false, there is a propagation of increased credibility to the remaining assertion, but it doesn't matter which one, since it is not a directed quantity flowing specifically from one member to another. An interpretation of these differences from the point of view of Fuzzy Logic is given by Yager (Yager, '79).

The formulae of Plausible Inference may be developed as modifications to the inference rules of formal logic. The relationships of formal logic (reading '->' as implies) are represented in Modus Ponens as:

A -> B and [TV(A)=T;C(A)=1],
Therefore [TV(B)=T;C(B)=1].

Modus Tollens reads as:

A -> B and [TV(B)=F;C(B)= -1],
Therefore [TV(A)=F;C(A)= -1].

Polya pointed out that in normal human reasoning two other rules of inference may be inferred. These rules are called here the Method of Confirmation and the Method of Denial.

The Method of Confirmation, in the form suggested by Polya, may be stated as:

A -> B and [TV(B)=T],
Therefore the credibility increment
propagated to A from B is 'more-likely'.

The Method of Denial in Polya's form is:

A -> B and [TV(A)=F],
Therefore the credibility decrement
propagated to B from A is 'less-likely'.

Note that these equations fill the gaps forbidden in formal logic, propagating credibility change and possibly truth value change where nothing is transmitted in formal logic.

We now introduce a number Δ to represent the change of credibility propagated. Δ is directed to the antecedent from the consequent (or vice versa) and represents the increment (or decrement) of credibility propagated between them. Our notation is:

$\Delta(A|B)$, the change of credibility propagated to A from B.
$\Delta(B|A)$, the change of credibility propagated to B from A.
$0 \leq \Delta \leq 1$.

Note also that the syntax of all four equations specifies whether what is transmitted is an increment or decrement. See Table I.

TABLE I, (A -> B)

CONFIRMATION	TV(B)=T	$\Delta(A\|B)$ is an increment
TOLLENS	TV(B)=F	$\Delta(A\|B)$ is a decrement
PONENS	TV(A)=T	$\Delta(B\|A)$ is an increment
DENIAL	TV(A)=F	$\Delta(B\|A)$ is a decrement

One further definition will permit us to state quantitative equations of Plausible Inference concisely.

Let $\Delta 1$ be the increment or decrement of credibility transmitted when there is either complete certainty or complete disbelief in either the antecedent or consequent; i. e., in confirmation, when there is certainty that the consequent is true, the increment of credibility transmitted to the antecedent, Δ, is the value of $\Delta 1$. Our notation for 'A -> B' is:

$\Delta 1(A|B|+)$, the value of $\Delta(A|B)$ when $C(B) = 1$.
$\Delta 1(A|B|-)$, the value of $\Delta(A|B)$ when $C(B) = -1$.
$\Delta 1(B|A|+)$, the value of $\Delta(B|A)$ when $C(A) = 1$.
$\Delta 1(B|A|-)$, the value of $\Delta(B|A)$ when $C(A) = -1$.

2.1 Quantitative Plausible Inference

PI operates with a dynamic ERIS knowledge base. Whenever a state-change takes place, causing either a change in the truth value or credibility of an assertion, a new PLANNER-like context layer is pushed. Once such a change or disturbance has taken place, propagation of beliefs occurs along lines of support for logically related assertions within the context layer until a state of equilibrium is reached.

In Plausible Inference, the values of $\Delta(A|B)$ or $\Delta(B|A)$ may be calculated by the following four equations for the inference 'A -> B'.

1) Confirmation

If in a new context layer, $[C(B) > 0]$ is asserted, when previously $C(B)$ had been zero, then

$$\Delta(A|B) = \Delta 1(A|B|+) * C(B).$$

Suppose that A represents the assertion 'The robot hand is grasping object1', and B represents the assertion 'The hand touch sensors are on'.

Experience may have shown that if we are certain the touch sensors are on, our confidence in the robot grasping something makes $\Delta 1 = 0.7$. If, however, those touch sensors are noisy, the robot may have to assign a credibility less than one to 'touch sensors are on'. This decrease in certainty of touch proportionately reduces the credibility increment transmitted to the grasp assertion below 0.7 .

2) Graded Tollens

If in a new context layer $[C(B) < 0]$ is asserted, when previously $C(B)$ had been zero, then

$$\Delta(A|B) = \Delta 1(A|B|-) * |C(B)|.$$

Unlike the situation in formal logic, the decrease in the credibility of A may be limited to a quantity much less than $C(B)$ if B is false, as in the implication: 'Most professional basket-ball players are more than 2 meters tall' -> 'John Doakes, the basket-ball player, is more than 2 meters tall'. There is no difficulty in representing this in Plausible Inference by a $\Delta 1(A|B|-)$ less than one.

If $C(B) = -1$ and $\Delta 1(A|B|-) = 1$, then $\Delta(A|B) = 1$, and $C(A) = -1$ regardless of other support. (Unless there is support for $C(A) = 1$, which would indicate a contradiction).

3) Graded Ponens

If in a new context layer [C(A) > 0] is asserted, when previously C(A) had been zero, then

$$\Delta(B|A) = \Delta 1(B|A|+) * C(A).$$

If C(A) = 1 and Δ1(B|A|+) = 1, then Δ(B|A) = 1, and C(B) = 1 regardless of other support. (Unless there is support for C(B) = -1, which would indicate a contradiction).

4) Denial

If in a new context layer [C(A) < 0] is asserted, when previously C(A) had been zero then

$$\Delta(B|A) = \Delta 1(B|A|-) * |C(A)|.$$

Note that graded Tollens and Ponens become identical to their counterparts in formal logic when C(B) = -1 or C(A) = 1 respectively, with Δ1(A|B|-) = Δ1(B|A|+) = 1. To get full formal logic requires in addition that Δ1(A|B|+) = Δ1(B|A|-) = 0 for equations 1 and 4, thus blocking transmission of belief in the forbidden directions. While running in the PI system, it is possible to temporarily reset the Δ1's for a selected subset of assertions, if there is a need to restrict inferencing to the rules of formal logic for that subset. (The reader who is familiar with S and B's work can translate our PI notation into S and B's notation by noting that our C(A) is their CF(A,B), our Δ(A|B) is their MB(A,B) or MD(A,B), and Δ1(A|B|+), is MB'(A,B). S and B's equations for strength of evidence are equivalent to our equation 1 for Confirmation.)

3. Combining Credibilities

We have just considered belief propagation in an implication with a single antecedent and consequent. When dealing with more complex logical expressions with multiple support for antecedents or consequents, S and B's methods for combining beliefs propagated from consequents can be extended to beliefs propagated from antecedents. The following rules include both cases. (Rules marked with a ** do not appear in S and B.)

3.1 Defining Criteria

(1) Notation

A is usually an antecedent or a hypothesis.
B is usually a consequent or evidence.

I(A) is defined as the combination of increments of credibility to A. $[I(A) \geq 0]$.

D(A) is defined as the combination of decrements of credibility to A. $[D(A) \geq 0]$.

$C(A) = I(A) - D(A)$.

B+ is defined as the set of confirming evidence
B- is defined as the set of disconfirming evidence
A+ is defined as the set of hypotheses that imply the evidence.
A- is defined as the set of hypotheses that deny the evidence.

$\wedge$ is defined as logical AND
$\vee$ is defined as logical OR
~ is defined as logical NOT

(2) Relation between belief and disbelief

$I(A) = D(\sim A)$.

In words, the belief in evidence for any proposition A equals the disbelief in evidence against it.

(3) Absolute Confirmation or Disconfirmation

(A) If $\Delta(A|B+) = 1$, then $C(A) = 1$, regardless of B-.
(B) If $\Delta(A|B-) = 1$, then $C(A) = -1$,regardless of B+.
(C) If $\Delta(A|B-) = \Delta(A|B+) = 1$, this is contradictory and C(A) is undefined.

Similar considerations hold for $\Delta(B|A-)$ and $\Delta(B|A+)$. **

(4) Commutativity

If B1 $\wedge$ B2 indicates an ordered observation of evidence, first B1, then B2

(A) $\Delta(A|B1 \wedge B2) = \Delta(A|B2 \wedge B1)$
[C(A) from B1 $\wedge$ B2] = [C(A) from B2 $\wedge$ B1].

If A1 $\vee$ A2 indicates an ordered formulation of hypotheses, first A1, then A2

(B) $\Delta(B|A1 \vee A2) = \Delta(B|A2 \vee A1)$. **
[C(B) from A1 $\vee$ A2] = [C(B) from A2 $\vee$ A1]. **

(5) Missing Information

If [TV(B2)=U]
Then $\Delta(A|B1 \wedge B2) = \Delta(A|B1)$.
[C(A) from B1 $\wedge$ B2] = [C(A) from B1]

If [TV(A2)=U]
Then $\Delta(B|A1 \vee A2) = \Delta(B|A1)$. **
[C(B) from A1 $\vee$ A2] = [C(B) from A1]. **

3.2 Combining Functions

When combining the increments of credibility of multiply supported assertions, first the increments and decrements of credibility (confirming and disconfirming) are combined by rules given below, then the difference between confirming and disconfirming beliefs is taken to give the resultant credibility. In what follows, C(A) or C(B) need not be +1 or -1 unless explicitly noted.

(1) Convergence of Evidence to a Given Hypothesis

(COMB1) Assume A -> (B1 $\wedge$ B2)
~A -> (B3 $\wedge$ B4).

B1, B2 are confirming evidence for A; B3, B4 are disconfirming evidence.

Suppose B1,B2,B3,B4 are all true.

Then

I[A] = Δ[A|B1] + Δ[A|B2](1 - Δ[A|B1]),

D[A] = Δ[A|B3] + Δ[A|B4](1 - Δ[A|B3]),

C[A] = I[A] - D[A].

(2) Convergence of Hypotheses on Given Evidence **

(COMB2) Assume (A1 V A2) -> ~B

(A3 V A4) -> B

Let all antecedents be false.

Then

I[B] = Δ[B|A1] + Δ[B|A2](1 - Δ[B|A1]) **

D[B] = Δ[B|A3] + Δ[B|A4](1 - Δ[B|A3]) **

C[B] = I[B] - D[B]. **

(3) Fanout of Evidence to Conjunctions of Hypotheses

(COMB3a) Δ[(A1 Λ A2)|B] = min(Δ[A1|B], Δ[A2|B])

(COMB3b) Δ[~(A1 Λ A2)|B] = max(Δ[~A1|B], Δ[~A2|B])

(4) Fanout of Evidence to Disjunctions of Hypotheses

(COMB4a) Δ[(A1 V A2)|B] = max(Δ[A1|B], Δ[A2|B])

(COMB4b) Δ[~(A1 V A2)|B] = min(Δ[~A1|B, Δ[~A2|B])

(5) Fanout of an Hypothesis to Conjunctions of Evidence **

(COMB5a) Δ[(B1 Λ B2)|A] = min(Δ[B1|A], Δ[B2|A]) **

(COMB5b) Δ[~(B3 Λ B4)|A] = max(Δ[~B3|A], Δ[~B4|A]) **

(6) Fanout of an Hypothesis to Disjunctions of Evidence **

(COMB6a) Δ[(B1 V B2)|A] = max(Δ[B1|A], Δ[B2|A]) **

(COMB6b) Δ[~(B3 V B4)|A] = min(Δ[~B3|A], Δ[~B4|A]) **

(7) Credibility transmitted to an Implication from its members

(COMB7a) for IMPLIES true, Δ[IMPLIES A B] = max[C(~A), C(B)].**

(COMB7b) for IMPLIES false, Δ[IMPLIES A B] = min[C(A), C(~B)].**

An example may serve to make the application of these rules clearer.

Assume the following assertions are in the data base:

(A) The robot hand is attached to object1.
(B1) The robot hand touch sensors are on.
(B2) Finger spread is close to a visually measured dimension D of object1.
(B3) Object1 moves with the hand, or
(B4) Torque or force on the hand saturates when the hand tries to move, and all velocities remain zero.

and $A \rightarrow (B1 \wedge B2 \wedge [B3 \vee B4])$.

Let c.m. stand for confirming measurements, d.m. for disconfirming measurements, m1 for measurement 1, etc. An example of a confirming measurement might be an above threshold voltage on the touch sensor line. A disconfirming measurement would be its absence.

$$\begin{aligned} \text{Let } I(Z) &= \Delta(B1 \wedge B2 \wedge [B3 \vee B4] \mid \text{c.m.}) \\ &= \min(\Delta[B1 \mid m1], \Delta[B2 \mid m2], \Delta[B3 \vee B4 \mid \text{c.m.}]) \\ &= \min(\Delta[B1 \mid m1], \Delta[B2 \mid m2], \max(\Delta[B3 \mid m3], \Delta[B4 \mid m4])). \end{aligned}$$

$D(Z) = \Delta(B1 \wedge B2 \wedge [B3 \vee B4] \mid \text{d.m.})$ is calculated similarly.

4. Comparisons with Other Plausibility Systems

MYCIN's plausibility deduction machinery is limited to the Confirmation rule of inference for implications going from evidence to hypotheses. Even Modus Ponens and Tollens are missing. The representation also is restricted to single implications whose evidence statements are conjuncts. Embedded disjuncts are allowed, but evidence statements that are disjuncts must be represented by separate statements for each term of the disjunct. Implications embedded within implications are not allowed. None of these restrictions applies to ERIS or PI.

The PROSPECTOR system uses plausibility measures for propagation of belief that are based on a modified subjective Bayesian probability, while utilizing the same Zadeh minimum maximum choice employed in PI for conjuncts and disjuncts (Duda, Hart, Nilsson and Sutherland, '77; Duda, Hart, Konolige, and Reboh, '79). Problems have emerged with their probability measures in the special cases when statistical independence cannot be assumed. Also, as in MYCIN, changes of belief are propagated only along the direction from evidence to hypothesis. Thus, for example, if the user begins by indicating strong suspicion of the presence of a porphyry copper deposit, the program should change its expectations for such things as quartz-sulfide veinlets (Modus Ponens). Originally it did not because the system could not reason from hypothesis to evidence; however, a fix has been made that now permits this by introducing an approximation to an exact Baysian formulation. This enforces constraints on the user. To use the fix, he must supply probability knowledge he may not have.

Both MYCIN and PROSPECTOR use a static system of rules that cannot be changed during a working session. The PI system works with a dynamic base that can infer new rules and introduce outside information that may change the truth value of assertions in successive PLANNER-like context layers. The rules COMB1 and COMB2 given above assume that credibility is being calculated from conditionally independent lines of support in one context layer. Thus, if in a new context layer there is a very small change in credibility to a non-zero belief, essentially the old belief will be propagated to any assertions linked to it and would be added to the previously propagated value as if independent. This error is eliminated in the PI system by checking to see if a new increment of credibility comes from the same source as the previous increment. If it has, the contribution of the previous increment is removed, unless it was 1 (certainty). An attempt to change certainty results in a contradiction. Note that if Credibilities are added that are not conditionally independent, the calculation is in error.

The PI system lacks the powerful techniques of the probability calculus available in PROSPECTOR, but provides the user with the ability to enter his degrees-of-belief in both directions of reasoning even when he does not know exact conditional probabilities. The PI system is capable of solving the same kinds of static problems handled in these systems, and in addition, can model dynamically changing world-states. Because this flexibility is available, it appears feasible to add other features to the PI system enabling it to imitate more closely human belief behavior.

5. Next Steps

Several studies provide rules that may be used to model human performance in weighing evidence and guessing the credibility of assertions in unfamiliar contexts. Polya suggested a number of qualitative rules describing how we revise already formed opinions in the light of new evidence for certain special circumstances. These rules were based on his own introspection and examination of the rational behavior of others, and can be implemented by modifying the combination rules given above. The extensive investigations by Collins of human protocols provide other rules for binding set variables involved in deduction, analogy, induction, generalization, and confirmation (Collins '78).

5.1 Quantitative Opinion Revision Rules

Polya stated his rules in the following form:

(1) If B is unlikely to be true without A, and B changes to true, there is more than a normal increase in strength of belief in A.

(2) If B is supported by many other propositions, B becoming true supports C(A) only weakly.

(3) If A is the only justification known for B, and A becomes false, C(B) is greatly decreased.

These qualitative rules may be given a quantitative form such as the following three rules.

(REV1) Assume in context 1, (A1 V A2 V .... V An) -> B
and TV(A1)=U,
C(A1)=0,
A2, ...,An are false.

C(B) = -D(B), computed by COMB2 and MISSING INFO;
TV(B)=F by C-TV Consistency.

If TV(B)=T in context 2,
C[A1] = I[A1] = Δ[A1|B] + |C[B]ctx1|(1 - Δ[A1|B])

(REV2) Assume in context 1, $(A1 \vee A2 \vee \ldots \vee An) \rightarrow B$

$TV(Ai) = U; i = 1,2,..n.$

$C(Ai) = 0; i = 1,2,..n.$

This state of beliefs leaves TV(B)=U.

If TV(B)=T in context 2,

then $\Delta(A1|B) = (1/n)\Delta1(A1|B|+) * C(B)$.

(REV3) Assume in context 1, A -> B,

and no other support, present or potential,

exists for B.

Let TV(A)=T; C[A] > 0

In context 2, let TV(A)=F. Then

$$C[B] = -D[B] = -[\Delta[B|A] + C[A]ctx1 * (1 - \Delta[B|A])].$$

The loss of belief in consequence B is proportional to previous confidence in A.

5.2 Estimating Unknown Credibilities

We are constantly estimating the credibility of assertions that are not directly inferrable by the methods described above. Collins' analogy example is a vivid one: 'This car model is like another car that Chrysler makes which has a particular problem with the automatic choke, so perhaps that is what is wrong with this car.' The rules suggested by Polya and Collins appear to be implementable as a control structure to direct the matching operations that must take place. This control structure could be represented as a knowledge base of assertions exactly like any other domain, so it could infer its conclusions for control using the present PI system. If the problems of representation and implementation can be solved, this may greatly extend the power of the PI system.

6. Applications of Inexact Reasoning

Polya's thesis in Patterns of Plausible Inference was that even in mathematics, a field where certainties are possible, the mathematician constantly uses plausible reasoning to suggest likely answers before attempting to prove them, and that it is employed in all human reasoning. The PI system, employing all four modes of reasoning, has been used to trouble-shoot plans generated by the ERIS planner. Such plans are models of the normal operation of the space craft. If a command sequence collected from the plan fails when executed by the space craft computer, the PI system can use evidence from tests of asserted functions suspected to be false to narrow down the possible candidates that may have caused the problem. A forthcoming paper describes the details. Other uses of PI may be in combination with formal logic to suggest most promising paths of search for proofs, and as indicated earlier, in the control structure of ERIS to reason about what to do next.

One final comment is in order. When we quoted Polya at the beginning of this paper, we had a double motive. Of course we are underscoring the manifold developments that may now permit a machine to engage in a form of plausible inference, but in addition, we wish to point up his insight into the essential ingredient, a proper background. The recent development of representations, truth-maintenance, and credibility theory have all contributed to giving the computer that necessary background.

References

Bellman and Zadeh '70, 'Decision-Making in a Fuzzy Environment', Management Science, vol. 17, 141-164.

Collins '78, <u>Studies of Plausible Reasoning</u>, Final Report, October 1976 - February 1978, Vol. I: 'Human Plausible Reasoning,' Bolt Beranek and Newman Report No. 3810.

Doyle '78, 'Truth Maintenance Systems for Problem Solving', M.I.T., AI-TR-419, Jan. 1978

Duda, Hart, Nilsson, and Sutherland '77, 'Semantic Network Representations in Rule-Based Inference Systems', S.R.I., AI Center, TN-136, 1977

Duda, Hart, Konolige, and Reboh '79, 'A Computer-Based Consusltant for Mineral Exploration', Final Report, September 1979, S.R.I., AI Center, Menlo Park, CA. 94025

Fikes '75, 'Deductive Retrieval Mechanisms for State Description Models', IJCAI4, Tbilisi, USSR, Sept. 1975, pp. 99-106

London '78, 'Dependency Networks as a Representation for Modelling in General Problem

Solvers', Ph. D. Thesis, U. of Maryland, Computer Science Dept., TR-698, Sept. 1978

Polya '54, 'Patterns of Plausible Inference', Princeton University Press, Princeton, N.J., 1954

Shortliffe and Buchanan '75, 'A Model of Inexact Reasoning in Medicine', Math. Biosci. 23, pp. 351-379, 1975

Shortliffe '76, 'Computer-Based Medical Consultations: MYCIN', Elsevier, New York, 1976

Stallman and Sussman '77, 'Forward Reasoning and Dependency-Directed Backtracking in a System for Computer-Aided Circuit Analysis', Artificial Intelligence, vol. 9, no. 2, pp. 135-196

Thompson '80, 'Logical Support in a Time-Varying Model', These Proceedings.

Yager '79, Measurement-Informational Discussion of Fuzzy Union and Intersection', Int. J. Man-Machine Studies, vol. 11. 189-200

Logical Support in a Time-Varying Model[1]

Alan M. Thompson
Information Systems Division
Jet Propulsion Laboratory
California Institute of Technology
Pasadena, California 91103 USA

Abstract

Techniques using logical dependencies as a means of monitoring the justifications of beliefs in a database have proved to be a valuable tool for automatic deduction and planning. A new technique for determining logical support dynamically within existing structure eliminates the need for declaring dependencies in clause form. Procedural specialists define the behavior of logical connectives and allow domain axioms to be expressed in a declarative form. The system uses dependency relationships in the world model to determine the consequences of actions and assumptions. The use of context layers in addition to dependencies remedies certain difficulties encountered in previous modelling schemes, resulting in improved model accuracy following a state change.

I. Introduction

Early automatic problem solving systems required a programmer to associate with each action description a complete representation of the changes that should be made to the world model when the execution of the action was simulated. In highly coupled problem domains where a single action may alter the truth values of a large number of statements in the model, this requirement can lead to a difficult programming problem. In recently developed knowledge-based problem solving systems, this problem is avoided by the use of logical dependencies in the model database as a means of determining the consequences of assumptions and actions. [Doyle, London, Thompson80]

As the scope of the task of problem solving has expanded to include inference from logical relations as well as the planning of action sequences, the need has grown for the world-modelling component to operate on the same logical relationships used by the inference process. Thus, if some simulated action

---

1 This paper presents the results of one phase of research carried out at the Jet Propulsion Laboratory, California Institute of Technology, under Contract NAS 7-100, sponsored by the National Aeronautics and Space Administration.

changes a state[2] that had been deduced from other states, then (at least some of) the antecedent states must also be changed. Similarly, an action (or the introduction/retraction of an assumption) may provide or remove logical support for other facts. This led to the notion of logical dependency as a means of maintaining the justifications of derived facts and of a process of "Truth Maintenance" (TM) as a means of monitoring the belief or disbelief in statements as a consequence of changes to the database. [Doyle] The TM process guarantees that every statement that is "believed" is well-founded, i.e., its justification is not circular. All other statements are regarded as having an unknown truth value.

The "ERIS" system, currently under development at JPL, is designed to generalize the TM process in several respects. Whereas the systems of Doyle and London required logical dependencies between model statements to be explicitly declared in clause form by a separate inference process, in the ERIS system TM functions are combined with inference functions in such a way that support can be determined within existing structure without the need for applying rewrite rules. This is accomplished through the use of a graph labelling method which operates on expressions using the full range of logical connectives.

A more complete description of the ERIS system is available in other references. [Thompson79, Thompson80] This paper discusses the techniques used by ERIS to determine the changes to be made to a model following changes to the truth values of model statements. In particular, its behavior following a state change caused by the simulated execution of an action will be illustrated.

II. Pertinent System Components

II.1 The Database

States and relations are expressed as nested n-tuples in an indexed database. Every expression has a unique representation, called its node. The database is fully invertible -- every expression has pointers to the nodes of its subexpressions (if it is non-atomic) and to those of its super-expressions, if any. The links point only to the expressions immediately above or below, so that paths from a top-level expression to its constituent atoms must be followed one node level at a time, and vice versa.

---

[2] The following definitions apply to this discussion: A statement is represented by an "expression" (normally a Predicate Calculus formula). Expressions are stored in an indexed (associative) database as "nodes". The term "state" refers to a node with a particular truth state (represented by a "truth value" (TV), such as true, false, etc.). "Axioms", or "rules", are statements of known truth value whose state may not be changed. Statement "connectives" include the normal truth-functional connectives (AND, OR, etc.) plus connectives for representing knowledge about causality.

Truth Values and other properties of nodes (statements) are context dependent, using a PLANNER-style context layer scheme. Temporal contexts provide a valuable method of capturing the history of a process and facilitating hypothetical reasoning, but it is important to emphasize that sequential "temporal" backtracking is not used by the system. Instead, when retraction of assumptions is found necessary, "relevant" backtracking is initiated as in similar dependency-based systems. [Stallman, London].

A key to the functioning of the system is the definition of a particular subcontext called the "Consistency Context" (CC). Truth Values determined by logical support within this context may not be refuted except in the attempt to resolve contradictions. During normal (static) deduction the CC is identical with the entire modelling context, but in a time-varying model the CC contains only the context layers for the various sets of axioms used by the system and the consequences of the most recent action. Other (older) support may be refuted by the modelling process as needed to maintain the consistency of the model. Logical support within the CC is described as "current" support.

II.2. Connective Specialists

The truth-functional connectives are defined to the system by associating with the name of the connective a procedural specialist responsible for the logical behavior of expressions linked by the connective. When a new TV is asserted for a connected expression or a component of a connected expression, the specialist examines the TV's of the expression and its arguments. If a node has current well-founded support for a truth value of TRUE or FALSE, then the state of that node may not be changed. Knowledge of the truth table for a connective is coded into its specialist and is used to check the consistency of the expression. If the new TV was assigned to the expression, then the TV's of the arguments may have to be changed; if the new TV was assigned to one of the arguments, then the TV of the expression may change unless it has current support, in which case the TV of some other argument may have to change. Connectives used to represent knowledge of causality are defined to the system similarly, by providing an appropriate procedural specialist that determines the logical behavior of relations using the connective.

The other important function of the connective specialists is to detect the case where the change in truth value of one argument results in the loss of a line of support for belief in the state of another argument. If there is no alternate well-founded (non-circular) support for the other state, then it must be marked as uncertain. The support checking process used is a generalization of Doyle's, in that any expression using a truth-functional connective can behave as an inference rule if it has appropriate logical support. Each connective specialist has encoded knowledge of the circumstances of this case and of the ways to check for lost support of its arguments. If a node loses its primary justification, lost support checks propagate through the network to all nodes dependendent upon it, using the same mechanism as normal truth value support propagation. This allows uniform treatment of all knowledge of the behavior of a connective, concentrated in a single specialist.

III. Network Truth Maintenance in a Time-Varying Model

III.1 Refuting Support for Changed States

Basic modelling behavior is illustrated in Figure 1. In the initial model, shown in (a) A was proved false by Modus Tollens using the axiom (IMPLIES A B) and other support (not shown) for B:False. A causality axiom (also not shown) states that "A may be made true by action K." When the action is added to the planned sequence, the causality specialist creates a new context (by adding a context layer to the list that defines the current context), and the CC is set to the new layer (plus the layers containing the domain axioms). State A:True is asserted in the new layer. Since A and the implication are true in the CC, the IMPLIES specialist concludes that B must be made true, which is permissible since the support for B:False is not in the CC and therefore dates from before the execution of the action. This starts a chain of assertions that refutes all support for B:False. Figure 1(b) shows the state of the model after the network stabilizes.

There are several problems associated with the use of logical dependencies for

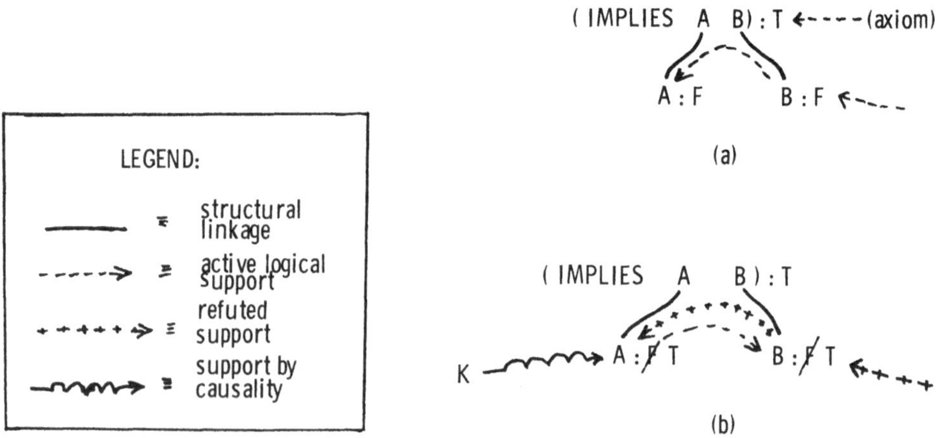

Figure I. Basic Model Revision

world-modelling, all relating to the requirement for refuting logical support for a changed state. These fall into three categories: local uncertainty, global uncertainty, and the need for adequate domain knowledge. An example of local uncertainty is shown in Figure 2. Suppose that C was originally proved true by the truth of A, B, and the axiom (IMPLIES (AND A B) C), as shown in (a). If C is made false by a planned action (K), it is necessary to refute the support provided by the conjunction. Since both A and B are true (with support outside the CC), it is indeterminate whether one or both should be changed. London outlines two approaches to the problem of local indeterminacy: the introduction of uncertainty into the network by allowing the propagation of a TV of "Unknown" (indicated in Figure 2(b) by a "?") to refute the support for the antecedents, or else by requiring domain knowledge to be sufficient to determine with certainty which one(s) of the antecedents have changed. The latter alternative is in general unacceptable, but the method discussed here provides an effective combination of these two approaches unavailable in the earlier system.

London pointed out that, in the case of refuting conjunctive support, the spread of uncertainty can in some cases be mitigated by new lines of support emanating from the state originally changed, as would occur in the case of "exhaustive support" with complete domain knowledge [London]. Since with incomplete knowledge this new support is not guaranteed to propagate to the refuted conjunction, some uncertainty may still remain in the net. A useful heuristic is to assume that a node will not change truth value unless either support is found for a new truth value or else a labelling of Unknown is required to maintain consistency.[3] By using a multi-level agenda mechanism, ERIS gives the propagation of uncertainty a lower priority than that for truth values of True or False. Thus, the system may find a line of support that eliminates the uncertainty without having to mark nodes as Unknown unnecessarily.

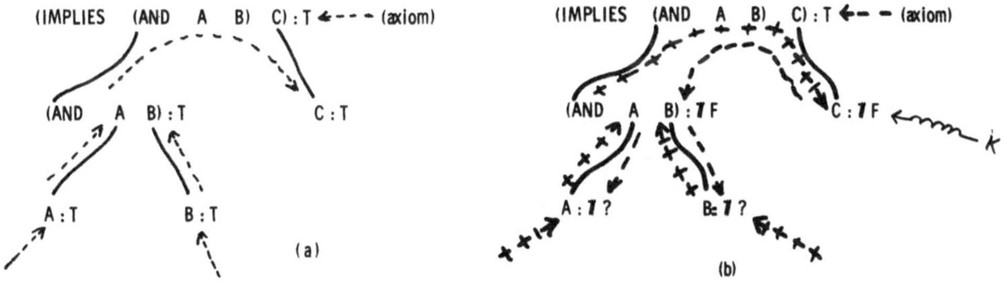

Figure 2. Local Uncertainty

[3] This is equivalent to the "commonsense" notion that things stay the same unless we have a reason for believing that they have been changed. Obviously this assumption can result in erroneously believed states which should be refuted by observation. Most commonsense knowledge of causality could be acquired in this way.

Furthermore, the CC method provides a way to undo the labelling of nodes marked Unknown in the event that the labelling is subsequently proved incorrect or unnecessary. If the spread of uncertainty attempts to refute the current support for some state, the new support may provide justification for a node previously marked Unknown. Alternatively, if, in refuting support for a conjunction, uncertainty was spread to the argument states and subsequently one of the states receives a TV of False due to another line of support, the refutation of the other states should be undone. Part of this behavior is accomplished by allowing "well-founded" TVs (True and False) to override a TV of Unknown even within the CC.

In Figure 3, an initial network is shown in (a), with states F:True and G:True supporting the truth of B and C, respectively, (through relationships not shown in the figure), and the conjunction in turn supporting A. Additional rules link C through D to E, with TVs as shown. D is providing alternate support for C in this case. Suppose that, as a result of an action, node A is made false and E is made true. Since D is already true, no further propagation from E would occur, but E now provides current support for D. A process to refute the conjunction would be scheduled with a lower priority, and (assuming that no other line of support is found for either B or C in the interim) both would be given TVs of Unknown. As a result, all support for both B and C would be refuted, propagating Unknown to F, G, and D, as shown in (b).

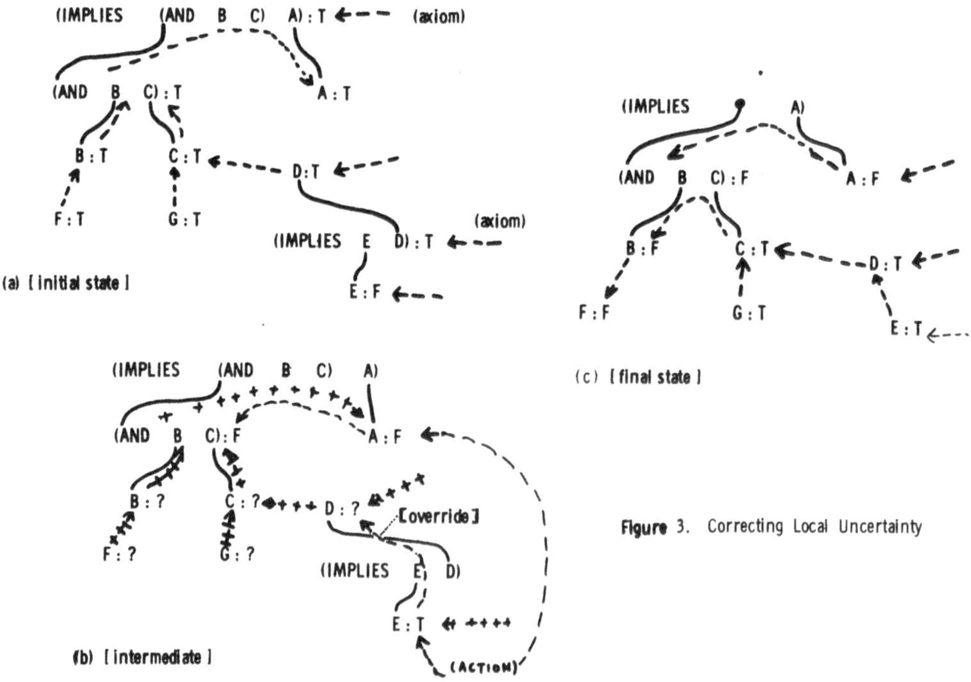

Figure 3. Correcting Local Uncertainty

At this point the specialist for IMPLIES looks at the relation (IMPLIES E D) and, noting that E:T is providing support for D:T, would attempt to refute the support for E:T. However, since E is true within the CC, it is a consequence of the most recent action and may not be refuted. The only other alternative is to restore support to D:T and consequently to C:T. Since a record of the previous justifications for each node is kept, the refutation of the original support for C:T (and D:T) may be retraced and the support restored. Given that C is now known to be true, the AND specialist would have no alternative but to label B false, since the conjunction must be kept false in the CC. This would result in retracing the refutation of B:T and changing the refuted support from Unknown to False, resulting in the final model shown in (c).

As another example of recovering from the unnecessary introduction of uncertainty, suppose that only A had been affected by the action and not E, and that, after the propagation of uncertainty from the conjunction, current support for C:F had been propagated through other relationships (not shown in the figure).[4] In this case the propagation of uncertainty caused by the earlier refutation of C:T would be retraced and changed from Unknown to FALSE. Then, since C makes the conjunction false, the earlier refutation of B:T would have been unnecessary, according to the "unchanged states" heuristic mentioned above, and the refutation of B:T would be retraced and undone, restoring the original support.

Another feature implemented using the CC method is the detection of global indeterminacy in the network following a state change. In the network of Figure 4(a) the two conjunctions are both initially false and are not providing support. Suppose that an action causes A to become true and therefore, by the axiom (IMPLIES A C), C becomes true as well. Since both B and D were true before the action, it would appear that both conjunctions would become true. If this is so, then both E and (NOT E) would have current support for a TV of True, and a contradiction would result.[5] The inconsistency could be avoided, however, by forcing a state change on either B or D, but since there is more than one choice, the net is globally indeterminate after the change, as in shown in (b).

This global indeterminacy could be reduced to local uncertainty by introducing uncertainty whenever states existing prior to the action are used as evidence during consequence modelling, e.g., after A is made True, if no current support for

---

[4] Even though uncertainty propagates at a lower priority, the discovery of support for C:F after refuting support for the conjunction could occur as a result of resolving uncertainty elsewhere in the network.

[5] The detection of a contradiction normally indicates an inconsistent set of assumptions. The assumptions supporting the contradiction are identified by tracing the conflicting justifications for the node. [Stallman] In the absence of assumptions, current support for a contradiction indicates an inconsistent set of domain axioms and the consequent uncertainty of all deductions using them.

Figure 4. Global Indeterminacy

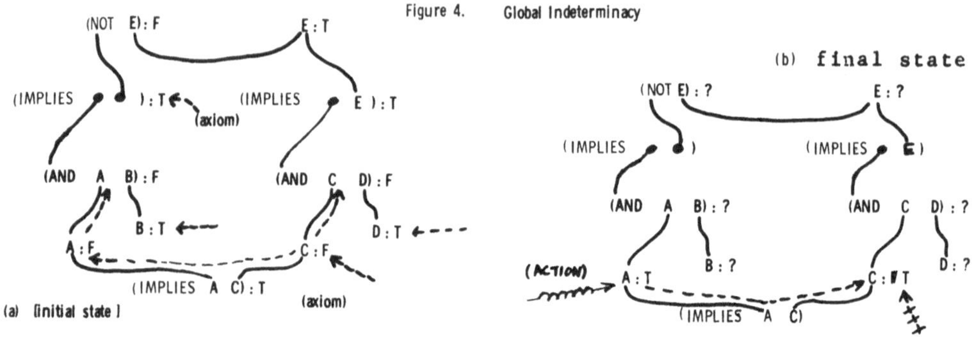

the TV of B or of (AND A B) is found, then mark them both as Unknown. This is undesirable from the standpoint of the "unchanged states" heuristic. Using the "relevant backtracking" approach, however, it is possible to implement the heuristic by regarding as assumptions each instance of using states as evidence that are not in the CC. In the example, using this rule enables the "unchanged state" assumptions of B:T and D:T to be identified, and, by refuting the conjunction of the assumptions, the desired result (Figure 4(b)) is obtained. Note that if the assumption set has only a single element, its refutation results in a completely determined network. This approach allows the "unchanged states" assumption to hold unless proved invalid, but it is justified only by assuming that the set of rules is consistent unless proved inconsistent by the inability to find a consistent labelling for nodes in the net.

The behavior of the model revision mechanism following a change in truth value to a node may also be understood by considering the following analogy: Suppose that before the state change is made to the node, the structure of each of its separate justifications is analyzed. The nature of the analysis is to trace the justifications backwards to identify for each one the set of nodes that, together with their corresponding model values, represent the system's assumptions about the world state and, with the given domain axioms, provide the ultimate logical support for the state of the node to be changed. We may call this node set the "foundations" of the justification [Doyle], since they have no logical support in themselves, but are obtained by observation or hypothesis.

After the state change is made, the new node state may provide support for other states in the model by using the domain axioms as inference rules. Presumably, this set of extended consequences of the change would include some of the foundations of the justifications of the old node state, thereby invalidating each justification that depends upon them. Uncertainty in the model results when there is a non-singular foundation set for some justification in which none of the foundations were changed by the inference following the change. Since by definition there is no way of determining which of the nodes to change, all must be marked as unknown, thereby invalidating the justification. Using the techniques outlined in the examples above, the behavior of the model revision process in ERIS accomplishes this result without the need for a preliminary support analysis. The necessary revision occurs automatically as a result of the inference following a state change.

The remaining difficulty encountered in modelling is the problem of capturing all the consequences of a state change. The examples above describe modelling behavior as it pertains to instantiated structure with initially known truth values, but it is often necessary to invoke deduction to determine whether nodes of unknown truth value that could possibly have been providing support were in fact providing support for the state of a node before it was changed. If such nodes were supporting a state that is changed by an action (or assumption), then their support must also be refuted. Failure to determine the support status of indeterminate nodes with respect to a changed state may cause some of the consequences of an action to go undetected.

This is the problem of "exhaustive support," as described by London, and is a major problem in dependency-based modelling. It is intimately related to the issue of the completeness of the domain knowledge and of the deductive search process, since the basic problem is that of proving a universal, namely the completeness of the set of consequences (or the non-existence of a line of support from the action consequent to any supposedly unaffected state). The problem of the completeness of domain knowledge is outside the scope of the system (although indeterminacy following a state change may allow it to detect incompleteness). Deducing the TVs of previously undecided states to determine their support status may place large demands on the deductive component. The approach planned for this system is that if the search for action consequences grows too expensive it will be abandoned. If this occurs during the course of planning an action sequence, then checkpoints to measure the continued validity of critical states (most notably, protected subgoals) will be inserted into the planned sequence. This raises the possibility that the detection of an unexpected condition during the plan execution presents an opportunity for learning.

III.2 Different Notions of "Lost Support"

In the model revision systems of Doyle and London, dependencies are used to determine the consequences of assertions introduced or retracted by an inference process. In both these systems, the model always reflects the current "state of the world," without keeping a history of the changes made. When reasoning about actions, however, this record becomes essential when the justification for certain assertions is retracted. Again, the combination of temporal context layers with logical dependencies provides an effective method of retaining and using this knowledge.

Figure 5 shows the two primary modes of "lost support" functioning employed by the ERIS system. In the initial model, given the axiom (IMPLIES A B):T and the state B:F, the IMPLIES connective specialist establishes support for A:F by Modus Tollens (a). In (b), the model shows the result of assuming the execution of an action that makes A true. If support for A:T is subsequently refuted or retracted, such as is shown in (c) where A is made false by a different action, support for B:T is lost, and the model revision process must either find an alternate well-founded justification for B:T or else cease to believe in the state as shown (unless B:F is also a consequence of the second action).

A different behavior is required if the loss in support for A:T occurs due to retracting the assumption that the action which made it true had ever been executed in the context of the model, i.e., if we pose the question, "What if K1 had never been done?" In this case, we would want the support structure refuted by the action assumption to be reconstructed, as shown in (d). This is not as simple as "undoing" a temporal context layer, however, as the demand for efficient planning requires that this question be answered in an arbitrary context without the expensive sequential temporal backtracking used in "PLANNER"-style systems. [London]

This can be done by an analysis of the network for alternate non-circular justifications, as in the normal lost-support procedure (see [Doyle], for example), but which allows the node to revert to its old state if no alternate support is found. Even if alternate support is found, it may date from a later context than the old support, in which case the node reverts only in the context prior to that in which the new support is valid. This means, effectively, that the node remains unchanged until a later action assumption (the one causing the alternate support) occurs. ERIS employs procedures that find the earliest alternate support, as well as allowing the retraction of assumptions made in earlier context, whether the assumptions are about events or hyphotheses. [Thompson80]

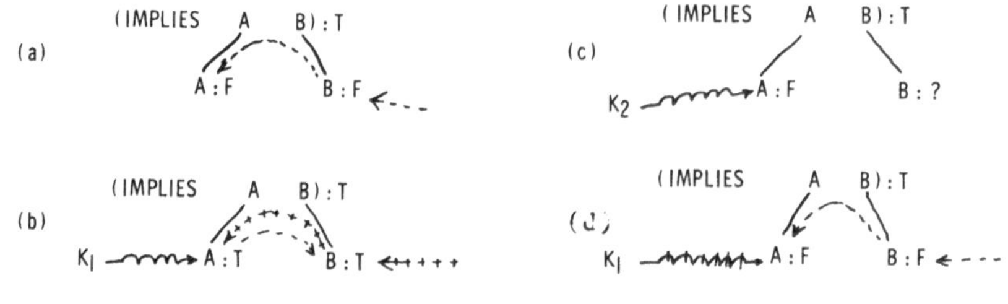

Figure 5. Lost Support

IV. Concluding Remarks

By allowing logical support to be determined within existing structure, the restriction that dependencies be explicitly declared by the deductive process in clause form is eliminated. This is accomplished by connective specialists that use graph labelling to maintain consistency in the network, greatly reducing the overhead for symbol manipulation required by the deduction and model revision processes, as rewrite rules are also eliminated (except for the initial simplification of expressions). The creation of new structure is limited to the instantiation of rules during deduction. By avoiding the structural manipulation of expressions, the network becomes, in effect, a "compiled" knowledge representation upon which the procedural specialists operate merely by the propagation of logical support and control "signals."

The potential efficiency of the connective specialist approach seems worth pursuing, and application efforts currently in progress (in the field of spacecraft command sequence generation) seem promising. The preceeding discussion was restricted to the system's behavior as implemented for propositional relations (no unbound object variables). At the present state of the implementation, first-order rules must be fully instantiated before they participate in deduction and TM. Although efficiency estimates are not available, recent work indicates that the propositional behavior described here extends smoothly into first-order by having the connective specialists examine binding information contained in the truth value property of the expressions.

References:

[Doyle] Jon Doyle, "Truth Maintenance Systems for Problem Solving", A.I. Laboratory, MIT. AI-TR-419.

[London] Philip E. London, "Dependency Networks as a Representation for Modelling in General Problem Solvers", Dept. of Computer Science, U. Maryland. TR-698.

[Stallman] Richard Stallman and Gerald Sussman, "Forward Reasoning and Dependency-Directed Backtracking in a System for Computer-Aided Circuit Analysis", MIT AI Memo 380, also in AI Journal, Vol 9, No 2 (October 1977).

[Thompson79] Alan Thompson, "Network Truth Maintenance for Deduction and Modelling," Proc. IJCAI6, Tokyo (1979).

[Thompson80] Alan Thompson, "Network-Based Reasoning in ERIS--First Report," Jet Propulsion Laboratory Publication (Pending), Pasadena, CA.

AN EXPERIMENT WITH THE BOYER-MOORE THEOREM PROVER:

A PROOF OF THE CORRECTNESS OF A SIMPLE PARSER OF EXPRESSIONS

Paul Y Gloess[1]
International Fellow, SRI International
Menlo Park, California 94025, U.S.A.
and
Boursier de Recherche I.R.I.A., 78150 Le Chesnay, France

ABSTRACT

The objective of this report is to convey the essential idea of a proof by the Boyer-Moore theorem prover of the correctness of a parser. The proof required a total of 147 functions and lemmas --all of which have been listed in the appendix of [4].

Included in the following text are a description of the original problem submitted to the theorem prover and a sketch of the resultant proof, together with a discussion of the reasons that induced us to introduce some auxiliary functions. The report also contains the computer-generated proof of one of the main lemmas: INIT.SEG. The complete proof is available from the author.

We conclude with some remarks on our experiment and comments on the use of the theorem prover.

1. INTRODUCTION

It seemed to us that a good way to learn something about the Boyer-Moore theorem prover was to submit an example to it. Although the specific nature of the example was of secondary importance to us in our selection, it was apparent from the outset that the example actually chosen presented potential difficulties. Our impression was indeed confirmed subsequently as we proceeded with the proof.

Of principal interest are the very particular conditions of our experiment:

- We knew almost nothing about the Boyer-Moore theorem prover [2] when we submitted our problem to it: therefore we did not tailor our example for this system.

- We had no interest in facilitating the task of the theorem prover: we did not alter our specification of the problem in the course of the proofs to suit the theorem prover; of course we did not alter the theorem prover to suit our needs.

Such ideal conditions are seldom met in practice. The fact that we were able to obtain a mechanical proof of our theorem with this system illustrates its effectiveness.

The problem was to prove the correctness of a parser. We did not spend much time in the design of our functions, especially the parser. This made the problem even more difficult than it could have been with simpler and more efficient parsing algorithms, —such as the one we were recently proposed (see Section 2).

To solve our problem, we introduced several auxiliary functions and proved a lot

---

[1]Supported by a grant from I.R.I.A. and ONR contract N00014-75-C-0816.

of lemmas, all listed in [4]. There may conceivably exist a shorter path that could be utilized by the system to solve exactly the same problem, that is:

> "Have the system accept the functions NLISTP, NULLP, LEFTPAR, LEFTPARP, ATOMP, CONNECTP, TOPOFTREE, UNARY, SUBTREE1, BINARY, SUBTREE2, TREEP, CHARCHAIN, CONCATCH, RIGHTPAR, RIGHTPARP, PRINT, FIRSTCH, REMAIN, G, ONEORTWO, LASTCH, EVALEXPR as total and prove the theorem EVALEXPR.IS.CORRECT without adding any axiom (see Appendix of [4] for the text of these functions and theorem); a version of the theorem prover without the builtin "linear arithmetic" (see [2]) should be used."

2. THE PROBLEM[2]

The parser we are concerned with is a parser of expressions such as:

((A + B)*(- C)), (A + (B + C)), A ,...

Note that our operators are either unary or binary, and that we require full bracketing and no more. For the parser, these are nonexpressions:

A + B, (A + B + C), - A, ((A + B)).

The function of the parser, called EVALEXPR, is to distinguish valid expressions from invalid, and to return a tree for every valid expression. For example a representation of the tree corresponding to the above first valid expression might be:

```
      *
     / \
    +   -
   / \ /
  A  B C
```

To specify the correctness of the parser, we define a function PRINT (similar to the one used by INTERLISP to print lists). For any unary-binary, tree such as the one above PRINT returns an expression. For the above tree the printed expression is the one already mentioned:

((A + B)*(- C)).

What we mean by correctness of the parser is this: "parsing the expression printed from every unary-binary tree restores back the original tree". In a more formal way:

(TREEP X) => (EVALEXPR "NIL" (PRINT X)) = X.

This is the precise statement of the theorem we have proved for EVALEXPR. (As we can see EVALEXPR takes 2 arguments: the first one plays only an internal role and when it is set to "NIL" the second one is the text to parse.)

One can object that the above theorem does not signify the correctness of the parser, preferring the observation that "for every expression, the parser returns a tree that prints like the original expression". Alternatively, one might require the conjunction of both theorems.

---

[2]All the definitions and lemmas involved are listed in [4].

However, it was our intention to prove only the first theorem. After doing so, we decided to name it "EVALEXPR.IS.CORRECT".

We shall now give our representation of trees and expressions and explain the definitions of TREEP, PRINT and EVALEXPR.

LISP Representation of Trees and Expressions

It is perhaps not yet clear to the reader that any significant difference exists between trees and expressions. If this were true there would be no real problem! In such examples as the proof of an optimizing compiler [1], trees and expressions are used synonymously, since one is only interested in their semantics. In the present context, however, we shall insist on a representation that differentiates between them.

We used the existing shell CONS to represent trees and expressions, so that all our functions could be tested in Lisp.

For example the tree given above (see previous page) is represented by the S-expression ((*) ((+) A B) ((-) C)) .

Note that we have adopted the convention that a tree should be a P-list (the CDR of a tree is "NIL" or a list).[3] We also have adopted the convention of representing operators (such as "+", "*", ...) by lists of one element. Operators are thus distinguished from the terminals "A", "B", "C", ... represented by atoms. Although not absolutely necessary, this convention is consistent with our definition of expressions.

We also use lists to represent expressions. But here the length of the list is arbitrary. First of all, to avoid any confusion between the brackets of expressions, and the lisp-brackets, we use "<" and ">" instead of "(" and ")". For example, the expression ((A + B) * (- C)) is represented by the S-expression (< < A (+) B > (*) < (-) C > >) .

The Theorem Prover told us to exclude "<" from atoms. How did it do so? By finding and displaying the following counterexample: the atom "<" would be a tree but prints as ("<") which in any case is a nonexpression. Such bugs are frequent because they correspond to special cases which seem relatively unimportant: the theorem prover finds them immediately while solving the base case of the induction.

On the other hand "NIL" and even ">" are perfectly valid atoms. Thus (("+") "NIL" ">") is a valid tree and the corresponding expression ("<" "NIL" ("+") ">" ">") is valid and well parsed by EVALEXPR, whereas ("<" "<" ("+") "A" ">") is recognized as a nonexpression.

---

[3] It is actually the Theorem Prover that made us alter our first definition of TREEP by showing us a counterexample (the parser returning only P-list trees).

Finally our predicate ATOMP for recognizing atoms is defined by
(DEFN ATOMP(X) (NLISTP X)&(X ≠ "<")).

Our predicate CONNECTP recognizes operators and is defined by
(DEFN CONNECTP(X) (LISTP X)).
Thus an operator might be any list, since we do not verify that it contains only one atom. We found that this test led to longer proofs, by multiplication of otherwise identical cases, and therefore decided to make exception to our principle of not changing our initial definitions.

The Functions TREEP and PRINT

This predicate recognizes trees and should be obvious from the previous discussion.

The PRINT function is also very simple. If the input X is a nonlist, PRINT returns (CONS X "NIL"). If X is a unary tree, PRINT returns the concatenation ("<")¦¦(CAR X)¦¦(PRINT (CADR X))¦¦(">") , where the "¦¦" operator stands for CONCATCH (our APPEND function). If X is binary the result is ("<")¦¦(PRINT(CADR X))¦¦(CAR X)¦¦(PRINT(CADDR X))¦¦(">") .

The Parser EVALEXPR

EVALEXPR is the most complicated function in this problem. It takes two arguments L (Left) and R (Right). L is always set to "NIL" in an external call to EVALEXPR and R contains the text to parse.

The algorithm we use is quite simple, but very inefficient. The main work of EVALEXPR is to try to decompose the input. Of course, when the text to parse is reduced to a list of one atom, nothing needs to be done. Otherwise the task of decomposition consists in finding the main operator of the expression, according to the pattern:
("<")¦¦left_sub_exp_or_empty¦¦main_op¦¦right_sub_exp¦¦(">").

When the element immediately following the initial "<" is an operator, decomposition has already been accomplished.

Otherwise EVALEXPR skips one element (moving to the right) by adding the first element of R to the end of L, as signified by the recursive call

```
(EVALEXPR (APPEND L (CONS (CAR R) "NIL"))
          (CDR R))
```

Then EVALEXPR will continue skipping until L and R are such that (CDR L) --ie. left_sub_exp_or_empty-- is an expression and (CAR R) is an operator --ie. main_op--. The search will eventually cease. For example, if (CDR L) happens to be an expression whereas (CAR R) is not an operator: EVALEXPR will return "<" to mean that

L and R do not comprise an expression.[4]

Once the decomposition is obtained, the left part L starts with a "<" followed by either an empty list or an expression (called left_sub_exp_or_empty above); R starts with an operator (called main_op). So all EVALEXPR needs to do now is to:

- check that the last element of R is a closing bracket, i.e:
  (LASTCH R) = ">";

- check that the elements between main_op and the last element of R form an expression, i.e:
  [EVALEXPR "NIL" (BEGIN (CDR R))] ≠ "<";

- return either "<" or the adequate parse-tree obtained by CONSing the main operator, the parse-tree resulting from the left subexpression (if any), the one resulting from the right subexpression (which is always present), and "NIL" (to form a P-list).

EVALEXPR contains three different recursive calls. The first of these is

```
(EVALEXPR (APPEND L (CONS (CAR R) "NIL"))
          (CDR R)),
```

which corresponds to skipping one element; the second is

```
(EVALEXPR "NIL" (CDR L)),
```

for the left operand; the third call is

```
(EVALEXPR "NIL" (BEGIN (CDR R)))     ,
```

corresponding to the right operand.

The totality of EVALEXPR stems from the fact that either the sum of the lengths of L and R decreases as in the second recursive call, or remains unchanged as in the first call; in the latter case the length of R decreases in both the first and third recursive calls.

Note that LASTCH (which returns the last element of a list) and BEGIN (which returns the list of all but the last elements of a list) are recursive fonctions. Therefore, some lemmas dealing with the manipulation of objects resembling character strings will be necessary.

Why this Definition of EVALEXPR?

The reader might wonder if we did not pick this particular definition of the parser to facilitate the task of the theorem prover. The answer is no! Our definition was actually more difficult to handle than more elaborate ones. Having from the beginning excluded a definition suggested by an LL(1) grammar of expressions, because it involved mutually-recursive functions (mutual-recursivity is

---

[4]Until very late in our proofs, we had chosen "NIL" as a returnvalue for nonexpressions; when we introduced the auxiliary predicate EXPRP and submitted a lemma relating EXPRP to EVALEXPR, the theorem prover found that ("NIL") was a valid expression according to EXPRP, but not with respect to EVALEXPR. We had unwantedly exluded "NIL" from available atoms within an expression. So we switched to "<" which is the only nonlist excluded by ATOMP.

not implemented in the theorem prover), we derived our definition from the very natural formulation

```
s is an expression
  iff  s is a string of one atom
     or ∃ op:operator. ∃ s1:expression. s = '<'¦¦op¦¦s1¦¦'>'
     or ∃ op:operator. ∃ s1,s2:expression. s = '<'¦¦s1¦¦op¦¦s2¦¦'>'.
```

We just replaced existential quantifiers by a loop scanning the input string. We did optimize our definition by stopping the scanner, in the binary case, as soon as s1 has been found. Note that such an optimization requires some knowledge of the theory of parsing: we implicitly use the fact that "a proper initial segment of an expression is not an expression". This fact is precisely the key-lemma of our proof (see Chapter 4). We did this simple optimization because of our slight concern for efficiency as an implementer, not as a prover.

A very elegant parser has recently been offered to us. Lack of space does not permit us to include it here, but the idea is the following. This parser, say EV, returns a pair consisting of a tree of the first sub-expression of the input s, and the remainder of s. The theorem to prove is then

```
∀ t:tree. ∀ s:string.  (EV (PRINT t)¦¦s) = (t . s)    .
```

We checked by hand that this theorem can be proved by a simple induction on the tree t. This formulation of EV and the correctness seem to have the idea of the initial-segment "wired in" so that the corresponding lemma is no longer necessary. However EV contains two embedded recursive calls whose outermost is "(EV (CDR (CDR (EV (CDR s)))))". Such a recursion does not presently allow the theorem prover to prove the totality of EV. J Moore shows in [10] how to prove the totality of the TAK function which has a similar recursion (the method will be implemented in a future version of the theorem prover). To prove the totality of EV using the same method would require the introduction of an auxiliary parser with a simple recursion (e.g EVALEXPR), and then prove that this auxiliary parser satisfies the schema of EV. The proof of EV's totality might well be as difficult as our proofs of EVALEXPR's correctness.

Proving that EV's definition implies EV's correctness does not require proving EV's totality, because no EV-induction is necessary; proving EV's totality is nevertheless essential to claiming the existence of EV as a "real" function.

3. SKETCH OF THE PROOF

The theorem we want to prove is:

```
EVALEXPR.IS.CORRECT:
    (TREEP X) => (EVALEXPR "NIL" (PRINT X)) = X    .
```

This conjecture suggests an induction on the variable X according to the schema of

TREEP, which is what the system does. Three interesting cases are to be considered:[5]

- The base case: the tree X is reduced to a leaf,

- The unary case: X has only one branch, so that (PRINT X) is the list

```
[APPEND (CONS "<" "NIL")
        (APPEND (CONS (CAR X) "NIL")
                (APPEND (PRINT (CADR X))
                        (CONS ">" "NIL"],
```

  constituting an expression of the form < operator operand > , which is easily parsed by EVALEXPR. The "pointer" moves only once to the right and the expression is immediately decomposed into its left part ("<") and right part operator||operand||(">"). The induction hypothesis
  (EVALEXPR "NIL" (PRINT (CADR X))) = (CADR X)
  and the expansion of (EVALEXPR "NIL" (PRINT X)) resolve this case.

- The binary case: here, of course, resides the entire difficulty of the problem. A simple expansion of EVALEXPR is insufficient, since the "pointer" must skip a variable number (n+1) of symbols if n is the length of the subexpression that commences after the initial bracket. Precisely, (PRINT X) is here the list

```
[APPEND (CONS "<" "NIL")
        (APPEND (PRINT (CADR X))
                (APPEND (CONS (CAR X) "NIL")
                        (APPEND (PRINT (CADDR X))
                                (CONS ">" "NIL"].
```

  Therefore, the proof consists in showing that the "pointer" must skip until past (PRINT (CADR X)) (which is an expression by virtue of induction hypothesis and the fact that X, being a tree, cannot be "<"). In other words, (PRINT (CADR X)) must be the smallest subexpression starting after the initial "<".

Finally the proof of the whole problem is mainly based on the following lemma:

"Any proper initial segment of an expression is not itself an expression."

Let us call this the "initial-segment lemma," the computer proof of which is presented in Chapter 5.

This kind of generalization requires the creation of auxiliary functions. Consequently, it is far beyond the system's capacity.

4. THE HANDWRITTEN AUXILIARY FUNCTIONS

After several unsuccessful attempts it was evident that all the lemmas in the world could not solve our problem.

Without a significant generalization (involving new functions), the intrinsic schema of EVALEXPR could never be used. The general induction principle of Boyer and Moore [3] requires that all changing measured arguments of a function reference whose schema contributes to the induction be variables.

---

[5]The Theorem Prover considers five cases, whereas the schema itself contains only two cases --the base case and the general case, which includes two induction hypotheses (corresponding to CADR and CADDR)--. The general case splits into four cases because the conjecture is an implication.

Furthermore, the function EVALEXPR appeared especially difficult for two reasons. First, as we saw in Section 2, it contains three different recursive calls, but one of these --namely the shift of one position to the right-- seems to precede the other two chronologically. The second negative aspect is that some of the recursive calls "govern" (see [3]) themselves.

All these reasons led us to introduce the function CUT, then INCLUP and, finally, EXPRP. Although this idea did not emerge easily, at least for the first function, it appears now that these functions have a common characteristic: each of them reflects a specific part of the complex schema of EVALEXPR and is thus "simpler" than EVALEXPR.

Decomposition of an Expression by the Function CUT

The goal achieved by this function is to move the "pointer" to the right until it splits the input expression into two segments,

```
< left right,
      ^
```

whose CONS is returned as value of CUT. The "pointer" stops (in the general case where the expression begins with a "<") as soon as "left" is the empty list and "right" begins with an operator (unary case), or "left" is an expression (binary case). Note that CUT does not check the right part, except to stop the "pointer" in the unary case.

The only recursive call contained in CUT's definition is

```
(CUT (APPEND L (CAR R))   (CDR R))
```

where L and R are the arguments.

The most interesting property of CUT is

```
CUT.HELPS.EVALEXPR: (EQUAL (EVALEXPR (CAR(CUT L R)) (CDR(CUT L R)))
                           (EVALEXPR L R))
```

The proof of EVALEXPR.IS.CORRECT is thus reduced to the proof of

```
[EQUAL [CUT "NIL"
            (APPEND (CONS "<" "NIL")
                    (APPEND EXP1
                            (APPEND (CONS OP "NIL")
                                    (APPEND EXP2
                                            (CONS ">" "NIL"]
       [CONS (CONS "<" EXP1)
             (CONS OP (APPEND EXP2 (CONS ">" "NIL"]],
```

where EXP1 and EXP2 are expressions and OP is an operator.

The Ordering Relation INCLUP

This function takes the four arguments L1, R1, L2 and R2 and checks a logical relation between the ordered pair <L1,R1> and the ordered pair <L2,R2>, as best expressed by the following schema:

```
     L1          R1
   |----|-----------------|

   |-------|--------------|
      L2          R2
```

L1, R1, L2, R2 are lists. The relation (INCLUP L1 R1 L2 R2) holds when (APPEND L1 R1) = (APPEND L2 R2) and L1 is an initial segment of L2. However, INCLUP is not defined that way; it uses the same recursive call as CUT.

Of course, the INCLUP relation holds between CUT's arguments and CUT's result, as signified by the lemma

```
INCLUP.LR.CUTLR: (INCLUP L R
                         (CAR (CUT L R)) (CDR (CUT L R)))  .
```

This lemma, joined to the preceding ones, reduces the proof of EVALEXPR.IS.CORRECT to that of the central lemma we formulated in Chapter 3: "Any proper initial segment of an expression is not itself an expression."

The Initial Segment Lemma and the Function EXPRP

Let us remind the reader that EVALEXPR assigns to nonexpressions only the value "<". Hence, a possible statement of the initial-segment lemma is:

```
[IMPLIES (AND (NOT (EQUAL (EVALEXPR L (APPEND R S)) "<"))
              (LISTP S))
         (EQUAL (EVALEXPR L R) "<"].
```

This is actually a slightly more generalized form of the lemma, since "NIL" has been replaced by the variable L. Although it allows an induction using the schema of EVALEXPR, the proof fails. The essential reason for this is that EVALEXPR does not suggest the right induction --at least the one we had in mind. Furthermore, it seems that the argument L, however necessary from the induction viewpoint, is misleading. So we decided to define another function: EXPRP. As we shall see, EXPRP makes obsolete the first argument, L.

The proof we conceived of and subsequently checked with the system is as follows. Suppose that S is a list, and R and R¦¦S are both expressions. It is obvious that R has to start with a "<". It can be proved that each decomposition --of R and R¦¦S-- will have the same left part i.e, the lemma CUT.CONCATCH. Since R is an expression, the search for the main operator will stop at least two elements before the end (see lemma TWO.ELEMENTS), because the right part must start with an operator and finish with a ">". Hence, since we know that this search does not depend on the right part (except, perhaps, for the first element which is the main operator [see CUT, Section 4]), the search will certainly not depend on S, which is after the final ">" of R. So we have the situation:

```
 R:  < left_part operator right_sub_exp >
R¦¦S: < left_part operator right_sub_exp_and_more >.
```

right_sub_exp_and_more must be an expression, but this is not possible (by induction hypothesis), since right_sub_exp is an expression and a proper initial segment of right_sub_exp_and_more.

Note that the variable L plays no role in this proof. At first glance the induction

```
  (p right_sub_exp S) => (p R S),
```

on R only, seems sufficient.

A second more careful look shows that the variable S of the induction hypothesis needs to be instantiated, according to the schema:

```
(p right_sub_exp (CONS ">" (BEGIN S))) => (p R S),
```

so as to take into account the final bracket of R, but omit that of R¦¦S.

As the system does not instantiate those variables that do not participate in the induction, we used an artifice to force the instantiation of S in the induction hypothesis: the adjunction of the dummy argument DUMMY to the auxiliary function EXPRP. Otherwise R would have been its only argument.

The only recursive call contained in the definition of EXPRP is

```
(EXPRP [BEGIN (CDR (CDR (CUT "NIL" R]
       [CONS ">" (BEGIN DUMMY]),
```

where the first argument represents the right subexpression of R when R is an expression. Note that EXPRP does not refer to the left subexpression of R, whereas EVALEXPR does. Also note that EXPRP does not depend on DUMMY (see appendix of [4]).

Expressed in terms of EXPRP, the initial-segment lemma becomes

```
[IMPLIES (AND (EXPRP (APPEND R S) S)
              (LISTP S))
         (NOT (EXPRP R S].
```

Naturally we had to prove that EXPRP recognizes the same set of expressions as EVALEXPR, i.e, the lemma:

```
EXPRP.EQUAL.EVALEXPR: [EQUAL (EXPRP R DUMMY)
                             (NOT (EQUAL (EVALEXPR "NIL" R) "<"]   .
```

5. COMPUTER-PROOF OF THE INITIAL-SEGMENT LEMMA

The system has proved a contrapositive form of the initial-segment lemma:

```
INIT.SEG: [IMPLIES (AND (LISTP B)
                        (EXPRP A B))
                   (NOT (EXPRP (CONCATCH A B) B].
```

The beginning of the proof shows what induction schema was chosen from the function EXPRP. The schema contains five base cases and one general case. In the remainder of the proof the five base cases split into six, numbered from eight to three. The general case bisects into cases no. 2 and no. 1, because the induction hypothesis is an implication. Case 2 is comparatively barren. Case 1 is the interesting one of the pair.

From this point, the system first expands the hypothesis "(EXPRP A B)", which is not useful and necessitates the additional lemma REBUILD.EXPRP.FROM.EXPANSION.

Then the hypothesis resulting from the induction (i.e, the conclusion of the induction hypothesis) is transformed into

```
[NOT (EXPRP (CONCATCH (CDR (CDR (CUT "NIL" A))) (BEGIN B))
            (CONS ">" (BEGIN B],
```

owing to the four linked technical lemmas CONCATCH.BEGIN.CONS.LASTCH, TWO.ELEMENTS,

LASTCH.CUT and LISTP.CDR.CUT. For example, LASTCH.CUT asserts that the last element of "(CUT L R)" is the last element of R if R is a list.

The next step is the elimination, as usual, of (CAR A) and (CDR A).

Finally, the function EXPRP in the conclusion expands by virtue of the lemma EXPRP.CONCATCH.EXPANDS. The other lemmas cited contribute to the new form of the conclusion,

```
  [OR conclusion_of_induction_hypothesis
      (NOT (EQUAL (LASTCH B) ">"))],
```

which is not visible since this conjecture is immediately reduced to (TRUE).

Note that the lemma CUT.CONCATCH (already mentioned in 4), although not cited in the present proof, plays an important role in the transformation of the conclusion. This lemma affirms the equality of the left subexpressions of expressions A and (CONCATCH A B), i.e, in a slightly stronger form:

```
CUT.CONCATCH : [IMPLIES (NOT (EQUAL (EVALEXPR U A) "<"))
                        (EQUAL (CUT U (CONCATCH A B))
                               (CONS (CAR (CUT U A))
                                     (CONCATCH (CDR (CUT U A)) B]
```

The computer proof of INIT.SEG starts below. All cited lemmas (except for the basic axioms dealing with the shell [CONS, CAR, CDR, LISTP]) and functions are listed in the appendix of [4].

```
PROVE.LEMMA(INIT.SEG (REWRITE)
         (IMPLIES (AND (LISTP B) (EXPRP A B))
                  (NOT (EXPRP (CONCATCH A B) B)))
         NIL NIL)
Name this *1.

Of course, when all else fails (and it has), induct on
something!

There are two plausible inductions. They don't merge. And,
neither is clean! And they appear equally likely. So the
one which will probably lead to eliminating the nastiest
expression was picked. We will induct according to the
following schema:
     (AND (IMPLIES (NOT (LISTP (CDR (CUT "NIL" A))))
                   (p A B))
          (IMPLIES (NOT (LISTP A)) (p A B))
          (IMPLIES (NOT (LISTP (CDR A)))
                   (p A B))
          (IMPLIES (NOT (EQUAL (CAR A) "<"))
                   (p A B))
          (IMPLIES (NOT (LISTP (CAR (CDR (CUT "NIL" A)))))
                   (p A B))
          (IMPLIES (AND (LISTP (CDR (CUT "NIL" A)))
                        (LISTP A)
                        (LISTP (CDR A))
                        (EQUAL (CAR A) "<")
                        (LISTP (CAR (CDR (CUT "NIL" A))))
                        (p (BEGIN (CDR (CDR (CUT "NIL" A))))
                           (CONS ">" (BEGIN B))))
```

```
                    (p A B))).
```

The inequalities BEGIN.CDR.LNG.LESSP, CDR.CUT.LESSEQP and EXPRP.IS.TOTAL establish that the measure (LNG A) decreases according to the well-founded function LESSP in the induction step of the schema. This produces eight new conjectures:

```
 8¦(IMPLIES (AND (NOT (LISTP (CDR (CUT "NIL" A))))
  ¦             (LISTP B)
  ¦             (EXPRP A B))
  ¦        (NOT (EXPRP (CONCATCH A B) B))),
  ¦which simplifies, using the lemmas CAR.NLISTP,
  ¦CAR.CONCATCH.LISTP and CDR.CONCATCH.LISTP, and unfolding
  ¦EXPRP and CONCATCH, to:

   *¦(TRUE).
```

... <u>Author's</u> <u>Note</u>: we skip Cases 7 through 2 for lack of space.

```
 1¦(IMPLIES
  ¦ (AND
  ¦  (LISTP (CDR (CUT "NIL" A)))
  ¦  (LISTP A)
  ¦  (LISTP (CDR A))
  ¦  (EQUAL (CAR A) "<")
  ¦  (LISTP (CAR (CDR (CUT "NIL" A))))
  ¦  (NOT
  ¦      (EXPRP (CONCATCH (BEGIN (CDR (CDR (CUT "NIL" A))))
  ¦                       (CONS ">" (BEGIN B)))
  ¦             (CONS ">" (BEGIN B))))
  ¦  (LISTP B)
  ¦  (EXPRP A B))
  ¦ (NOT (EXPRP (CONCATCH A B) B))),
  ¦which we can simplify, rewriting with LISTP.CDR.CUT and
  ¦LASTCH.CDR.LISTP.ELIM, and expanding the definitions of
  ¦LASTCH, EXPRP and CONCATCH, to:

   1¦(IMPLIES
    ¦ (AND
    ¦  (LISTP A)
    ¦  (LISTP (CDR A))
    ¦  (EQUAL (CAR A) "<")
    ¦  (LISTP (CAR (CDR (CUT "NIL" A))))
    ¦  (NOT
    ¦    (EXPRP (CONCATCH (BEGIN (CDR (CDR (CUT "NIL" A))))
    ¦                     (CONS ">" (BEGIN B)))
    ¦           (CONS ">" (BEGIN B))))
    ¦  (LISTP B)
    ¦  (EXPRP (BEGIN (CDR (CDR (CUT "NIL" A))))
    ¦         (CONS ">" (BEGIN B)))
    ¦  (EQUAL (LASTCH (CDR A)) ">"))
    ¦ (NOT (EXPRP (CONS (CAR A) (CONCATCH (CDR A) B))
    ¦             B))),
    ¦which we can again simplify, applying
    ¦CONCATCH.BEGIN.CONS.LASTCH, TWO.ELEMENTS, LASTCH.CUT,
    ¦LISTP.CDR.CUT and LASTCH.CDR.LISTP.ELIM, and expanding
    ¦LASTCH, to:

     1¦(IMPLIES
      ¦ (AND
      ¦     (LISTP A)
      ¦     (LISTP (CDR A))
      ¦     (EQUAL (CAR A) "<")
```

```
|     (LISTP (CAR (CDR (CUT "NIL" A))))
|     (NOT (EXPRP (CONCATCH (CDR (CDR (CUT "NIL" A)))
|                           (BEGIN B))
|                 (CONS ">" (BEGIN B))))
|     (LISTP B)
|     (EXPRP (BEGIN (CDR (CDR (CUT "NIL" A))))
|            (CONS ">" (BEGIN B)))
|     (EQUAL (LASTCH (CDR A)) ">"))
| (NOT (EXPRP (CONS (CAR A) (CONCATCH (CDR A) B))
|             B))).
|Eliminate the undesirable expressions (CDR A) and
|(CAR A), by using the lemma CAR/CDR.ELIM backwards
|and then generalizing.  We thus obtain:

 1|(IMPLIES
  | (AND
  |  (LISTP (CONS Z X))
  |  (LISTP X)
  |  (EQUAL Z "<")
  |  (LISTP (CAR (CDR (CUT "NIL" (CONS Z X)))))
  |  (NOT
  |   (EXPRP
  |      (CONCATCH (CDR (CDR (CUT "NIL" (CONS Z X))))
  |                (BEGIN B))
  |      (CONS ">" (BEGIN B))))
  |  (LISTP B)
  |  (EXPRP
  |        (BEGIN (CDR (CDR (CUT "NIL" (CONS Z X)))))
  |        (CONS ">" (BEGIN B)))
  |  (EQUAL (LASTCH X) ">"))
  | (NOT (EXPRP (CONS Z (CONCATCH X B)) B))).
  |This can be further simplified, applying the
  |lemmas CDR.NLISTP, CAR.CONS, CDR.CONS,
  |EXPRP.CONCATCH.EXPANDS,
  |REBUILD.EXPRP.FROM.EXPANSION,
  |LISTP.CAR.CDRCUT.CONCATCH and
  |EXPRP.CONCATCH.EQUAL.COMMON, and expanding the
  |functions CONCATCH, EVALEXPR and CUT, to:

  *|(TRUE).
```

Ha! That finishes the proof of *1.

Q.E.D.

Load average during proof: 2.855174
Elapsed time: 271.84 seconds
CPU time (devoted to theorem proving): 99.956 seconds
IO time: 2.802 seconds
CONSes consumed: 122298

PROVED

6. REMARKS OF A USER

We present some historical remarks on our experiment. We then give more general comments on the interaction with the system.

Our Experiment

We started the proofs with two principles:

1. Never modify the initial definitions, unless a bug is found.

2. Give the theorem prover as much autonomy as possible.

We never departed from the first principle. We followed the second one during the first two thirds of our experiment, while we had no clear idea of what the computer-proof would be (if it be), although we introduced some auxiliary functions at this time. Then, we had gained such an intimate knowledge of the system that we undoubtedly could not refrain from being more persuasive: we introduced EXPRP and INIT.SEG with the firm intention of imposing our own proof.

Only then did we discover the last bug in our parser EVALEXPR! Even more surprising was the immediate proof that we obtained for a lemma that we had set aside for a long time after unsuccessfull attempts to prove it. The system had learned from the lemmas we had added to the data-base for other purposes.

Interaction with the System

- How do we give hints to the system?

- When and why do we interrupt a proof?

- When and why do we break up a conjecture into lemmas?

We give here our own answers to these questions.

Suppose we are sitting at a terminal, with a particular theorem to prove. We let the proof run. The following situations are typical:

- The proof terminates with a message of failure, no generalization[6] has been performed, and no induction has been attempted at the time of the failure: we are sure that our conjecture was wrong. We scrutinize the last steps of the proof: they usually suggest some missing hypothesis or some "wrong" definition of a function.

- The proof fails but a generalization has been performed: it is probably too strong; we carefully examine the formula prior to the generalization and infer some less drastic generalization which yields a new lemma to work upon.

- The proof fails because the conjecture has nothing to induct upon: the conjecture needs some non trivial generalization. Auxiliary functions may need to be introduced.

---

[6] A generalization usually consists in the replacement of some occurrences of a common subterm with a new variable; hypotheses about this variable are eventually added. The CAR/CDR elimination is a trivial example which we do not consider as it preserves intersatifyability.

- The proof goes on and on ..., without any generalization. There are two cases. The number of generated subgoals (as a result of some induction) is so big that we cancel the proof for aesthetic or practical reasons. We try to reformulate the conjecture so as to reduce the number of hypotheses, or simplify the definition of a function by reducing the number of tests (all the dependent events must be reascertained). Alternatively, one of the subgoals requires an induction, which generates new subgoals ..., suggesting that the proof is looping (by the fact that the number of variables increases, typically through CAR/CDR elimination). Here as above, a remedy may not be easy to find.

Once the main lemmas have been proved, and seem to "logically" imply some conjecture (without induction or generalization), some real difficulties often arise in practice. Irrelevant lemmas prevent the good ones from applying and are difficult to track. Special lemmas, traditionnally called bridge-lemmas need to be created.[7] The order in which lemmas are proved may be critical and the use of MOVE.LEMMA, which permits temporary deactivation of lemmas, becomes necessary.

These difficulties may be the price to pay for not having to specify what lemma must be invoked and where, in thousands of occasions (like proof-checkers or low-level theorem provers require it). They are closely related to the existence of critical pairs among lemmas. They concern the theory of rewrite-rules and equality, see [5, 6, 7, 8, 9, 11, 12].

ACKNOWLEDGEMENTS

We would like to thank Robert S. Boyer and J Strother Moore for their invaluable advice (which concerned their theorem prover in general but not our specific problem).

Our appreciation should also be expressed to the Theorem Prover for its very accurate proofs and incredible patience!

REFERENCES

1. Robert S. Boyer, J Strother Moore. A Computer Proof of the Correctness of a Simple Optimizing Compiler of Expressions. Technical Report N00014-75-C-0816-SRI-4079, SRI International, Menlo-Park, California 94025, January, 1977.
2. Robert S. Boyer, J Strother Moore. A Theorem Prover for Recursive Functions: a User's Manual. CSL-91, NR 049-378, SRI International, Menlo-Park, California 94025, June, 1979.
3. R. S. Boyer and J S. Moore. A Computational Logic. Academic Press Inc., New York, 1979.
4. P. Y Gloess. A Proof of the Correctness of a Simple Parser of Expressions by the Boyer-Moore System. Technical Report N00014-75-C-0816-SRI-7, SRI International, Menlo Park, California 94025, August, 1978.
5. Ps, J-P Laurent. Adding Dynamic Paramodulation to Rewrite Algorithms. Technical Report CSL-102, SRI International, Menlo Park, CA94025, December, 1979.

---

[7] Many of our bridge-lemmas were introduced for a different reason: the necessity of instantiating free variables of hypotheses of lemmas such as the transitivity of "<"; they would now be superfluous since an arithmetic-package has been built in a new version of the theorem prover, see [2].

6. G. Huet. Confluent Reductions: Abstract Properties and Applications to Term Rewriting Systems. Rapport Laboria n°250, IRIA-LABORIA, Domaine de Voluceau, 78150 Le Chesnay, France, August, 1977.

7. G. Huet. Equations and Rewrite Rules: a Survey. CSL-111, SRI International, Menlo Park, California 94025, January, 1980.

8. D. E. Knuth and P. G. Bendix. Simple Word Problems in Universal Algebras. In J. Leech, Ed., Computational Problems in Abstract Algebra, Pergamon Press, New York, 1970, pp. 263-297.

9. D. S. Lankford. Canonical Inference. Automatic Theorem Proving Project Report ATP-32, University of Texas, December, 1975.

10. J S. Moore. A Mechanical Proof of Takeuchi's Function. Information Processing Letters 9, 4 (November 1980), 176-181.

11. D. R. Musser. Convergent Sets of Rewrite Rules for Abstract Data Types. USC Information Sciences Institute, 4676 Admiralty Way, Marina del Rey, California 90291, December, 1978. Extended Abstract

12. R. E. Shostak. An Algorithm for Reasoning About Equality. Communications of the ACM 21, 7 (July 1978), 583-585.

AN EXPERIMENT WITH "EDINBURGH LCF"

Jacek Leszczylowski
Institute of Computer Science
Polish Academy of Sciences
P.O.BOX 22
00-901 Warszawa, PKiN
POLAND

1.ORGANISATION OF THE PAPER

EDINBURGH LCF is a machine implemented system whose purpose is to support a user in expressing problems concerning algorithms, programs and programming languages as well as in proving theorems describing facts about them.

In this paper we shall describe one of the possible ways (and we hope that it is "the most natural" one) of stating and proving the following fact: "the" normalization function of if-expressions terminates. We chose this example because it seemed to be interesting enough to present the main features of EDINBURGH LCF, as well as small and easy enough to give an almost fully detailed example of the application of the system. The same methodology is also useful for dealing with bigger proofs.

In sec.2 and sec.3 we shall present a quick overview of two main parts of EDINBURGH LCF: PPLAMBDA and ML, respectively. While developing the example (the process is described in sec.4-10), we shall give the necessary definitions or precise specifications of the tools being used; ML-programs defining functions used in the paper will be given in APPENDIX 1-3. The final comments are given in sec.11.

2.PPLAMBDA

This is a Polymorphic Predicate LAMBDA-calculus - a deductive calculus in which one can deal with:

1.domains - in the sense of the theory developed by D.Scott. There are two elementary domains: a domain with one element and an other (called: tr) of the truth values, consisting of true, false and the undefined element. There are also domain constructors including: sum of domains (denoted: +), Cartesian product (denoted: #), constructor of the domain of continous functions (denoted: ->) . Using these constructors one can define new domains from domains defined previously, as well as domains constructors defined by expressions in which variables occur.

2.functions - there are some standard functions such as: TT and FF - truth value constants, FIX - the least fixed point operator, the functions associated with domain constructors like FST and SND which allow one to access the first and the second element of an element of a Cartesian product. It is an important feature that constants have so called "generic" types. This means that they can be used as function symbols in many different contexts. For example, the constant UU denotes the undefined element in every domain. Based on the already defined functions one can introduce new ones either using constructors like conditional or composition or by describing some properties of the function by means of new axioms.

3.theorems and proofs - theorems are expressed and proofs conducted in a Gentzen-style logic where inference rules play the dominant role.

More formally:

1.each type is interpreted as a domain which is a complete partial order.

2.the terms of PPLAMBDA are those of the typed lambda-calculus together with the fixed point operator.

3.Formulae are built up from terms by the binary relation symbols: == (equality) and << (the partial order) or by the usual operators of the predicate calculus: & - conjunction, IMP - implication, ! - universal quantification.

3.META LANGUAGE - ML

ML is a general purpose programming language. It is designed as the main tool for supporting PPLAMBDA users and its characteristic features are:

1.a polymorphic type discipline providing the flexibility of typeless programming languages and the security of compile-time typechecking; this is what ensures that well-typed programms cannot perform faulty proofs

2.data types - one can define one's own data types together with their primitive functions. Some of the predefined data types correspond to syntactic classes of PPLAMBDA. For example, type to the class of PPLAMBDA types, form to the class of formulae, type thm to the class of theorems. Using functions defined on those types one can "manipulate" PPLAMBDA. For example, if f:thm->thm is a ML-function then by applying f to some t:thm one can produce f(t) which is a theorem. In this way the inference rules of PPLAMBDA are represented in the system; one can also define one's own derived rules by programming them in ML.

3.functions are first class values - ML is fully higher order. This allows users to define functionals by programming them. For instance: tacticals - see sec.10.

The other important aspects of EDINBURGH LCF are:

1.structuring of information - let us call the PPLAMBDA-kernel all objects which are given as primitives within the calculus. Each activity such as type or constant definitions or the introduction of new axioms will cause an extension of the kernel. One can "freeze" this development at any moment and give a name to the defined calculus. Then we can use the types, constants and theorems from the calculus while building other ones. Retrieving theorems from files and other similar actions can be done by means of ML.

2.security - if we assume that the PPLAMBDA-kernel is consistent then it is ensured that one cannot derive false theorems; the ML type-checker ensures that the only way to generate theorems is by using the inference rules. Of course, one can introduce inconsistency into the system by introducing inconsistent axioms, but it is possible to avoid the danger sometimes; for instance, one avoids the possibility of inconsistency when one defines new functions using only the functions of the PPLAMBDA-kernel.

4.INTRODUCING A NEW TYPE IN LCF

To introduce a new type in LCF one has to declare a new type operator using the ML-function: newtype. For any n:int and a:token, the ML-expression: (newtype n a) will introduce the identifier: a of the defined type operator whose arity is n. We shall define an if-expression either as an atom or as an expression beginning with IF. Thus atom is the only parameter of the definition. In the case of non-parametric types the arity is 0.

Let us remember that each type is interpreted as a domain which is a complete partial order. From this fact it follows that for each type t there exists the undefined element UU:t which is less than any other element of the type with respect to the relation <<. Notice that if a problem we want to consider in LCF is originally formulated in set-theoretic terms without an ordering structure then the following problem arises: how does one represent a set as a domain? It seems that the most natural way to represent a set is as so called flat domain - each element is greater than UU and none of them is greater than any other element besides UU.

There are two ways of introducing the primitive functions of the newtype. The first one is the following: let A be a type which is going to be defined and let B be a "representing" type which is built up from other types (possibly A as well) using the domain constructors . B is to be isomorphic with the type A (to ensure this we have to introduce special constants and axioms). We define primitive functions of the type A using these special constants and the functions associated with the domain constructors used in the definition of the type B. This the secure way we mentioned in point 2 of sec.4. To ensure that the domain A is flat it is enough to use only flatness-preserving constructors while building the domain B.

When we are sure we shall not introduce any inconsistency to the system, we can use "abstract" way of introducing primitive functions of a type. In this case we shall not use any representing type and the axioms characterizing the primitive functions will concern only functions over the type. To ensure that the domain is flat we have to add additional axioms. We shall choose this way of introducing primitive functions of the if-expressions type.

To present the information to the system we have to fulfil some syntactic requirements. There are two Ml-functions which allow us to declare constants of PPLAMBDA and their properties: newconstant and newaxiom. For any a:token and t:type , the ML-expression: newconstant(a,t) will introduce constant: a with its generic type: t. For any a:token and f:form, the ML-expression: newaxiom(a,f) will introduce the theorem:]-!X.f to the system and the theorem name: a (A]-w is the shape of the theorem in the system, where A represents its assumptions (hypotheses) and w conclusion, X represents all free variables of the formula f). There is also another ML-function: map which for any function f and list l produces a list of results of applying f to each element of l. Comments can be included in ML-programs between %...%. The ML prompt is # .Thus everything which occurs in a line beginning with # and ending with ";;" will be user's contributions and others will be outputs of the system.

5.IF-EXPRESSIONS

We shall use the function "DEF" which is standard in PPLAMBDA; its generic type is ":* -> tr" (* stands for any type) and if DEF x == TT for some x then it means: x is not equivalent to the undefined element.

Having in mind the definition of if-expressions we mentioned at the beginning of sec.4, we shall present the full session with the system in which we introduce the type. We shall assume that the subsequent sessions are the continuations of their predecessors and we shall omit the responses of the system while presenting them.

```
===============================================================================

# newtype 1 `ifexp` ;;
() : .

# map newconstant

     [ `ISAT` , ": * ifexp -> tr " ;
       `IF`   , ": * ifexp -> * ifexp -> * ifexp -> * ifexp " ]  ;;

[();()] : (. list)

# "X:* ifexp","Y:* ifexp","Z:* ifexp" ;;
"X","Y","Z" : (term # (term # term))
```

```
# map newaxiom

     [ `ISATax` , " ISAT X == TT   IMP   DEF X == TT " ;

       `DEFax`  , " DEF X == TT & DEF Y == TT & DEF Z == TT
                          IMP  DEF(IF X Y Z) == TT "       ]  ;;

[ ]-"!X.ISAT X == TT IMP DEF X == TT";
  ]-"!X.!Y.!Z.DEF X == TT & DEF Y == TT & DEF Z == TT
           IMP  DEF(IF X Y Z) == TT"                ] : (thm list)
```

===

Notice that we had introduced the types of varibales which were used later while introducing the axioms. The system has quantified all free variables in the formulae ; in this form all axioms and proved theorems are stored. The two constants: "ISAT" for testing whether an expression is an atomic one, and "IF", a constructor of new expressions; "ifexp" is a postfixed operator thus "* ifexp" is a parametrized type.

6.FORMULATION OF THE PROBLEM

If we have a flat domain A and function F ranging over A then for a given argument X the following fact: F terminates for X, can be expressed as follows: DEF(F X) = TT. We now define constants and axioms which will ensure that the domain of if-expressions will be flat ; we also define a normalization function NORM. The following constants: FI1,FI2,FI3 will be "inverse" functions with respect to IF and operator: ...=>...|... is the conditional operator.

===

```
# map newconstant

    [ `FI1`  , ": * ifexp -> * ifexp " ;
      `FI2`  , ": * ifexp -> * ifexp " ;
      `FI3`  , ": * ifexp -> * ifexp " ;
      `IND`  , ": * ifexp -> * ifexp " ;
      `NORM` , ": * ifexp -> * ifexp "  ] ;;

# "A:* ifexp","B:* ifexp","X:* ifexp","Y:* ifexp","Z:* ifexp",
  "G: * ifexp -> * ifexp"  ;;

# map newaxiom

    [ `Xax` , "X == UU  IMP  IF X Y Z == UU" ;
      `Yax` , "Y == UU  IMP  IF X Y Z == UU" ;
      `Zax` , "Z == UU  IMP  IF X Y Z == UU" ;
      `EQax`, "ISAT X == TT  &  X << Y   IMP   X == Y" ;
      `INVa`, "DEF(IF X Y Z) == TT   IMP  ISAT(IF X Y Z) == FF  &
       FI1(IF X Y Z)==X  &  FI2(IF X Y Z)==Y  &  FI3(IF X Y Z)==Z";
      `INDd`, "IND G X == ISAT X => X |
                        IF (G(FI1 X)) (G(FI2 X)) (G(FI3 X))" ;
      `INDa`, "FIX IND X == X" ;
      `N1` , "ISAT X == TT  IMP  NORM X == X" ;
      `N2` , "ISAT X == TT   IMP
                 NORM(IF X Y Z) == IF X (NORM Y) (NORM Z)" ;
      `N3` , "NORM(IF (IF X Y Z) A B) ==
                    NORM(IF X (IF Y A B) (IF Z A B))"   ]  ;;

% Thus the formula which is going to be proved can be introduced
  to the system in the following way: %
```

```
# let   F = "!X. DEF X == TT  IMP  DEF(NORM X) == TT" ;;
```

===

7.COMPUTATIONAL AND STRUCTURAL INDUCTION

It seems to be natural to use the structure of an object while proving a fact about it. However, the only tool of this kind which is elementary in LCF is the simultaneous computational induction rule: INDUCT - a primitive ML-function. One can specify INDUCT as follows:

```
INDUCT l w (th1,th2) = A1 u A2 ]- w[...FIX(funi)...]
```

where:

```
l = [...(funi,fi)...] - list of pairs of terms
w[...fi...] - a formula
th1 = A1 ]- w[...UU...]
th2 = A2 u {w} ]- w[...funi(fi)...]
```

In the above specification "funi" are functions or functionals of the induction. For more details see Bird's book [2] where Scott's induction rule is decribed.

We are able to derive structural induction for if-expressions from INDTAC because we have already introduced the necessary axioms when we wanted to ensure that the domain was flat. The most important one is `INDa`. This way of asserting the well-foundedness is shown in full details in APPENDIX 1 of [10]. We shall call our structural induction rule: STRUCIND ; its definition represented by a ML-program is given in APPENDIX 1. Its specification is the following:

```
STRUCIND "X" w (TH1,TH2,TH3) = A]- !X.w[X]
```

where:

```
"X" - a variable
w[X]  - a formula, with free variable X
TH1 = A]- w[UU]
TH2 = A]- !X. ISAT X == TT IMP w[X]
TH3 = A]- !X.!Y.!Z.w[X] & w[Y] & w[Z] IMP w[IF X Y Z]
```

8.GOALS AND TACTICS

The formula F presented at the end of the sec.6 does not determine our aim - the theorem we want to prove; one should also specify the fact that we want to prove it without any assumptions.

There is a predefined type in ML which can be used to specify our goal completely. Its definition is the following:

```
goal = form # simpset # form list
```

The first element of the Cartesian product is for stating the formula which is going to be proved, the third one for listing assumptions, the second one is (from the user's point of view) an abtract type consisting of simplification rules - these are (possibly conditional) equivalences of terms to be used as left-to-right rewriting rules. Now we show how we can define our goal; we shall use the following tools to define our simplification set: BASICSS - constant of the type simpset (including some basic axioms from PPLAMBDA-kernel) , ssadd:thm->simpset->simpset - function adding a theorem to a given simplification set, itlist - a primitive Ml-functional iterating a given function over a given list and argument).

```
================================================================================

# let  ss = itlist ssadd
         (map (AXIOM `IFEXP`) ``ISATax DEFax N1 N2 N3``) BASICSS ;;

% The result of the above instruction is that the variable ss is of
  the type simpset and its value is the extension of BASICSS by
  adding the listed above axioms to it. %


# let  g = (F,ss,nil) ;;

% This is our goal; nil means - we start with no assumptions %

================================================================================
```

We shall base our proof on partial subgoaling methods (called: tactics). The system can support this kind of proofs via the predefined types: tactic and proof defined as follows:

proof = thm list -> thm

tactic = goal -> (goal list # proof)

We shall explain now the use of the tool. Let us define when a theorem A']-f' achieves a goal (f,ss,A) - this is the case when up to renaming of bound variables: f' is identical with f and each member of A' is either in A or is in the hypothesis list of a member of ss. Then a theorem list achieves a goal list if the first element achieves the first in the goal list etc. Thus, a tactic: t will work "properly" if for any goal: g and any goal list: gl and any proof: p such that t(g) = gl,p ,if we have a theorem list: thml which achieves a goal list: gl then p(thml) will achieve goal: g. Whenever gl is empty then p(nil) will be the theorem which we wanted to prove while stating the goal g.

9.PLANNING THE PROOF ; TACTICALS

We now design our structural induction tactic: INDTAC producing subgoals and the proof. Its definition is given in APPENDIX 2 and its specification is the following:

```
INDTAC (!X.f,ss,A) = [(f[UU],ss,A);(f,ss1,w.A);
                      (f[IF X Y Z],ss2,w1 @ A)] , p
```

where:

```
f - of the shape: !V.(f' IMP f1 & ...& fn)
    with: !V = !X1....!Xm
w = "ISAT X == TT"
ss1 = ssadd ( w ]- w ) ss
. and @  - list cons and append, respectively
w1 = [ f[X] ; f[Y] ; f[Z] ] - inductive assumptions
ss2 - ss extended by adding all the theorems of the form:
f ]- !V.(f' IMP f1) ,..., f ]- !V.(f' IMP fn)
where : X,Y Z are instantiated for the induction variable
p uses the STRUCIND rule.
```

If we apply INDTAC to the goal g then we get three subgoals; let us call them: g1,g2,g3. We shall analyse how to prove them. Let :

```
f1 = " DEF UU == TT IMP DEF(NORM UU) == TT "
f2 = " DEF X == TT IMP DEF(NORM X) == TT "
f3 = " DEF(IF X Y Z) == TT IMP DEF(NORM(IF X Y Z)) == TT "
```

be identical to the first elements of the subgoals up to renaming variables, respectively. f1 should be true because the antecedent is false - see the PPLAMBDA axiom asserting that "DEF UU == UU" and the fact that the domain tr consists of three different elements. f2 should be true because the consequent is true - see axioms N1 and ISATax as well as the assumption "ISAT X == TT".

Let us analyse now how to prove f3. Observe that one can easily prove (by simple reasoning by contradiction) that if "IF X Y Z" is defined then so are X, Y and Z. Let us assume that we already proved these facts and named them respectively: \`1\`,\`2\`,\`3\`. They are available via the Ml-function FACT. For example,

```
=================================================================================

# FACT `IFEXP` `2` ;;
  ]- "!X.!Y.!Z.DEF(IF X Y Z) == TT IMP DEF Y == TT" : thm

% Notice now that using these theorems and the inductive assumptions
  we can prove f3 if we have proved the following formula: %

# let f = "!X.!Y.!Z.DEF X== TT & DEF(NORM Y)== TT & DEF(NORM Z)== TT
                     IMP  DEF(NORM(IF X Y Z)) == TT" ;;

=================================================================================
```

The system allows us to prove the formulae: F and f simultaneously but for clarity reasons we shall prove the formula f as a lemma and use the lemma ,in turn, to prove our theorem.

Now the last tool necessary for designing the proof: tacticals - mechanisms for combining tactics to form larger ones. We shall use only one of the standard tacticals of the system, which is called: THEN . For any tactic: T1 and T2, the composed tactic; T1 THEN T2 applies T1 to the goal and then applies T2 to all resulting subgoals produced by T1; the proof function returned is the composition of the proof functions of T1 and T2.

10. THE PROOF

If we specify our lemma by the following goal:

```
=================================================================================

# let  g = (f,ss,nil) ;;

=================================================================================
```

and apply tactic INDTAC to it then we get three subgoals.

By a similar analysis to the one we demostrated in sec.9 we can convince ourselves that the first goal can be solved by using the rewriting rules from the simplification set ss. There is a standard tacitic: SIMPTAC which can be used for this purpose; applied to any goal: (w,ss,A) SIMPTAC produces list of goals [(w',ss,A)] and proof p where w' is the simplification of w by ss and p justifies all the simplifications made. In the case where w' is a tautology then the goal list is null.

Let us take a look at the formula occurring in the second of subgoals; it will be the formula identical to f if we omit the quantifier. Notice that the consequent of the implication can be proved by using the axioms N2 and DEFax as well as conjuncts of the antecedent. Notice also that the third subgoal can be solved by using facts `1`,`2`,`3`. Unfortunately, we cannot add these facts to the simplification set because the simplification would loop. On the other hand, one can use this axiom after instantiating the variables to the ones which are necessary at this particular moment of the proof. To overcome the problem we shall design a function which will perform the required task. This function is of general use and describes an implication proving method. Its definition is given in APPENDIX 3 and its specification is the following:

```
IMPTAC thml (!X.f1&...&fn IMP f' ,ss,A) = [(f'',ss',A)] , p
```

where:

```
X - is a vector of variables
f'' is f' up to renaming variables
ss' - an extension of ss created by adding all theorems
      of the form:

      fi ]- fi       and
      [!V.(wi IMP vi) ; f] ]- v'

      where:
      ]-!V.(wi IMP vi) - element of the theorem list thml
      wi is f , up to renaming variables
      v' - the result of the substitution:
           if [...(uj,xj)...] - list of such pairs
           of variables that  wi[...uj/xj...] = f
           then  vi[...uj/xj...] = v'
```

```
================================================================================

# let T = INDTAC THEN SIMPTAC THEN
        (IMPTAC (map (FACT `IFEXP`) ``1 2 3``)) THEN SIMPTAC ;;

% We defined our tactic %

# let gl,p = T g ;;
gl = []: (goal list)
p  = - : proof

% Since the empty list of theorems achieves empty list of goals
  then we may apply the produced proof to it proving the lemma: %

# let lemma = p[] ;;

# let ss = ssadd lemma ss ;;

# let gl,p =  T  (F,ss,nil) ;;

# p[] ;;

]-"!X.DEF X == TT IMP DEF(NORM X) == TT" : thm

================================================================================
```

11.A FEW REMARKS

As we mentioned in sec.1 the aim of the paper was to present an application of EDINBURGH LCF to an interesting problem. But the aim of any application itself is to find a general tactic which can be used in totally different examples. Why? The answer to the first question is the following: because LCF is essentially a tool placed somewhere between a theorem prover and a proof checker; this is why we are looking for general proof strategies which when found can be programmed in ML and be used to tackle other goals.

We chose the example deliberately to be able to compare our solution with the solution presented in chapter 4 in [4]. There, more things were done automaticaly and a user assisted only in suggesting auxiliary lemmas; the proof itself was much bigger because the theorem prover had to prove at the very beginning that the definition of the NORM function was acceptable - this means that some measures decrease according to a well-founded function in every recursive call of NORM. In EDINBURGH LCF we could use much higher level approach - we took axioms `INDd` and `INDa` to ensure well foundedness of the type of if-expressions. This was done in a very natural way due to explicit presence of the logic in the system; this makes the system "human-oriented" and easy to extend. The Boyer-Moore theorem prover is very efficient in its use of built-in strategies, but difficult to extend; by contrast the need to conduct all inference through the basic inference rules (as ML procedures), which appears necessary if we wish to allow users to extend the system reliably by programming, leads to some inefficiency in LCF. This we have found tolerable and indeed it can be reduced significantly by direct implementation of ML (which at present is compiled into LISP). For a nice general comparison of these systems see [9]. The LCF style of solving the problems by programming proof strategies seems to be natural (it took the author 2 months to be able to work with the system) and well fitted especially to doing large proofs with machine assistance. By this we mean neither that a large proof is submitted step by step and merely checked by the machine (as in [11]), nor that the system discovers the large proof by itself, but that the problem may be split into parts, each of which is tackled semiautomatically by a subgoaling method.

Was it a successful experiment? To answer this question we have to emphasize that the tactic "T" was used twice in the example (we found a strategy which worked in solving different goals). Moreover, it turned out that the strategy was quite general. Namely the sequence of tactics: structural induction and IMPTAC was used by the author in several other proofs; for example, while proving correctness of the primitive functions: DELETE , INSERT, SEARCH defined over the type of binary search trees. Structural induction over lists as well as SIMPTAC were the main tactics used in proving the theorems given by J.Backus in [1] . The system allowed us to formulate them in more general form; hence the proofs were straightforward. (see [13]).

Notice that the theorem was proved for any function NORM satisfying the axioms. If one wanted to have NORM defined as the least fixed point of a functional, then instead of proving the theorem one could prove a general metatheorem concerning the shape of the functional. Namely, that if all elementary components of the functional terminates then the least fixed point terminates. We chose the kind of the proof we presented here because it seemed to us that the price we have to pay for generality is too high - we would lose the simplicity and clarity of the proof.

ACKNOWLEDGEMENTS

I wish to thank Avra Cohn, Mike Gordon, Robin Milner and Chris Wadsworth for their friendly help during my "fights" with EDINBURGHLCF and while preparing the draft version of the paper.

BIBLIOGRAPHY

[1] J.Backus, "Can programming be liberated form the von Neumann style? A functional programming style and its algebra of programs", Comm ACM 21,8 (1978).
[2] R.Bird, "Programs and Machines; and introduction to the theory of computation",Wiley (1976).
[3] A.Blikle, "Assertion Programming", Proc. 8th MFCS Symposium, Olomouc, Czechoslovakia, (1979).
[4] R.S.Boyer, J S.Moore, "A Computational Logic", Academic Press, New York (1979).
[5] W.H.Burge, "Recursive Programming Techniques", Addison-Wesley, Reading, Mass. (1975).
[6] R.M.Burstall, J.A.Goguen, "Putting theories together to make specification",Proc. 5th Intern.Joint Conference on Artificial Intelligence, Cambridge Mass., published by Dept.of Comp. Sci., Carnegie Mellon, (1977).
[7] C.Cartwright, private communication.
[8] A.Cohn, "High level proof in LCF", Proc. 4th Workshop on Automa ted deduction, Austin, Texas (1979).
[9] A.Cohn, "Remarks on Machine Proof", 1980 , manuscript
[10] M.Gordon, R.Milner, C.Wadsworth, "Edinburgh LCF", Springer- Verlag, (1979).
[11] L.S.van Benthem Jutting, "Checking Landau's `GRUNDLAGEN` in the AUTOMATH system", thesis, Tech.Hogeschule, Eidhoven, The Netherlands (1977).
[12] F.W.von Henke, "Recursive data types and program structures", Internal report, GMD, Bonn (1976).
[13] J.Leszczylowski, "Theory of FP systems in EDINBURGH LCF", Internal Report,Comp.Sci.Dep.,Univ.of Edinburgh, (1980)
[14] R.Milner, "A theory of type polymorphism in programming", Journal of Computer and System Sciences 17 (1978).
[15] R.Milner, "LCF: a way of doing proofs with a machine", Proc.8th MFCS Symposium, Olomouc, Czechoslovakia (1979).
[16] R.Milner, "Implementation and application of Scott's logic for computable functions", Proc. ACM Conference on Prov ing Assertions about Programs, SIGPLAN notices (1972).
[17] D.Scott, "Lattice theoretic models for various type-free calcu li", Proc. 4th International Congress in Logic, Me thodology and Philosophy of Sciences, Bucharest (1972).
[18] R.A.Wulf, R.L.London, M.Shaw, "Abstraction and verification in ALPHARD: introduction to language and methodology", Carnegie Mellon University, (1976).
[19] S.Zilles, "Algebraic specification of data types", Computation Structures Group Memo 119. M.I.T. (1974).

APPENDIX 1

```
let  C t t' =  ( TRANS(SYM(CONDCONV(lhs(concl th))),th)
                 where  th =
                 SUBSOCCS [[1],ASSUME "^(fst(destcond t')) == ^t" ]
                          (REFL t')  )  ;;

let   CC x w t (th1,th2,th3)  =

        CASES  (fst(destcond t))
               ( (SUBST [(C "TT" t)     ,x] w th1),
                 (SUBST [(C "FF" t)     ,x] w th2),
                 (SUBST [(C "UU:tr" t)  ,x] w th3)  )  ;;

let   STRUCIND X w (Z1,Z2,Z3) =

     let [t] = snd(desttype(typeof X))
     in
     let f = ":^t ifexp -> ^t ifexp"
     in
     let b = GEN X (SUBST [SYM(MINAP "(UU:^f) (^X)") ,X] w Z1)
```

```
and F = genvar f
in
let w' = "!^X.^(substinform ["^F ^X",X] w)"
in
let th1 = MP (SPEC X Z2) (ASSUME "ISAT ^X == TT")
and th2 = ASSUME w'
and Y,Z = let f = destquant(snd(destquant(concl Z3)))
          in
          fst f , (fst(destquant(snd f)))
in
let t1,t2,t3 = (SPEC X th2),(SPEC Y th2),(SPEC Z th2)
in
let th3 = CONJ(t1,(CONJ(t2,t3)))
and th4 = SPEC "^F ^Z"(SPEC "^F ^Y"(SPEC "^F ^X" Z3))
in
let th5 = MP th4 th3
in
let th6 = INST ["FI1 ^X",X;"FI2 ^X",Y;"FI3 ^X",Z] th5
and t' = "ISAT ^X => ^X |
               IF (^F (FI1 ^X)) (^F (FI2 ^X)) (^F (FI3 ^X))"
in
let th7 = CC X w t' (th1,th6,Z1)
and th8 = INSTTYPE [t,":*"] (AXIOM `IFEXP` `INDdef`)
in
let s = GEN X (SUBST [SYM(SPEC X (SPEC F th8)),X] w th7)
in
let th9 = INDUCT ["IND:^f -> ^f",F] w' (b,s)
and th10 = INSTTYPE [t,":*"] (AXIOM `IFEXP` `INDax`)
in
GEN X (SUBST [(SPEC X th10),X] w (SPEC X th9))  ;;
```

APPENDIX 2

```
letrec  sf x =

        (let t = destquant(snd x)
         in
         sf((fst t).(fst x)  , snd t)) ? x ;;


letrec S t = S(SPEC (fst(destquant(concl t))) t)? t ;;



let INDTAC (w,ss,wl) =

    let L,w' = destquant w
    in
    let [t] = snd(desttype(typeof L))
    and Y,Z = let t = variant(L,union(formlvars wl,formvars w))
              in
              t,(variant(L,union(formlvars wl,(t.(formvars w)))))
    in
    let w1 = substinform [Y,L] w'
    and w2 = substinform [Z,L] w'
    in
    let Z1 = substinform ["UU:^t ifexp",L] w'
    and Z2 = substinform ["IF ^L ^Y ^Z",L] w'
    and ss1 = ssadd (ASSUME "ISAT ^L==TT") ss
    and l = map ASSUME [w';w1;w2]
    and g x y l =  revitlist GEN l (DISCH y x)
```

```
in
letrec f z = let x = S z
             and l = fst(sf([],(concl z)))
             in
             let w,w'=destimp(concl x)
             in
             let th = MP x (ASSUME w)
             in
             (if phylumofform w' = `conj`
                 then (g (SEL1 th) w l).(f(DISCH w (SEL2 th)))
                 else [g th w l])
in

let [al;bl;cl] = map f l
in
let ss2 = itlist ssadd (al @ bl @ cl) ss
in
let p = \[tl;t2;t3].STRUCIND L w' (tl,
        (GEN L (DISCH "ISAT ^L==TT" t2)),
        (itlist GEN [L;Y;Z] (DISCH w' (DISCH wl (DISCH w2 t3)))))
in
([(Z1,ss,wl);(w',ss1,"ISAT ^L==TT".wl);
  (Z2,ss2,([w';wl;w2] @ wl))],p) ;;
```

APPENDIX 3

```
letrec  rdestconj f =  if isconj f
                    then (fst(destconj f)).(rdestconj(snd(destconj f)))
                        else [f] ;;


let  IMPTAC thl  ((w,ss,wl):goal) =

     let [w,ss,wl],p = (REPEAT GENTAC) (w,ss,wl)
     in
     let w',w'' = destimp w
     and l = map concl thl
     in
     let wl = rdestconj w'
     in
     let F = \x.(filter
                ((can (\y.(formmatch y x))) o fst o destimp o concl)
                 (map (S o ASSUME) l))
     in
     let t = combine(wl,(map F wl))
     in
     let thml = let f = fst o destimp o concl
                in
                flat(map ((\x,y.(map (\z.(MP z (ASSUME x))) y)) o
                (\x,y.(x,(map (\z.(INST (formmatch (f z) x) z)) y)))) t)
     in
     let ss' = itlist ssadd (thml @ (map ASSUME wl)) ss
     and p = p o (\x.[x]) o (revitlist (\x y.MP y x) thl) o (itlist DISCH l)
     in
     [(w'',ss',wl)],(p o (itlist DISCH wl) o hd) ;;
```

AN APPROACH TO THEOREM PROVING ON THE BASIS OF A TYPED LAMBDA-CALCULUS

R.P. Nederpelt (*)

Eindhoven University of Technology
Department of Technology

P.O. Box 513, 5600 MB Eindhoven
The Netherlands

Abstract and introduction

This paper describes a system of typed lambda-calculus suited to representing mathematical texts, and discusses some theorem proving aspects of the system. In part 1 a formal exposé is given of the system, with comments on the principles chosen. A natural manner of rendering mathematical texts in the system will be explained in part 2. Finally, in part 3 the system will be investigated as regards its potentials for (partial) theorem proving.

An idea for a "completely" formalized, yet natural language for rendering mathematical texts was conceived by N.G. de Bruijn. In 1968 he developed the mathematical language Automath. Automath has since been extensively applied and tested on numerous mathematical topics, language theory has been developed for establishing its computational soundness, and a computer programme has been produced for checking the formalized texts. For Automath and applications, cf. references [1], [2], [5] and [8].

Automath is essentially a typed lambda-calculus presented in a modified form in order to make it more accessible to the customer. The underlying typed lambda-calculus, which we call Λ, is relatively simple and in a sense natural. In contrast to types in the usual typed systems, types in Λ behave like terms, and do not look different from terms. The system Λ obeys the same "nice properties" as does Automath, such as the Church-Rosser property, unique (strong) normalization and closure (see section 1.6). For language theory of Automath and Λ, see [3], [4] and [6].

Since Λ has a simple, transparent structure, it would seem promising to investigate its practical theorem proving aspects. However, there has not been much experience in this direction. Some considerations and comments, arising from a Λ-text example presented in this paper, may nevertheless be helpful for practical applications to theorem proving.

(*) The author wishes to thank A.V. Zimmermann for remarks concerning the use of the English language.

1. A CONCISE DEFINITION OF THE SYSTEM Λ

1.1. Notations

Like Automath, Λ uses notations that deviate from lambda-calculus conventions. For $\lambda_x B$, x having type A, we write $(A\lambda_x)B$. Instead of (BA), i.e. "function" B applied to "argument" A, we write $(A\delta)B$. Here δ is the signal for a functional application. There are two main reasons for inverting the order of function and argument: (1) there is an analogy between the dual operations "abstraction" (leading from B to $(A\lambda_x)B$) and "application" (leading from B to $(A\delta)B$); (2) the inversion is very practical when rendering mathematical texts by means of Λ (see e.g. sections 2.6 and 2.7).

1.2. Terms

We introduce the set T^o of *terms* by the following recursive definition; the alphabet consists of variables, brackets and the symbols λ and δ.

(1) The empty term is a term of T^o; each variable is a term of T^o.

(2) If A and B $\in T^o$ and if x is a variable, then $(A\lambda_x)B$ and $(A\delta)B$ are terms of T^o.

The empty term is rendered invisibly; examples of other terms are: x; $((\delta)\lambda_x)y$.

The set T^c is defined as the closed fragment of T^o, i.e. the set of all closed terms present in T^o; a *closed* term is a term without free variables. We say that G is a *subterm* of F, if F and G are terms and G occurs in F. In fact, we obtain more subterms than is usual in lambda-calculus. For example, (λ_y) is a subterm of $(\lambda_y)x$ in our sense, but not in ordinary lambda-calculus. Those subterms that resemble lambda-calculus subterms, we call *genuine* subterms, characterized by: a genuine subterm G of F is a subterm of F for which the first symbol (if any) occurring after G in F is δ or λ. So z, $(\lambda_y)x$ and the entire term are examples of genuine subterms of $((\lambda_x)(\lambda_y)x\delta)z$, but (λ_y) is a subterm that is not genuine.)

1.3. Typing

As concerns types, the present system (and Automath) follow a point of view different to that in ordinary typed lambda-calculuses: the types are not given beforehand as a separate set. "Types" and "terms" can both be found in T^o, and it depends on the circumstances whether a given T^o-term is considered a "type" or a "term". Formally, one can only say that, for A and B $\in T^o$, the relation "B is a type of A" may hold. If so, one often finds the structure of A closely reflected in B. Every term may act as a type, so that one may construct chains of terms: $A_1, A_2, \ldots, A_n$, each A_{i+1} being a type for A_i. (In this context we note that a term ending in the empty term, if present in such a chain, must be the final entry; such terms act as a kind of "final types".)

For an intuitive understanding of the connection between "terms" and "types", we refer to sections 2.2 and 2.3. For the moment we only mention that we need types, as usually, to enable us to attach "classes" to "objects": if the term A represents an object, then the type of A represents the class of that object. But in our system it is a matter of interpretation whether a term is considered an object or a class. Moreover, we shall extend both notions: "object" and "class", in a manner deviating from the set-theoretical tradition.

Formally, we define types as follows. Firstly, we define the type of a variable. If $(B\lambda_x)$ is a subterm of a term A, and x occurs as a variable in A that is bound by the λ_x mentioned, then the type of the bound x, relative to A, is B. Next, we define the type of a term ending in a variable. If C is a subterm of $A \in T^O$ and C ends in a variable x (so $C \equiv Dx$ for some term D), where x is bound in A, then the type of C, relative to A, is the (adapted) concatenation of D and the type of x. (We say *adapted* concatenation, because one possibly has to rename variables in order to prevent undesirable bindings, resulting from a so-called "clash of variables".)

We denote the type of C, relative to A - if constructed in this manner - by $Typ_A[C]$, or, when no confusion is possible, by Typ[C]. Example:
$Typ[((\lambda_z)y\lambda_x)(x\delta)x] \equiv ((\lambda_z)y\lambda_x)(x\delta)(\lambda_z)y$.

For a variable that is free in A, there is no term in A that can naturally act as its type. It is also natural that the empty term, acting as a final type, has no type. Hence it is imaginable that, for $A \in T^O$, Typ_A is only defined for subterms of A ending in a variable bound in A.

The iterates of Typ are denoted, as usual, by Typ^2, Typ^3, etc.; $Typ^0[C] \equiv C$.

1.4. Reductions

We provide T^O with the usual reduction relations, called α-, β- and η-reduction. Since α-reduction is a mere renaming of variables, we prefer to consider T^O as being the set of all α-equivalence classes, terms being α-equivalent if and only if one term reduces to the other by means of α-reduction.

The essential reduction for T^O is *β-reduction* (symbol: $>_\beta$), which is the formalization of the application of a function to an argument. It is induced by the rule: $(A\delta)(B\lambda_x)C >_\beta {}^x_A C$, the latter term being the result of (simultaneously) substituting A for all x's free in C. (Here, again, one should prevent "clash of variables".) We also have *η-reduction*, induced by: $(A\lambda_x)(x\delta)C >_\eta C$ if x does not occur freely in C.

As usual, *conversion* is the equivalence relation generated by $>_\beta$ and $>_\eta$. We denote conversion by the symbol ~. In the interpretation, we do not distinguish between terms that are mutually convertible: if $B \sim C$, then B and C have the same "meaning". A consequence is, that, if $Typ[A] \equiv B$ and $B \sim C$, then both B and C may be regarded as types of A.

1.5. Strong functionality

In systems of natural reasoning there is a natural desire to restrict the functional applicability. In words: $(A\delta)B$ is only permitted as a subterm of a given F, if B has a domain, say C, and if A fits in domain C. We shall express "having a domain" and "fitting in" by means of the function Typ.

When term B has the shape $(C\lambda_y)D$ for certain terms C and D, or converts to a term of this shape, then it is obvious that one may attach domain C to term B. It may, however, be the case that B does not obey the above property, but that $\mathrm{Typ}^n[B]$ does, for some natural n. Then we say nevertheless that B has a domain, being the same as the domain of $\mathrm{Typ}^n[B]$.

For n = 1, this can easily be understood: if x denotes a function from naturals ($\mathbb{N}$) to reals ($\mathbb{R}$), then the term x itself does not show its domain; however, this domain can be extracted from Typ[x], which denotes $\mathbb{R}^{\mathbb{N}}$. For higher values of n, a similar argument holds. (It can be shown that the domain for a term of Λ, if existent, is unique but for conversions.)

Formally, we define: $F \in T^c$ is *strongly functional* if and only if the following condition applies: for all genuine subterms of the form $(A\delta)B$ it holds that there exist a non-negative number n, a variable y and terms C and D such that

(i) $\mathrm{Typ}^n[B]$ exists and $\mathrm{Typ}^n[B] \sim (C\lambda_y)D$,

(ii) Typ[A] exists and $\mathrm{Typ}[A] \sim C$.

We define Λ as the set of all closed terms that are strongly functional.

Note: For theoretical purposes, a much weaker form of functionality is sufficient to ensure the validity of the "nice properties" (see the next section). One may define "weakly functional" terms, which have a functional structure comparable to that in usual typed lambda-calculuses. The weak system is helpful, proofs of (strong) normalization for Λ being given with the aid of the analogous proofs for the weaker system, which are relatively easy. For details, see [4] and [6].

1.6. Properties of Λ

(1) The *Church-Rosser* (or *diamond*) *property*, i.e.: if $A \in \Lambda$ reduces to B, and A reduces to C, then B and C have a common reduct.

(2) *Closure*, i.e.: if $F \in \Lambda$, then $\mathrm{Typ}[F] \in \Lambda$; moreover, if F reduces to G, then $G \in \Lambda$.

(3) (Unique) *normalization*, i.e.: if $F \in \Lambda$, then there is a unique normal term H such that F reduces to H. (H is *normal* when there is no G such that $H >_\beta G$ or $H >_\eta G$.)

(4) *Strong normalization*, i.e.: each reduction sequence starting from a strongly functional term, terminates.

For proofs of these theorems, see [4] and [6].

2. EXPRESSING MATHEMATICS IN Λ

2.1. Translation of a text

There is a standard way, in a sense natural, for translating mathematical texts into Λ. We shall comment on the principles of this manner of translating. The idea is, that a mathematical text transforms into a longdrawn term of Λ. Not only mathematical entities such as sets and functions, present in the original text, become subterms of this term, but also text units such as theorems and assumptions have their direct counterparts in subterms. The order of the text units in the original reasoning is generally maintained in the translation.

For obtaining a term of Λ, having no free variables, one should in principle have a text that is complete in a double meaning: the text should not have gaps in the reasoning or argumentation, and all foreknowledge (axioms, theorems, definitions used in the text) must be explicitly given. In practice one only translates a portion of text when all foreknowledge is accessible in translated form, so that the text under consideration becomes, after translation, a mere extension of an already existing (possibly very long) Λ-term.

We shall now discuss a possible way of translating some mathematical notions.

2.2. Sets and propositions

Our (long) Λ-term opens with two subterms: (λ_τ) and (λ_π). We think of τ as being the class of all sets, and π as being the class of all propositions.

If we wish to express, somewhere in the translation, that variable s must denote a set, we write $(\tau\lambda_s)$ in our Λ-term. Then $\mathrm{Typ}[s] \equiv \tau$, in correspondence with our interpretation of τ and of Typ. Analogously, if we wish to regard variable p as a proposition, we write $(\pi\lambda_p)$.

An element x of set s may now be introduced by embodying the subterm $(s\lambda_x)$ in our term. For the analogous subterm $(p\lambda_t)$, where p is a proposition, there is a nice and practical interpretation: t is a *proof* of p. (This so-called *propositions-as-types notion* has fairly recently been introduced by several investigators, among others De Bruijn; for comment, see [4].)

Let us state that a subterm X belongs to the n-grade if $\mathrm{Typ}^n[X]$ is a final type, i.e. a term ending in the empty term. Then we have obtained interpretations for terms of four different grades. The empty term belongs to the 0-grade. It acts as a type for the 1-grade terms τ and π, which have the above interpretations. Terms interpretable as sets or propositions, belong to the 2-grade. Finally, there are terms in the 3-grade that represent elements of sets or proofs of propositions.

Hence, if X is an element (or proof) in the 3-grade, then Typ[X] is a set (or proposition) in the 2-grade, $\mathrm{Typ}^2[X]$ is τ (or π respectively) and $\mathrm{Typ}^3[X]$ is the empty term.

It is striking that we only need these four grades for representing a large section of mathematics, although Λ has possibilities for arbitrary n-grades (n being a non-negative number).

2.3. Functions

It is convenient to use the functional structure of lambda-calculus in describing functions. For example, the identity function on A can obtain the term $(A\lambda_x)x$ as its counterpart in Λ. We take the term $(A\lambda_x)A$ as type of this function, usually written A^A. This interpretation of $(A\lambda_x)A$ is not self-evident, but such a *type-valued function* is, again, very practical in use. We note that this policy corresponds with the formal identity $\mathrm{Typ}[(A\lambda_x)x] \equiv (A\lambda_x)A$. For further explanation, see [3].

Following the above convention concerning type-valued functions, there is a plausible interpretation for the term $(p\lambda_t)q$, where p and q are propositions, viz.: $p \Rightarrow q$. This can be understood as follows. If u is a proof of q (so $\mathrm{Typ}[u] \sim q$ according to the propositions-as-types notion), then function $(p\lambda_t)u$ conveys any proof t of p into proof u of q. Hence $(p\lambda_t)u$ proves the implication $p \Rightarrow q$, so that the type of $(p\lambda_t)u$ must be $p \Rightarrow q$. But $\mathrm{Typ}[(p\lambda_t)u] \sim (p\lambda_t)q$, so the latter represents the implication $p \Rightarrow q$.

Analogously, term $(A\lambda_x)q$, where A embodies a set and q a proposition, represents: $\forall_{x\in A}[q]$. Here q is a term that may contain the free variable x.

2.4. Assumptions and introductions

The text unit "Let $x \in A$" introduces a variable x of type A. In translation this becomes $(A\lambda_x)$. Analogously, the assumption "Assume p" can be translated by $(p\lambda_t)$. Note that the latter mode of translation is in accordance with the propositions-as-types notion: the subterm $(p\lambda_t)$ can be read as: "Let t be a proof of proposition p".

2.5. Axioms, axiomatic notions

Axioms and axiomatic notions may be regarded as introductions (or assumptions) with an unbound validity range. For example, the primitive notion "natural number" can be introduced by means of the subterm $(\tau\lambda_{\mathbb{N}})$. The first Peano axiom, "1 is a natural number", reads: $(\mathbb{N}\lambda_{\mathrm{one}})$, and so on.

2.6. Definitions

When object α of class β is abbreviated by variable x, then it is to be understood that each occurrence of x "means" α. This is essentially what the definition $x := \alpha$ does. Let A and B be translations into Λ of α and β. Then we can write the definition in translation as $(A\delta)(B\lambda_x)$, since β-reduction enables us to again replace by A every x bound by this λ. Moreover, by strong functionality both A and x must have type (convertible to) B. These observations imply that the effect of the insertion of $(A\delta)(B\lambda_x)$ is that x "means" A.

2.7. Theorems, lemmas and intermediate results

In translating theorems, we lean heavily on the propositions-as-types notion. Let B be the translation of a proposition that we regard as a theorem, and let A be the translation of its proof. Then we may insert the subterm $(A\delta)(B\lambda_t)$, expressing both the theorem and its proof. By strong functionality, Typ[A] ~ B, in accordance with "A proves B". Variable t may be regarded as a name of the proof A. Theorem B may later be applied by referring to its proof, which can be done by calling the name t of the proof.

Lemmas and intermediate results may be treated analogously.

2.8. Deduction rules and logic

We shall briefly comment on the way in which logic can be incorporated. By introducing an axiomatic notion "contradiction": $(\pi\lambda_{cd})$, we can express the negation $\neg$ p of proposition p as p $\Rightarrow$ cd, or, in translation: $(p\lambda_t)cd$. The logical connectives $\wedge$, $\vee$ etc. now can be expressed by means of the implication and the negation.

The universal quantifier is already "present" in Λ, as we saw in 2.3. The existential quantifier $\exists$ then can be easily expressed as $\neg\,\forall\,\neg$.

The elimination and introduction rules of natural deduction now are implicitly present in the system. They are a result of the natural language structure, and need not be introduced as primitive rules or axioms. See also [7].

When wishing to apply classical logic, one adds the double negation rule: $\neg\neg p \Rightarrow p$ as an axiom.

2.9. Remarks on some translation difficulties

There are a number of peculiarities that hamper the translation of a mathematical text into Λ. We mention a few. (For more extensive comments on these topics, see [4] and [5]).

(1) The system Λ has "uniqueness of types". That is to say: if A converts to B, then Typ[A] converts to Typ[B]. This presents practical difficulties as to the hierarchy of types. For example, if x is a natural number, then x is not automatically a real number as well, since $\mathbb{N}$ and $\mathbb{R}$ are obviously non-convertible. A way out is to write in Λ a mechanism of embedding and "exbedding", to enable us to deal with sets and subsets.

(2) Two proofs of a certain statement are in principle different. This gives undesirable effects in the cases in which only the existence of a proof matters, not its

nature. For example, the natural logarithm ℓn will have two arguments in Λ: a number x, and a proof s that this number is positive. So in fact we should not write ℓn x, but $\ell n(x,s)$ (in Λ : $(s\delta)(x\delta)\ell n$). If s and t are two different proofs of the positiveness of x, however, then nevertheless $\ell n(x,s)$ and $\ell n(x,t)$ should be "equal". One can write in Λ an axiom yielding such an "irrelevance of proofs" in these cases.

(3) In Λ there is no primitive equality, apart from conversion. So some forms of equality (e.g. between sets, and between numbers) have to be expressed axiomatically. This treatment of equality is in principle feasible, but in practice somewhat cumbersome.

(4) When Λ is used in the form as described above, it gives rise to numerous repetitions inside the Λ-term. See the example in section 3.2. Front parts of subterms are often repeated; they are subterms themselves, but since they end in the empty term, they cannot be abbreviated as is done with definitions (cf. section 2.6). It is not hard, however, to extend Λ in such a manner that the abbreviations meant can be carried out.

3. AN APPROACH TO THEOREM PROVING ON THE BASIS OF Λ

3.1. The shape of a translated mathematical text

If one follows the translation conventions discussed in part 2, one obtains a single Λ-term that may be considered a concatenation of *fragments*. Each fragment is a subterm ending in the empty term. There are three kinds of fragments:

1. the *initial* fragments, which stand at the heading of the term, namely (λ_τ) and (λ_π),
2. *primitive* fragments of the form $(A\lambda_x)$, A being a 1- or 2-grade term (see section 2.2),
3. *stating* fragments of the form $(A\delta)(B\lambda_x)$, B being a 1- or 2-grade term.

The role of the initial fragments will be clear. The primitive fragments are the translations of axioms and axiomatic notions. The stating fragments are translations of theorems, lemmas, intermediate results, but also of definitions.

3.2. Example of a text in Λ

As an example we render the first few lines of Jutting's complete translation of Landau's "Grundlagen" (see [5]). Jutting's translation uses one of the Automath languages. We give the Λ-version. As a matter of fact, we use a slightly adapted version of Λ, being the counterpart of Jutting's Automath-version. The only difference with Λ as defined in part 1, is that the following rule of *type inclusion* is added: Let σ abbreviate either τ or π; if a term has type $(B_1\lambda_{x_1})(B_2\lambda_{x_2})\ldots(B_n\lambda_{x_n})\sigma$, then it also has type σ or $(B_1\lambda_{x_1})(B_2\lambda_{x_2})\ldots(B_m\lambda_{x_m})\sigma$ for any $1 \le m \le n$. The use of this rule has practical advantages, although it slightly detracts from the natural appearance of the system.

Since logic is not incorporated in Landau's book, Jutting had to write the preliminary part, dealing with logical notions, himself. The 11 fragments below mainly concern implication and negation; the content of each fragment will be explained afterwards.

For reasons of economy (cf. section 2.9, note (4)) we draw a line when a repetition is meant. E.g., the fourth line in the subjoined Λ-text should read $((\pi\lambda_a)(\pi\lambda_b)\pi\lambda_{imp})$. Numbers 1 to 11 are extra-textual, only meant for numbering the fragments. The Λ-text below, read uninterruptedly, yields a single Λ-term.

1. (λ_τ)
2. (λ_π)
3. $((\pi\lambda_a)(\pi\lambda_b)(a\lambda_x)b\delta)$

   $(\text{————}\ \pi\lambda_{imp})$

4. $((\pi\lambda_a)(\pi\lambda_b)(\pi\lambda_c)((b\delta)(a\delta)\mathrm{imp}\ \lambda_i)((c\delta)(b\delta)\mathrm{imp}\ \lambda_j)(a\lambda_x)((x\delta)i\delta)j\delta)$

$(\longrightarrow (c\delta)(a\delta)\mathrm{imp}\ \lambda_{trimp})$

5. $(\pi\lambda_{cd})$

6. $((\pi\lambda_a)(cd\delta)(a\delta)\mathrm{imp}\ \delta)$

$(\longrightarrow \pi\lambda_{not})$

7. $((\pi\lambda_a)((a\delta)\mathrm{not}\ \delta)\mathrm{not}\ \delta)$

$(\longrightarrow \pi\lambda_{nn})$

8. $((\pi\lambda_a)(a\lambda_{a1})((a\delta)\mathrm{not}\ \lambda_x)(a1\delta)x\delta)$

$(\longrightarrow (a\delta)\mathrm{nn}\ \lambda_{nni})$

9. $((\pi\lambda_a)((a\delta)\mathrm{nn}\lambda_w)a\lambda_{dn})$

10. $((\pi\lambda_a)(cd\lambda_{c1})(((a\delta)\mathrm{not}\ \lambda_x)c1\delta)(a\delta)dn\delta)$

$(\longrightarrow a\lambda_{dne})$

11. $((\pi\lambda_a)(\pi\lambda_b)((a\delta)\mathrm{not}\ \lambda_n)((cd\lambda_x)(x\delta)(b\delta)\mathrm{dne}\ \delta)(n\delta)(b\delta)(cd\delta)(a\delta)\mathrm{trimp}\ \delta)$

$(\longrightarrow (b\delta)(a\delta)\mathrm{imp}\ \lambda_{th})$

Fragments 1 and 2 are initial fragments, 5 and 9 are primitive fragments. The others are stating fragments, where 3, 6 and 7 concern definitions; 4, 8, 10 and 11 may be regarded as theorems.

Now we shall briefly comment on the content of these fragments. The class τ of all sets and the class π of all propositions are primitively introduced in fragments 1 and 2 (cf. section 2.2). In fragment 3, implication ("imp") is defined for each pair of propositions a and b, as a typed-valued function $(a\lambda_x)b$ (see also sections 2.3 and 2.7). Next, in fragment 4, the transitivity of implication is stated as a theorem (cf. section 2.7): if a, b and c are propositions, if i is any proof of implication a ➡ b, and j of b ➡ c, then $(a\lambda_x)((x\delta)i\delta)j$ proves a ➡ c. This is the case, because $\mathrm{Typ}[(a\lambda_x)((x\delta)i\delta)j]$ converts to $(a\lambda_x)c$ or to $(c\delta)(a\delta)\mathrm{imp}$, both representing a ➡ c.

In fragment 5, contradiction ("cd") is primitively introduced (cf. section 2.8). Negation ("not") is defined in fragment 6: not is by definition $(\pi\lambda_a)(cd\delta)(a\delta)\mathrm{imp}$. Hence, for any given proposition a', the term $(a'\delta)\mathrm{not}$ converts to $(cd\delta)(a'\delta)\mathrm{imp}$, which represents a' ➡ cd (or $\neg$a'). Therefore, term $(a'\delta)\mathrm{not}$ represents $\neg$a' as well. In fragment 7 the double negation is defined; $(a'\delta)\mathrm{nn}$ represents $\neg\neg$a'.

Fragment 8 contains the theorem: a ➡ $\neg\neg$a. For, given a proposition a and a proof a1 of this proposition a, $\mathrm{Typ}[((a\delta)\mathrm{not}\ \lambda_x)(a1\delta)x]$ converts to $(a\delta)\mathrm{nn}$, representing $\neg\neg$a, so $((a\delta)\mathrm{not}\ \lambda_x)(a1\delta)x$ proves $\neg\neg$a. In fragment 9, the double negation rule (i.e., for all propositions a it holds that $\neg\neg$a implies a) is primitively introduced as an axiom. Fragment 10 contains the falsum-principle (i.e., for all propositions a it holds that contradiction implies a) as a theorem. In fragment 11 the logical theorem: $\neg a \Rightarrow (a \Rightarrow b)$ is proved.

3.3. The construction of a proof

We ignore the proof given in the first part of fragment 11 (previous section), and try to construct a proof independently. In this construction we follow a strictly formal approach; we do not appeal to any mathematical insight. To begin with, we transform the text above into normal form. This is, of course, a crude and inefficient thing to do, especially for longer Λ-texts, but we obtain so doing a clearer view on the principles of proving.

Fragments 1 to 10 transform into the normal Λ-term:
$(\lambda_\tau)(\lambda_\pi)(\pi\lambda_{cd})((\pi\lambda_a)(((a\lambda_y)cd\lambda_x)cd\lambda_w)a\lambda_{dn})$, consisting of four primitive fragments. Now a proof of the theorem has to be a term with type (converting to) Y, where $Y \equiv (\pi\lambda_a)(\pi\lambda_b)((a\lambda_x)cd\lambda_n)(a\lambda_{x0})b$, which is the normal form obtained from the term expressing the theorem: $(\pi\lambda_a)(\pi\lambda_b)((a\delta)not\ \lambda_n)(b\delta)(a\delta)imp$.

We can describe the actual *proving state* as follows: on one hand we have a *stock* of variables and matching types, on the other hand there is a *target*, determined by one or more types. In our case, we have the following initial proving state: the stock consists of the "leading variables" of the four primitive fragments, namely τ, π, cd and dn, together with their types; the target is to find a term with type (converting to) Y.

In view of the shape of Y, it is appropriate to change the proving state: add a, b, n and x0 to the stock, with types as above, and change the target into a term X with type (converting to) b. None of the variables in the stock has a type that is b, converts to b, or ends in b. As to all stock-variables except dn, there is no way of changing the final variable of their types into b by reductions. Only dn can give us hope: the final a of its type is "internally bound", i.e.: bound by a λ that occurs inside the same fragment; so variable a can possibly be changed into b.

Hence we now direct our searching attempts at dn and, again, we change the proving state. We look for terms X_1 and X_2 (the new targets) such that $(X_2\delta)(X_1\delta)dn$ has a type (converting to) b. Then X_1 must have type π and X_2 must have type $((X_1\lambda_y)cd\lambda_x)cd$, according to strong functionality. Now $Typ[(X_2\delta)(X_1\delta)dn] \equiv X_1$, as can be easily computed, so X_1 must be b. The remaining target is an X_2 with type (converting to) $Y_2 \equiv ((b\lambda_y)cd\lambda_x)cd$.

In view of the shape of Y_2, we add x to the stock of variables, with its type: $(b\lambda_y)cd$. Now the target becomes a term X_{21} with type cd. The only possibilities are to use dn, n or x from the stock. We choose to use n and have to find an X_{211} such that $(X_{211}\delta)n$ has type cd. Then X_{211} must have type a, and, indeed, X_{211} := x0 does the job.

Thus we have reached the final proving state, in which no target is left. Recapitulating:

we found $X_{211} \equiv x0$; $X_{21} \equiv (X_{211}\delta)n \equiv (x0\delta)n$; $X_2 \equiv ((b\lambda_y)cd\lambda_x)X_{21}$; $X_1 \equiv b$ and $X \equiv (X_2\delta)(X_1\delta)dn$. The requested proof is $(\pi\lambda_a)(\pi\lambda_b)((a\lambda_x)cd\lambda_n)(a\lambda_{x0})X$. Inspection shows that we have found the "same" proof as given in fragment 11 of the example in section 3.2 when written in normal form.

There were only a few choices to be made in this simple proving problem. Yet, if we had chosen to use dn instead of n, when looking for X_{21}, we would have returned to a prior proving state. Hence, in principle we could have been caught in a loop.

3.4. Remarks on partial theorem proving on the basis of Λ

A general strategy for theorem proving on the basis of normal forms can easily be derived from the construction example above. This *normal strategy* does indeed work in uncomplicated cases, but it fails when major mathematical tools are needed, such as induction. In such cases there is, presumably, only hope for a mechanical theorem prover when it is built in an interactive way: the theorem prover must be able to react to hints from the human textwriter.

For general use one has to abandon the transformation into normal form. The normal proving strategy, as explained above, can however be adapted for non-normal Λ-texts, such as given in section 3.2. The strategy itself then becomes more complicated. It is, for instance, not sufficient to regard the final variable of a certain "type-term", but one also has to consider all variables that can possibly replace this variable when reductions are applied. This leads to the tracing of certain *chains* of variables. The comparison of variables, being a major activity in the normal strategy, then has to be replaced by a method of comparing variable chains.

Summarizing, there appear to be possibilities for *partial* theorem proving on the basis of Λ, in particular when small gaps have to be bridged. For exacting proofs, however, a form of interaction between man and machine appears indispensable.

REFERENCES

[1] *N.G. de Bruijn*, The mathematical language AUTOMATH, its usage and some of its extensions, Symposium on Automatic Demonstration, IRIA, Versailles, France, 1968. (Lecture notes in Mathematics, Vol. 125, pp. 29-61, Springer-Verlag, Berlin, 1970.)

[2] *N.G. de Bruijn*, AUTOMATH, a language for mathematics, Lecture Notes prepared by B. Fawcett, Les Presses de l'Université de Montréal, Canada, 1973.

[3] *D.T. van Daalen*, A description of AUTOMATH and some aspects of its language theory, Proceedings of the Symposium APLASM, Vol. I, ed. P. Braffort, Orsay, France, 1973.

[4] *D.T. van Daalen*, The language theory of Automath, doctoral dissertation, Technol. University Eindhoven, The Netherlands, 1980.

[5] *L.S. van Benthem Jutting*, Checking Landau's "Grundlagen" in the AUTOMATH system, doctoral dissertation, Technol. University Eindhoven, The Netherlands, 1977. (Mathematical Centre Tracts 83, Amsterdam, The Netherlands, 1979.)

[6] *R.P. Nederpelt*, Strong normalization in a typed lambda calculus with lambda structured types, doctoral dissertation, Technol. University Eindhoven, The Netherlands, 1973.

[7] *R.P. Nederpelt*, Presentation of natural deduction, Recueil des Travaux de l'Institut Mathématique, Nouvelle série, tome 2 (10), p. 115-126, Symposium: Set Theory, Foundations of Mathematics, Beograd, Jugo-Slavia, 1977.

[8] *J. Zucker*, Formalization of classical mathematics in AUTOMATH, Actes of the International Logic Colloquium, Clermont-Ferrand, France, 1975.

ADDING DYNAMIC PARAMODULATION TO REWRITE ALGORITHMS

Paul Y Gloess[1], Jean-Pierre H Laurent[2]

SRI International, Menlo Park, California 94025, U.S.A.

ABSTRACT

A practical and effective solution to a problem resulting from the existence of critical pairs in a set of rewrite rules is presented.

We show how to modify rewrite algorithms by introducing a dynamic paramodulation of rules. This can be done in different ways. The related issues are discussed.

A concrete example is given in the case of the "depth first" rewrite algorithm.

KEY WORDS AND PHRASES

critical pairs, Knuth-Bendix algorithm, paramodulation, rewrite algorithms, rewrite rules, theorem-proving.

1. INTRODUCTION

Rewrite algorithms often suffer from the existence of critical pairs among the set of rewrite rules. The application of a rule may inhibit further application of another rule that would otherwise apply.

One combinatorial way to avoid this inconvenience is to develop a tree of all possible derivations from the given term. However, for large sets of rules this is generally not a practical solution.

Another possible remedy is to try to complete the set of rules, using the Knuth-Bendix algorithm [7], but this is not an efficient solution in a theorem-proving environment such as the Boyer-Moore system [1, 2]. First, this algorithm does not always terminate. Second, to take each new lemma into account, it would be necessary to run it many times in the course of a proof.

We suggest introducing a dynamic strategy into the rewrite algorithms. Each rewrite algorithm can be modified so as to produce a small set of new rules while transforming a term. These new rules eliminate the difficulties caused by the critical pairs for the particular term under consideration.

---

[1]International Fellow at SRI, supported by a grant from I.R.I.A., 78150 Le Chesnay, France, and AFOSR contract F49620-79-C-0099.

[2]Maître Assistant, Université de Caen, 14032 Caen Cedex, France, supported by a NATO grant.

2. DYNAMIC PARAMODULATION

In this chapter we shall use Huet's notations [5] $t_{|u}$ to denote the subterm at occurrence u in term t and t[u <-- t'] to denote the term obtained by replacing this subterm with t'.

For example, let t be the term (f X (g Y)): then $t_{|NIL}$ is t, $t_{|(1)}$ is X, $t_{|(2)}$ is (g Y) and $t_{|(2\ 1)}$ is Y. There is no other occurrence in t besides NIL, (1), (2) and (2 1). "t[(2) <-- (h Z)]" is (f X (h Z)).

The Elementary Critical-Pair Problem

Let S be a set of rewrite rules L_i --> R_i. Consider the rules L_j --> R_j and L_k --> R_k. Suppose that there exists a subterm $(L_j)_{|u}$ that is not a variable and a most general unifier *s of L_k and $(L_j)_{|u}$. Then the two rules form a critical pair.

Let t be an input term and $t_{|v}$ a subterm that has the form of L_j. The rule L_j --> R_j is thus applicable. Suppose now that $t_{|v|u}$ (i.e, $(t_{|v})_{|u}$) has the form of L_k. Let *r be the unifying substitution such that *r(L_k) = $t_{|v|u}$. If rule L_k --> R_k is used first, the resulting subterm $t_{|v}$[u <-- *r(R_k)] may no longer have the form of L_j, so that L_j --> R_j becomes inapplicable.

For example if L_j --> R_j is (f (a X)) --> (g X) and L_k --> R_k is (a (b Y)) --> (c Y) , then $(f\ (a\ X))_{|(1)}$ and (a (b Y)) unify by the most general unifier *s = {<X, (b Y)>} . Hence these two rules form a critical pair. Let t be the term (f (a (b Z))). Both rules apply to this term. If rule (a (b Y)) --> (c Y) is used first, transforming t into (f (c Y)), rule (f (a X)) --> (g X) no longer applies.

In such a case, the minor replacement of $t_{|v|u}$ has been preferred to the replacement of $t_{|v}$, whereas the opposite choice is generally expected.

Paramodulation of a Rule, Fundamental Property

Assume in this section the same notations and hypotheses as in the previous section.

Definition 1: [3] The paramodulation of rule L_j --> R_j by rule L_k --> R_k consists in adding the "paramodulated rule" L'_j --> R'_j:
*s(L_j[u <-- R_k]) --> *s(R_j)

With the example of Section 2, the paramodulated rule is (f (c Y)) --> (g (b Y)).

Theorem 2: The term $t_{|v}$[u <-- *r(R_k)] has the form of *s(L_j[u <-- R_k)].

Hence, the paramodulated rule L'_j --> R'_j applies to the term $t_{|v}$[u <-- *r(R_k)]. If rule L_j --> R_j had become inapplicable after application of rule L_k --> R_k, rule L'_j --> R'_j could still be used, which solves our problem.

The proof of Theorem 2 can be found in our report [4].

---

[3] See Lankford [8] and Musser [9].

A More General Problem

In general, the problem posed by a critical pair is not as simple as described in the previous paragraph, where either L_k --> R_k or L_j --> R_j could apply.

The following situation is more troublesome:

- L_k --> R_k is applicable, but not L_j --> R_j.
- If L_k --> R_k is not applied, L_j --> R_j may apply after several intermediate rewrites.
- If L_k --> R_k is applied, L_j --> R_j will never be applicable.

Consider for example the rules

```
(a (b X)) --> (c X)                                    (1)
(d (d X)) --> X                                        (2)
(f (a (b X)) (e X)) --> (g X)     .                    (3)
```

Rules (1) and (3) form a critical pair. Let t be the term (f (a (b Z)) (d (d (e W)))) . Rule (2) must be used before Rule (3) can be applied. The latter will never apply if (1) is used. A "rewrite depth-first" strategy would produce the sequence

```
(f (a (b Z)) (d (d (e W)))) --> (f (c Z) (d (d (e W))))
--> (f (c Z) (e W))    .
```

The remainder of this report deals with this more general problem.

Knuth-Bendix Solution

A static solution is proposed by Knuth and Bendix. Before using the set S of rewrite rules, they modify it, with the hope of obtaining a convergent set S', by adding paramodulated rules corresponding to critical pairs and by simplifying rules whenever possible. Care must be taken in orienting new or modified rules to preserve termination. The algorithm iterates until all critical pairs reduce to identities. The convergence of S' implies that each term has a unique normal form. The order in which the rules are applied has no importance, and thus critical pairs are no longer a problem.

Unfortunately, the Knuth-Bendix algorithm has significant drawbacks. First, there is no general method to orient the new rules so as to achieve at each step the finite-termination property of the set of rules. Second, even if this property is always verified, the algorithm may not terminate. For example, starting with the two following rules taken from the queue data type described in [9],

```
1: (append Q1 (add Q2 I2)) --> (add (append Q1 Q2) I2)
2: (append (append Q1 Q2) Q3) --> (append Q1 (append Q2 Q3)) ,
```

the Knuth-Bendix algorithm might produce the infinite set of rules

```
3: (append (add (append X Y) Z) T)
   --> (append X (append (add Y Z) T))
```

```
4: (append (add (add (append X Y) Z) T) U)
   --> (append X (append (add (add Y Z) T) U))
...
n: (append (add (add ...  (add (append X Y) Z) ...  T) U) V)
   --> (append X (append (add (add ...  (add Y Z) ...  T) U) V))
...      ,
```

if it were using a different termination measure from the one proposed by Musser. The infinite set above has the finite-termination property.

In a lemma-based theorem-proving context (e.g, the Boyer-Moore system [1]), a proof of a theorem usually requires proving several lemmas[4]. The list of these lemmas is not known at the beginning, but is built up during the proof --sometimes mechanically, sometimes interactively. Hence one is not confronted by an invariable list of rules that could be processed once and for all. The Knuth-Bendix algorithm would have to be run at each step of the proof. In the case of an interactive proof, it would also become difficult for the user to distinguish his own lemmas from those introduced or modified by Knuth-Bendix. Furthermore, the proof would undoubtedly be slowed by the addition of superfluous lemmas.

For all these reasons, we believe that the Knuth-Bendix algorithm is not a practical solution to critical-pair problems in a theorem-proving context.

Dynamic Paramodulation

The idea we present in this report is to perform a dynamic paramodulation, as opposed to static methods, such as the one proposed by Knuth-Bendix.

Instead of modifying the set S of rewrite rules before using a rewrite algorithm A, we suggest introducing paramodulation within A itself. Applied to any input term t, the new algorithm A' will paramodulate rule L_k --> R_k into rule L_j --> R_j if, in the course of rewriting t, some conditions are verified, including the condition that L_k --> R_k has been applied previously.

Dynamic paramodulation may be introduced in different ways. Many algorithms A' can be conceived, depending on the chosen set of conditions for generating new rules and the range of these rules. We discuss this point in Chapter 4.

The following chapter describes an example of dynamic paramodulation inserted into the "depth first" rewrite algorithm.

3. ADDING CHRONOLOGY-DRIVEN PARAMODULATION TO THE DEPTH-FIRST ALGORITHM

Let us recall how the depth-first rewrite algorithm works. We name it RDP (REWRITE.DEPTH.FIRST in Appendix I).

Let t be an input term. If t is a variable, the result is t. Otherwise t is of the form (f arg_1 ... arg_n) where f is a function symbol and arg_1, ..., arg_n are

---

[4] The proof of each lemma normally requires at least one induction.

terms. RDP recursively rewrites the arguments arg_1, ..., arg_n to arg'_1, ..., arg'_n. Then it considers the term t' = (f arg'_1 ... arg'_n). If some rule can apply at the top level of t', the algorithm applies it, thus replacing t' by a new term t", and iterates from the beginning with t = t". Otherwise it returns t', which is of course irreducible.

We now explain how we add dynamic paramodulation to this algorithm. We call the resulting algorithm R&R (REWRITE.AND.REMEMBER in Appendix I). R&R eventually generates new rules. It does this in the course of rewriting --not a priori, but on the basis of some information built in parallel and called the "chronology." The chronology keeps track of each modification performed on the term by recording the location of the subterm that is being replaced, as well as the rule used. R&R returns not only a term, but also a chronology.

How does R&R use this chronology? To rewrite a term of the form (f arg_1 ... arg_n), it recursively rewrites the arguments arg_1, ..., arg_n and produces a corresponding chronology of the rewrites applied to these arguments. Let us call arg'_1, ..., arg'_n the rewritten arguments. The algorithm then tries to apply a rule at the top level of the term (f arg'_1 ... arg'_n): all rules whose left member starts with function symbol f are "candidates." Instead of trying each candidate as it is, the algorithm first "applies the chronology" to it. The candidate rule is thus eventually modified to reflect the changes already effected in the arguments.

These modifications (actually paramodulations) of the left member of the rule are introduced exactly where they occurred within the original term and in the order of the chronology.

<u>A Simple Example</u>

Consider the following rules:

```
(a X) --> (b X)                                            (4)
(b (c X)) --> (q X)                                        (5)
(f (m (b (c X)))) --> X                                    (6)
(m (a (c X))) --> (q X)      .                             (7)
```

Suppose we want to simplify the term (f (m (a (c Z)))).

We can see in Appendix I that RDP fails to perform the topmost transformation and rewrites the original term (f (m (a (c Z)))) to (f (m (q Z))). RDP applies rules (4) and (5), producing the following sequence of rewrites:

```
Step 1: (f (m (a (c Z)))) --> (f (m (b (c Z)))), by Rule (4) at
                                                  occurrence (1 1)    ,

Step 2: (f (m (b (c Z)))) --> (f (m (q Z))),      by Rule (5) at
                                                  occurrence (1 1)    .
```

Then Rule (6) is no longer applicable. RDP stops there and returns (f (m (q Z))) since none of the four rules applies to this term.

Our algorithm R&R does not stop there. Because there is no rule whose left member starts with q, the algorithm pops up to (m (q Z)) in the term. Rule (7) is a candidate at this level; before trying to apply it, however, we modify it, using the current chronology that remembers the replacements made by Steps 1 and 2. Since the current term being rewritten is at location (1) within the whole term and the locations refer to the current term, the current chronology[5] is specifically the list of two pairs

```
(<(1), (a X) --> (b X)>
,<(1), (b (c X)) --> (q X)>)     .
```

The first and second pairs describe the modifications performed in Step 1 and Step 2, respectively.

How do we "apply the chronology" to candidate Rule (7)? Let us rename X to X' in this rule to avoid any confusion, thus yielding (m (a (c X'))) --> (q X'). The first pair of the chronology suggests trying to use the rule (a X) --> (b X) at the level of the first argument (a (c X')) of the left member of (7). Terms (a (c X')) and (a X) unify with the obvious substitution {(X, (c X'))}. Paramodulation of (7) results in the rule

```
(m (b (c X'))) --> (q X'))     ,
```

which in turn is paramodulated according to the second pair of the chronology. The final rule is

```
(m (q X')) --> (q X')     .
```

Using the above rule, we can now rewrite the first argument of (f (m (q Z))) and obtain (f (q Z)) at the end of Step 3.

In Step 4, since (q Z) is irreducible with respect to the original rules, the algorithm pops up and considers (f (q Z)). The current chronology is now the list of three pairs

```
(<(1 1), (a X) --> (b X)>
 <(1 1), (b (c X)) --> (q X)>
 <(1), (m (q X')) --> (q X')>)     .
```

The only candidate rule for term (f (q Z)) is Rule (6), because it is the only one whose left member starts with f. Applying the above chronology successively yields

```
(f (m (b (c X)))) --> X   is left unchanged
```

---

[5] In general, a chronology is a sequence of pairs of the form <occurrence, rule>, where "occurrence" is a sequence of integers, as explained in Chapter 2, and "rule" is a rewrite rule (not necessarily a member of the original set).

```
(f (m (b (c X)))) --> X   gets   (f (m (q X))) --> X
(f (m (q X))) --> X       gets   (f (q X)) --> X   .
```

Now generated rule (f (q X)) --> X applies to term (f (q Z)) and reduces it to Z.

Rewriting with a Chronology: an Example in Which the Set of Rules is Irreducible

The previous example was a simple one, showing how R&R works. However, the set of rules was not irreducible (i.e, some rules could be simplified by other rules). Rule (7) can be simplified by (4) to

```
(m (b (c X))) --> (q X)    ,
```

which is simplified again to

```
(m (q X)) --> (q X)    ,
```

using Rule (5).

We could simplify the whole set to

```
(a X) --> (b X)
(b (c X)) --> (q X)
(f (q X)) --> X
(m (q X)) --> (q X)    .
```

With this new set of rules RDP as well as R&R return Z for (f (m (a (c Z)))) (see Appendix I). One could conclude that a preprocessing of the rules involving only one-way unifications (as opposed to the most general unification used by our algorithm when it paramodulates) suffices and thus renders our algorithm useless.

However, we now present another example. The set of five rules

```
(a (c X)) --> (b (c X))
(b (c (d X))) --> (q X)
(d (d X)) --> X
(f (m (b (c X)))) --> (d X)
(m (a X)) --> (q X)    ,
```

is irreducible, but, as shown in Appendix I, our algorithm R&R reduces (f (m (a (c (d Z))))) to Z; RDP returns (f (m (q Z))).

An Example Showing Generation of an Interesting Rule

We now present an example that is interesting for several reasons. First, it represents an actual occurrence during our use of the Boyer-Moore theorem prover [1] to prove the correctness of a simple parser of expressions [3]. Second, our algorithm generates a rule that is even more useful than the reduction performed. Third, we can clearly perceive the advantage of our approach over static methods such as the Knuth-Bendix algorithm.

The set consists of the two rules

(append X (nil)) --> X   ,   (8)

(inclup (append X A) R1 (append X B) R2)

--> (inclup A R1 B R2) . (9)

Although it is not essential for the present argument, let us recall the meaning of the predicate inclup: (inclup L1 R1 L2 R2) is true if and only if (append L1 R1) and (append L2 R2) are equal and L1 is an initial sublist of L2. nil is considered here as a function of zero arguments, as meant by the brackets.

In the course of our proof we encountered the term (inclup (append Z A) R1 (append Z (nil)) R2), hoping that it would simplify to (inclup A R1 (nil) R2). Instead of that, the strategy being the one of RDP, Rule (8) reduced the fourth argument, thus rewriting the original term to (inclup (append Z A) R1 Z R2). The further use of Rule (9) then became impossible.

As shown in Appendix I our chronology algorithm eliminates this difficulty by generating the rule

(inclup (append X A) R1 X R2) --> (inclup A R1 (nil) R2). (10)

It is interesting to note that in the course of reducing the term (inclup (append Z A) R1 (append Z (nil)) R2) to the term (inclup A R1 (nil) R2), it actually establishes and uses a simpler and more useful reduction, namely Rule (10).

Starting with the same two rules, (8) and (9), the Knuth-Bendix algorithm would produce three additional rules:

(inclup X R1 (append X B) R2) --> (inclup (nil) R1 B R2)
(inclup (append X A) R1 X R2) --> (inclup A R1 (nil) R2)
(inclup X R1 X R2) --> (inclup (nil) R1 (nil) R2) ,

instead of just one, i.e, (10). In its search for all critical pairs, it would also attempt several most general unifications, whereas R&R immediately picks the right one.

4. DYNAMIC PARAMODULATION STRATEGIES

The idea of dynamic paramodulation is to generate rules only in the course of rewriting the input term. However, two questions arise regarding application of this principle:

- Exactly when do we decide to generate a new rule?

- Which subterms will be matched against this new rule?

We discuss these points in the present chapter.

When to Generate a Rule

The paramodulated rule L'_j--> R'_j may be generated only if rule L_k--> R_k has previously been applied to the input term. This is the basic condition, but we may decide to generate the new rule at different moments after having applied rule L_k--> R_k.

One possible choice is to generate it as soon as rule L_k--> R_k has been applied. This implies either pointing out the critical pairs before rewriting the term, or verifying after each step whether the applied rule can paramodulate other rules.

Conversely, we may delay paramodulation until no rule from the initial set can be applied by the basic rewrite routine, i.e, when a normal form has been obtained. If paramodulation produces new rules and one of them applies, the basic rewrite routine starts again with the reduced term.

Between these extremes some intermediate method may be preferred. For instance, the algorithm R&R, described in Section 3, attempts paramodulations only immediately after it pops up to a subterm, i.e, when no rule applies to the arguments of this subterm. Only "candidate rules" (see Chapter 3) are tested for paramodulation, using the current chronology.

Of course, many other intermediate strategies might be devised.

Range of a Paramodulated Rule

Once a rule has been generated, where in the term is it available? Here too, several choices are possible.

One choice consists in matching the new rule against the current subterm only. Whether it applies or not, the rule is subsequently disabled. This is the simple strategy used in R&R.

Here is an example in which this strategy may seem inadequate.

Consider the set of two rules:

```
(a X) --> (b X)                                   (11)
(c (a Y)) --> Y    .                              (12)
```

Let us apply R&R to the term (f (c (a V)) (c (b W))). Rule (11) replaces (a V) with (b V). The paramodulated rule (c (b Z)) --> Z is created and replaces (c (b V)) with V. It is then disabled. The algorithm returns (f V (c (b W))) and not (f V W).

A radically different strategy could add all paramodulated rules to the initial set of rules, so as to render them available at any time for any subterm of the input term[6].

In the example above, this strategy would return (f V W). However, the general result may be a substantial increase in the number of rules (paramodulated rules may paramodulate rules...).

An intermediate strategy could take into account the level of the subterm

---

[6]One could even decide to keep these generated rules for future rewriting, as a means of enabling the rewrite routine to learn.

responsible for the generation of a rule. If it has been generated when rewriting the argument arg_i (itself, not a proper subterm) in (f arg_1 ... arg_n), a rule would be available for all arguments arg_1 ... arg_n.[7]

Given the term (f (c (a V)) (c (b V))) and the rules (11) and (12), we would obtain (f V W). However, the term (g (f U (c (a V))) (c (b W))) would rewrite to (g (f U V) (c (b W))) and not to (g (f U V) W).

As shown by these examples, the new rewrite algorithm is very dependent on the manner in which dynamic paramodulation is introduced.

5. CONCLUSION AND DISCUSSION

The dynamic paramodulation described in this report is an efficient and practical solution for avoiding difficulties caused by critical pairs.

However, further work is needed to determine strategies and, eventually, reasonable conditions on the set of rules such that, for example:

- The modified algorithm A^* yields some normal form, or at least is "closer to a normal form" than the original algorithm A (i.e, if A(t) and A(t') are identical, then A^*(t) and A^*(t') are identical.
- The modified algorithm A^* terminates. Note that even if the set of rules has the finite termination property, A^* may not be total (e.g R&R applied to (f (a (c))) with the set of rules {(f (a (b))) --> (f (a (c))), (a X) --> (d)}).

A possible way of preserving termination could be to restrict dynamic paramodulation in such a manner that the sequence of rules applied by A^* be equivalent to a sequence of rules taken from the initial set of rules. This idea lead us to the following lemma:[8]

Lemma 3: Suppose that

- rule R_k paramodulates rule R_j at redex v and R'_j is the resulting rule,
- t rewrites to t' by application of R_k at u,
- t' rewrites to t" by application of R'_j at u/v,
- the left member of R_j is linear (see [5, 6]) and contains the same variables as its right member,
- both members of R_k have the same variables,

Then t rewrites to t" by application of R_j at u/v.

---

[7]This may involve reprocessing of arguments.

[8]The proof of which will be included in a complete report.

I. APPENDIX: LISTING OF THE EXAMPLES MENTIONED IN 3

A Simple Example

```
_SHOW.DATA.BASE]

(a X) --> (b X)
(b (c X)) --> (q X)
(f (m (b (c X)))) --> X
(m (a (c X))) --> (q X)

NIL
_SHOW.REWRITE.AND.REMEMBER((f(m(a(c Z]

Your original term (f (m (a (c Z)))) finally rewrites to Z.

The chronology of the replacements is as follows:

1-  Term (f (m (a (c Z)))) rewrites to (f (m (b (c Z)))), by
 applying rule (a X) --> (b X)
 at occurrence (1 1).

2-  Term (f (m (b (c Z)))) rewrites to (f (m (q Z))), by
 applying rule (b (c X)) --> (q X)
 at occurrence (1 1).

3-  Term (f (m (q Z))) rewrites to (f (q Z)), by applying rule
 (m (q X)) --> (q X)
 at occurrence (1).

4-  Term (f (q Z)) rewrites to Z, by applying rule
 (f (q X)) --> X
 at occurrence NIL.

NIL
_REWRITE.DEPTH.FIRST((f(m(a(c Z]
(f (m (q Z)))
```

First Ireducible-Set Example

```
_SHOW.DATA.BASE]

(a X) --> (b X)
(b (c X)) --> (q X)
(f (q X)) --> X
(m (q X)) --> (q X)

NIL
_SHOW.REWRITE.AND.REMEMBER((f(m(a(c Z]

Your original term (f (m (a (c Z)))) finally rewrites to Z.

The chronology of the replacements is as follows:

1-  Term (f (m (a (c Z)))) rewrites to (f (m (b (c Z)))),
 by applying rule
 (a X) --> (b X)
 at occurrence (1 1).

2-  Term (f (m (b (c Z)))) rewrites to (f (m (q Z))),
```

```
 by applying rule
 (b (c X)) --> (q X)
 at occurrence (1 1).

3-  Term (f (m (q Z))) rewrites to (f (q Z)), by applying rule
 (m (q X)) --> (q X)
 at occurrence (1).

4-  Term (f (q Z)) rewrites to Z, by applying rule
 (f (q X)) --> X
 at occurrence NIL.

NIL
_REWRITE.DEPTH.FIRST((f(m(a(c Z]
Z
```

Second Irreducible-Set Example

```
_SHOW.DATA.BASE]

(a (c X)) --> (b (c X))
(b (c (d X))) --> (q X)
(d (d X)) --> X
(f (m (b (c X)))) --> (d X)
(m (a X)) --> (q X)

NIL
_SHOW.REWRITE.AND.REMEMBER((f(m(a(c(d Z]

Your original term (f (m (a (c (d Z))))) finally rewrites to Z.

The chronology of the replacements is as follows:

1-  Term (f (m (a (c (d Z))))) rewrites to (f (m (b (c (d Z))))),
 by applying rule
 (a (c X)) --> (b (c X))
 at occurrence (1 1).

2-  Term (f (m (b (c (d Z))))) rewrites to (f (m (q Z))),
 by applying rule
 (b (c (d X))) --> (q X)
 at occurrence (1 1).

3-  Term (f (m (q Z))) rewrites to (f (q (c (d Z)))),
 by applying rule
 (m (q X)) --> (q (c (d X)))
 at occurrence (1).

4-  Term (f (q (c (d Z)))) rewrites to (d (d Z)),
 by applying rule
 (f (q (c (d X)))) --> (d (d X))
 at occurrence NIL.

5-  Term (d (d Z)) rewrites to Z, by applying rule
 (d (d X)) --> X
 at occurrence NIL.

NIL
_REWRITE.DEPTH.FIRST((f(m(a(c(d Z]
(f (m (q Z)))
```

A Real Example

```
_SHOW.DATA.BASE]

(append Z (nil)) --> Z
(inclup (append Z A) R1 (append Z B) R2) --> (inclup A R1 B R2)

NIL
_SHOW.REWRITE.AND.REMEMBER(
                   (inclup (append Z A) R1 (append Z (nil)) R2]

Your original term (inclup (append Z A) R1 (append Z (nil)) R2)
finally rewrites to (inclup A R1 (nil) R2).

The chronology of the replacements is as follows:

1- Term (inclup (append Z A) R1 (append Z (nil)) R2)
 rewrites to (inclup (append Z A) R1 Z R2), by applying rule
 (append Z (nil)) --> Z
 at occurrence (3).

2- Term (inclup (append Z A) R1 Z R2) rewrites to
 (inclup A R1 (nil) R2), by applying rule
 (inclup (append X Y) Z X T) --> (inclup Y Z (nil) T)
 at occurrence NIL.

NIL
_REWRITE.DEPTH.FIRST((inclup (append Z A) R1 (append Z (nil)) R2]
(inclup (append Z A) R1 Z R2)
```

ACKNOWLEDGMENTS

We are grateful to Robert S. Boyer, Bernard Elspas, Gerard Huet, Jean-Marie Hullot, J Strother Moore, Robert E. Shostak and Richard J. Waldinger for their helpful comments and advice.

REFERENCES

1. R. S. Boyer and J S. Moore. A Computational Logic. Academic Press Inc., New York, 1979.
2. Robert S. Boyer, J Strother Moore. A Theorem Prover for Recursive Functions: a User's Manual. CSL-91, NR 049-378, SRI International, Menlo-Park, California 94025, June, 1979.
3. P. Y Gloess. A Proof of the Correctness of a Simple Parser of Expressions by the Boyer-Moore System. Technical Report N00014-75-C-0816-SRI-7, SRI International, Menlo Park, California 94025, August, 1978.
4. P. Y Gloess, J-P Laurent. Adding Dynamic Paramodulation to Rewrite Algorithms. Technical Report CSL-102, SRI International, Menlo Park, CA94025, December, 1979.
5. G. Huet. Confluent Reductions: Abstract Properties and Applications to Term Rewriting Systems. Rapport Laboria n°250, IRIA-LABORIA, Domaine de Voluceau, 78150 Le Chesnay, France, August, 1977.
6. G. Huet. Equations and Rewrite Rules: a Survey. CSL-111, SRI International, Menlo Park, California 94025, January, 1980.
7. D. E. Knuth and P. G. Bendix. Simple Word Problems in Universal Algebras. In J. Leech, Ed., Computational Problems in Abstract Algebra, Pergamon Press, New York, 1970, pp. 263-297.
8. D. S. Lankford. Canonical Inference. Automatic Theorem Proving Project Report ATP-32, University of Texas, December, 1975.
9. D. R. Musser. Convergent Sets of Rewrite Rules for Abstract Data Types. USC Information Sciences Institute, 4676 Admiralty Way, Marina del Rey, California 90291, December, 1978. Extended Abstract

HYPERPARAMODULATION: A REFINEMENT OF PARAMODULATION*

by

L. Wos, Argonne National Laboratory, Argonne, IL
R. Overbeek, Northern Illinois University, DeKalb, IL
L. Henschen, Northwestern University, Evanston, IL

Abstract

A refinement of paramodulation, called hyperparamodulation, is the focus of attention in this paper. Clauses obtained by the use of this inference rule are, in effect, the result of a sequence of paramodulations into one common nucleus. Among the interesting properties of hyperparamodulation are: first, clauses are chosen from among the input and designated as nuclei or "into" clauses for paramodulation; second, terms in the nucleus are starred to restrict the domain of generalized equality substitution; third, total control is thus iteratively established over all possible targets for paramodulation during the entire run of the theorem-proving program; and fourth, application of demodulation is suspended until the hyperparamodulant is completed. In contrast to these four properties which are reminiscent of the spirit of hyper-resolution, the following differences exist: first, the nucleus and the starred terms therein, which are analogous to negative literals, are determined by the user rather than by syntax; second, nuclei are not restricted to being mixed clauses; and third, while hyper-resolution requires inferred clauses to be positive, no corresponding requirement exists for clauses inferred by hyperparamodulation.

To illustrate the value of this refinement of paramodulation, we have chosen certain conjectures which arose during the study of Robbins algebra. A Robbins algebra is a set on which the functions o and n are defined such that o is both commutative and associative and such that for all x and y the following identity

$$n(o(n(o(x,y)),n(o(x,n(y))))) = x$$

holds. One may think of o as union and n as complement. The main interest in such algebras arises from the following open question: If S is a Robbins algebra, is S necessarily a Boolean algebra? The study of this open question entailed heavy use of an automated theorem-proving program to examine various conjectures. Certain computer proofs therein were obtained only after recourse to hyperparamodulation. (These lemmas were actually obtained prior to the work reported on here by Winker and Wos using a non-standard theorem-proving approach developed by Winker.)

Section 1. Introduction

In this paper we discuss the inference rule, hyperparamodulation, which is a refinement of paramodulation. We give some detail concerning its implementation, the motivation for its formulation, and the way it is used. We make comparisons of it to both paramodulation and hyper-resolution in terms of its properties and effectiveness. Since hyperparamodulation requires the user to select from among the input clauses those into which so-called equality

*This work was funded by the Applied Mathematical Sciences Program of the Office of Energy Research, U.S. Department of Energy and supported in part by the National Science Foundation under grants MCS-7903870 and MCS-7913252.

substitution is permitted, a heuristic is provided for making that choice. Similarly, since the user also designates within the chosen input clauses which terms are admissible for replacement, we suggest a heuristic for that purpose also.

We then turn in Section 4 to an example from algebra, specifically from Robbins algebra, for a detailed examination of the kind of proof most suited to the use of hyperparamodulation. Since this proof itself, the study from which it is taken, and the set of experiments underlying this paper all implicitly reflect certain views of the authors, we make those views explicit in Section 2. We also discuss there certain additional procedures and their effect on program and user complexity.

Although this study is a very preliminary study, the evidence appears to be of some significance. We believe that the appropriate purpose of workshops is to encourage the dissemination of such studies and experience. We welcome any evidence even remotely germane to any position taken herein.

Section 2. Tacit Assumptions

2.1. The Program as a Research Tool

One use of the automated theorem-proving program implemented at both Argonne National Laboratory and Northern Illinois University is that of attacking open questions in mathematics. An example of this is provided by our study of Robbins algebra. In attempting to answer the open question given in Section 4, a number of conjectures were made about possible lemmas. Each was submitted to the theorem-proving program with the proviso that no special programming features be added or employed, but also with the proviso that the user could adjust any or all of various input parameters to reflect his preference. We note first that, in most cases, the program did generate a proof using hyperparamodulation. Second, it was not necessary to change parameter settings from the defaults for any of the lemmas tried. Most significantly, the program was not able to get proofs using paramodulation alone. A number of these conjectures are still open, and in particular, the main question is still unanswered. No attempt was made to have the program produce a counter-example to any conjecture, although this capability is available in part.

At this point, it is natural for three questions to be asked. Was all of the research done with computer runs? If a proof was not found, wasn't one then in a quandary? If one can set various parameters at will, doesn't that give the set of experiments an ad hoc flavor?

Aside from the user making the conjectures about lemmas, the answer to the first question is, essentially, yes. If a proof was obtained from the first attempt, it was catalogued, and the next conjecture was tackled. If no proof was found, various parameters were changed to permit a wider and somewhat different search. In addition, sometimes lemmas the user judged worthwhile and which were found in previous runs were added. In most cases, but a few failures were sufficient to table the conjecture. Occasionally, some results obtained by hand were adjoined, somewhat in the spirit of using suggestions made by some other mathematician.

As for the second question, tabling of a conjecture leaves one no worse off than when the conjecture was made. Put another way, if a colleague fails to make any helpful suggestion, one still continues to seek his assistance with new conjectures.

As for the third, and perhaps most important, question, it should be noted that the various parameters have reasonable default values, and these were sufficient for most proofs generated. The user of the program is not required to have special knowledge about the problem or about the proof search in order to use the system. On the other hand, we must emphasize the belief that the user should be provided with the opportunity to use any intuition, special knowledge, or previous experience in the area under study. Any such advantage should be convertible easily into some biased use of the theorem-proving program. The various input parameters in part provide such a mechanism.

We believe more and more that flexible and efficient theorem provers are becoming useful tools. We cite our own experience with Robbins algebra, our success in answering the axiom dependence questions in ternary Boolean algebra [5], and the success in answering the previously open question concerning finite semigroups and involutions [6]. None of these required any additional programming. One might suspect a major reason for the achievements lies in our familiarity with the program and its use. As evidence to the contrary we cite the use to which our program was put by non-theorem-proving people to help design circuits in 3 valued logic [7]. Finally, in order to provide a flexible research tool for developing deductive techniques, we are at present willing to accept both an increase in complexity of the program and an increase in complexity of its use in exchange for the ability to imitate various types of reasoning under various restrictions and relaxations of different constraints. We would, of course, prefer a fully automated system, and we welcome any advance toward such.

2.2. Representation

Among the decisions to be made for the representation of problems in clause form, the most important one is that of choosing between emphasis and deemphasis of equality. Is it, for example, better to represent commutativity with

-P(x,y,z) P(y,x,z)

or with

EQUAL(f(x,y),f(y,x)) ?

For a variety of reasons, we clearly prefer an emphasis on equality. This, of course, presents us with various inherent problems such as the control of paramodulation. This inference rule in itself gives no hint for making good choices of clauses either as "from" or "into" operands. Similarly, it supplies no information to direct the choice of terms within the "into" clause to replace. Not nearly enough control is provided, for example, by the restrictions: block paramodulation from or into any argument which is itself a variable. We prefer that the mathematician, using intuition and/or knowledge of the field under study, be in a position to exercise much control over the set of generalized equality substitutions. (Recall that the unification performed in completing a paramodulation may force nontrivial instantiation on both the "from" and the "into" clause, which explains the recurrent phrase, "generalized equality substitution".) The inference rule given in Section 3 in part provides some of the missing control mechanism. In using it, one makes actual choices of possible into clauses and possible terms therein. Here again we are aware of adding complexity to both the program and its use. But, as will be seen in Section 4, the imitation of a valuable type of reasoning is now automated.

The following overly simplified summary lists some of the factors contributing to our preference for equality representations in spite of their complications. First, both the relevance and effectiveness of demodulation is heightened. This statement holds when demodulation is used to simplify or to canonicalize. Second, since the non-equality representation incurs artificial argument boundaries, redundancy of information results. For example, if one has appropriate demodulators to cause, among other things, right association, then the deduced fact that (a(b(cd))) = k will be retained in but one way in the equality representation. On the other hand, with the other often-used representation one will find

```
        P(a,f(b,f(c,d)),k)
    and P(f(a,b),f(c,d),k)
    and P(f(a,f(b,c)),d,k)
```

and the like. Third, the ability to attack a particular subterm immediately rather than building up to it by passing through various equality substitution axioms (as is required when using equality axioms instead of paramodulation) should, in our opinion, contribute to efficiency. And fourth, examination of the clauses produced during a run of the theorem-proving program is greatly facilitated because terms and formulas are easier and more natural to read. This is especially important in experimentation when one is trying to get the program to find a particular kind of proof.

To conclude this section, we first remark that the effort has obviously concentrated on equational systems. So no light is shed here on those problems which do not benefit from an equality representation. This concentration, in turn, implies for us the rejection of the use of inference rules such as hyper-resolution and UR-resolution [1,2,3] for problems dominated by equality. Much more experimentation in this regard would be valuable. But one should keep in mind, when experimenting, that techniques that are useful for inferences in one representation do not necessarily carry over to another. Conversely, certain problems will benefit from a particular representation, and this will strongly influence the choice of inference rule. Last, comparisons between clause representation coupled with so-called resolution-based systems and other representations with inferential procedures, such as natural deduction, should be made continually. Problems of the type cited in Section 4 are but one class on which comparison should occur. In this manner, we may be able to determine which techniques are effective and to which types of problems they apply.

2.3. Canonicalization

Because of our experiments, we are convinced of the need of canonicalization, especially in word problems, which are essentially equational in nature. For us, canonicalization is accomplished through demodulation. The most common and most successful set is exemplified by:

```
        o(x,y) = o(y,x)
        o(x,o(y,z) = o(y,o(x,z))
        o(o(x,y),z) = o(x,o(y,z))
```

where the function, o, can be thought of as union. Here, demodulators are used to rewrite terms when the rewritten version is "simpler" in a predefined lexical ordering of all terms. The middle demodulator deserves special note. It "cycles" terms and has proven extremely effective and efficient in the generation of canonical forms [4]. The user has control over the lexical ordering except, for the present, that variables are assumed to be the earliest lexical class.

We are painfully aware of the problems attendant to the use of canonicalization. For us, the marked gain in efficiency is sufficient compensation. Often, proofs would not be found if canonicalization were avoided. Rather than an approach based on the study of complete reductions, we have chosen to stay with our approach for reasons we cannot include here because of space limitations. This entails the addition of certain clauses in the input which are indicated by the user not to be canonicalized. For example, given that n(a) is lexically earlier than the constant, a, and given also that o(n(a),a) = n(a), we would probably adjoin

o(a,n(a)) = n(a)
o(n(a),o(a,x)) = o(n(a),x)
o(a,o(n(a),x)) = o(n(a),x)

to the input. (Such additions in part cope with certain refutation completeness problems.)

Here again we are cognisant of a possible ad hoc flavor to our approach. However, we believe we are developing a set of procedures and rules to help users decide among the various choices. See Section 3.5 for a sample of some of these rules.

Section 3. Hyperparamodulation

3.1. Motivation for its Formulation

For certain domains of reasoning, hyper-resolution is clearly superior to binary resolution. For comparison below we make the following points. First, hyper-resolution prevents many sets of clauses from interacting directly because of the positive-negative requirements. Next, a hyper-resolvent is often a more significant lemma or fact than a binary resolvent is. Third, if the hyper-resolvent is calculated by clashing away all the negative literals in one unification, then demodulation is not applicable until the hyper-resolvent is formed. In contrast, in a sequence of P1 steps, if demodulation were allowed after each step, certain paths of reasoning would be appreciably modified, and often the desired hyper-resolvent would not be generated. Fourth, no new nuclei are generated during the run. Fifth, the target literals (the negative literals of the nuclei) are all known at the start.

On the other hand, one might note the following disadvantages. Negative clauses have no value to the proof search other than termination. So some of the denial of a purported theorem is of no use in the proof search. Next, the rule is syntactically rather than semantically oriented. Finally, we note that certain problems heavily involving the notion of equality suffer greatly, in our opinion, from the problems mentioned in Section 2.2 when hyper-resolution is used.

So, with the foregoing in mind, and with the knowledge that paramodulation removes some of the stated disadvantages, it was natural to seek an inference rule which combined the good properties of paramodulation and of hyper-resolution. Such a rule would, intuitively, require the following: a single nucleus or "into" clause; a set of satellites or "from" clauses as correspondents to the positive clauses in a hyper-resolution; and a set of terms or subterms within the nucleus as correspondents to the negative literals. Then, just as hyper-resolution can be accomplished by an appropriate sequence of P1 inferences, the inference yielded by the new rule could be obtained by a sequence of paramodulations.

3.2. Formal Definition of Hyperparamodulation

We begin this section with an example and comments to illustrate certain complications in the definition below. Consider the clause

EQUAL(f(a,f(b,f(c,d))),k)

and let

EQUAL(f(c,d),c1)
EQUAL(f(b,c1),b1)
EQUAL(f(a,b1),a1)

also be given. Then there is but one sequence of paramodulations which yields EQUAL(a1,k) as an inference. The corresponding ordering problem does not exist in hyper-resolution.

Now consider paramodulation from a clause EQUAL(r,s) into a clause C(r') where r' is the term chosen to be replaced. Let u be the MGU of r and r'. Then unlike resolution, terms in the paramodulant (C(s))u are not necessarily instances of terms in C(r'). Those terms t(r') that contain the chosen r' are transformed to (t(s))u. Thus, in the above example, after the first paramodulation we have EQUAL(f(a,f(b,c1)),k). Here, c1 is not an instance of f(c,d), and f(b,c1) is not an instance of f(b,f(c,d)), etc. Thus, in the second paramodulation of that example, we must identify the into term of paramodulation as the transform of f(b,f(c,d)). In general, the transform of a term w = t(r') containing the chosen r' is (t(s))u; the transform of a term w not containing the chosen r' is just (w)u. A term w which is itself contained in the chosen r' in general will not have a corresponding transform.

In order to facilitate the definition, we therefore identify terms by position vectors. The position vector p1,p2,...,pm identifies the term occurring in literal p1+1, and within that literal occurring in argument position p2+1, etc. Thus, a in the clause

EQUAL(f(a,f(b,f(c,d))),k)

has position vector 0,0,0 because it occurs in the first literal, in the first argument position of that literal, in the first argument position of that term.

The important point here is that the position vector of the transform of a term t in a paramodulant is the same as the position vector of t itself (unless t disappears altogether because it was a subterm of the chosen r'). For example, in paramodulating EQUAL(f(c,d),c1) into EQUAL(f(a,f(b,f(c,d))),k), f(c,d) is transformed into c1, both of which have position vector 0,0,1,1 in their respective clauses. The term d with position vector 0,0,1,1,1 has no transform.

DEFINITION. Hyperparamodulation: Let a1,a2,...,ak be a set of positive equality unit clauses, and let n be a clause. Let t1,t2,...,tk be k distinct (at least as far as position) terms in n. Further, assume that, for all i,j with $1 \leq i < j \leq k$, ti does not contain tj as a subterm. Now let c1 be n. Define c2 to be the paramodulant, if such exists, of a1 with c1 into t1, and let v1 be the corresponding transformation of terms in c1. Then define c3 to be the paramodulant, if such exists, of a2 with c2 into v1(t2), and let v2 be the corresponding transformation of terms in c2. In general, define c(i+1) to be the paramodulant, if such exists, of ai with ci into the (compound) transform of ti. Then c(k+1) is called a hyperparamodulant. Equivalently, c(k+1) is said to be inferred by hyperparamodulation from a1,a2,...,ak and n. The clause, n, is called the nucleus, and the ai are the satellites.

To illustrate the necessity of ordering the terms and the corresponding satellites, consider first the example at the beginning of this section. There is only one correspondence of satellites to position vectors in the nucleus that yields a hyperparamodulant. Moreover, the sequence of steps must be carried out in the order inside out. For example, EQUAL(f(b,c1),c2) cannot be applied to f(b,f(c,d)) until after EQUAL(f(c,d),c1) has been used to "prepare the way." Now consider the two satellites

A1: EQUAL(f(a,b),f(c,d))
and A2: EQUAL(f(f(x,y),z),f(f(x,z),f(y,z)))

and the nucleus P(f(f(x,y),z)) with chosen terms t1:f(x,y) and t2:f(f(x,y),z). When the sequence of paramodulations is A1 into t1 followed by A2 into v(t2), the result is P(f(f(c,z),f(d,z))). If A2 is used into t2 first, the result is P(f(f(c,d),f(y,b))).

These examples illustrate why all the positions in the nucleus and their corresponding satellites must be chosen at the outset and fixed and why the paramodulations must be done in a particular order. We do note that the fundamental requirement is that the terms be processed inside out. Thus, one could partition the ti into smaller classes, each of which consists of an argument of a predicate or literal of the nucleus together with those of its subterms among the ti. Then the classes could be processed in any order, but within a class the inside-out rule must be observed. In fact, any ordering of the terms which obeys the partial order determined by containment is acceptable.

Because there is no straightforward syntactic way to determine allowable nuclei and the terms therein for paramodulation, we also make the following definition.

DEFINITION. A hyperparamodulation proof procedure is a proof procedure in which

1. the rules of inference are hyperparamodulation and possibly resolution or some refinement of resolution,

2. a method is given for determining which clauses are nuclei and which are satellites, and which are the chosen into terms,

and 3. a method is given for choosing the order in which clauses are used in making new inferences.

One can relax various restrictions in the above definition in the obvious way, for example, to allow non-unit satellites (although we have not needed to do this for our problem sets). There is also ample flexibility in the definition of a hyperparamodulation proof procedure for exploring heuristics for selecting nuclei, into terms, etc. Finally, space considerations preclude a discussion of questions concerning the effect of using hyperparamodulation with the choice of set of support, of various paramodulation strategies and the role of complete reductions.

3.3. Implementation and Use

Our hyperparamodulation proof procedure is based on the NIUTP system implemented at Argonne National Laboratory and Northern Illinois University [1,2,3]. The nuclei are chosen by the user from among the input clauses by attaching the dummy literal NUC(C) to each nucleus. Nuclei are those clauses that contain the NUC(C) literal and satellites those clauses without the NUC(C) literal. (Note, this requires us to include two copies of a clause that we want to be used as both a nucleus and a satellite, one with NUC(C) and one without.) The user also specifies by position vectors the terms that are to be paramodulated in each nucleus. These terms are said to be starred. Clauses are ordered for hyperparamodulation on the basis of weights assigned to the terms [2]. The program also makes extensive use of demodulation and canonicalization and performs both forward and backward subsumption on all satellites. Finally, new equality units that pass certain weight tests are dynamically added as demodulators and are used immediately to demodulate all existing clauses.

As discussed in Section 3.2, a hyperparamodulant must be calculated as the last in a sequence of paramodulants. Our program automatically chooses the terms in the nuclei in the inside-out order, regardless of how the user ordered the position vectors in the input.

Within a given position-vector level, terms are chosen left to right so as to avoid redundant paths. When a starred term is paramodulated by its corresponding satellite, the star is removed. The program will also at times paramodulate into starred terms which themselves contain starred subterms. In this case, the stars on the subterms are removed automatically. This corresponds to the situation in which the subterms had been associated with reflexive axioms (or in which the corresponding negative literals in a hyper-resolution system had been resolved away with closure). It is necessary for our program to handle this situation specially because 1. we do not include functional reflexive axioms in our input, and 2. we do not allow paramodulation from a variable so that EQUAL(x,x) cannot be used for this purpose.

When one paramodulation step is completed, two things may happen. First, if the result has no remaining starred terms, the clause is demodulated and canonicalized and tested for retention in the clause space. The NUC(C) literal is removed, and if the clause is kept, it becomes a satellite. If the new clause has starred terms, it is called an intermediate nucleus. In this case, two clauses are formed. The first is the intermediate nucleus itself. This is kept on a special list unless subsumed by some other intermediate nucleus. No demodulation or canonicalization is performed on intermediate nuclei. Second, a satellite is formed by deleting the NUC(C) literal and all remaining stars. As above, this corresponds to using closure on these terms. This satellite is then filtered through the normal tests for clause retention.

Once a starred term has been paramodulated, no more paramodulation can be done into that term or any of its subterms. For this reason it is necessary, in some cases, to include in the input various commuted and associated versions of some clauses. For example, suppose we had a satellite EQUAL(f(a,b),c) which we wanted to use with a starred term f(x,y), but with the constant, a, going in for y instead of x. With regular paramodulation we might be able to accomplish this by using commutativity into f(x,y) first. However, in hyperparamodulation such a step would remove the star, and the second step using f(a,b) would not be allowed. This is somewhat of an inconvenience because it adds another choice for the user. Our experience is that, on the whole, the benefits of hyperparamodulation far outweigh this and other inconveniences.

3.4. Comparison of Properties

For space considerations we limit ourselves to a brief comparison of hyperparamodulation to hyper-resolution and paramodulation. While hyper-resolution requires the nucleus to be a mixed clause and the inferred clauses to be positive, hyperparamodulation allows any clause to be a nucleus, and therefore places no such restriction on the inferred. Satellites for the former must be positive clauses, while in our experiments satellites have been positive units because our problems have been expressible as sets of equality units. In both, demodulation is suspended until the inference has been completed, thus avoiding blocking of certain potentially valuable paths. Hyperparamodulation easily allows for bidirectional search by including the denial of the theorem as a nucleus. This can lead to valuable negative inferences in a proof attempt. In contrast, hyper-resolution gains little from the denial of the theorem and is limited to forward reasoning.

As for paramodulation, its control is certainly a problem. For example, many are familiar with the explosion that can result from the generation of more and more positive equalities from the initial set of units. Little is known about strategies to screen and order the from and/or into candidates. Similarly, one knows little to do to direct the search profitably through the ensemble of terms and subterms which are admissible as into terms. In contrast, hyperparamodulation certainly attacks this question in that, for example, satellites are not allowed to interact. For which types of proof this is beneficial remains to be determined. And, of course, the user makes the decisions which affect the into space. Although such decision is arbitrary, we can shed some light on it and do so in the next subsection.

We make a few additional brief remarks about our hyperparamodulation experiments. In the sample problems from Robbins algebra, roughly 1/3 of the kept clauses in the paramodulation runs were from inferences in which the into clause did not correspond to a nucleus in the hyper runs. For these particular problems, those clauses do not correspond to steps in any proofs we know of, and so in this case anyway, they appear to be extraneous. The problem is even worse because considerable processor time is spent in attempted paramodulations with these clauses in which either the inference fails or the result is subsumed, i.e., redundant. These two factors keep the paramodulation program from generating enough useful clauses within the given resources. For example, we list statistics from two sample runs — one hyperparamodulation run and one paramodulation run with two different time limits:

RULE	TIME	SUCC. UNIF.	UNSUCC. UNIF.	GEN.	KEPT	PROOF
hyper	25 sec.	11,769	27,349	1857	510	yes
para	25 sec.	13,181	65,093	1118	280	no
para	110 sec.	42,628	272,192	3133	574	no

Note particularly the number of unsuccessful unifications and the ratio of kept to generated clauses.

We also make the following observations about our (admittedly limited, so far) experiments. When canonicalization was not used, large numbers of commuted and associated versions of a clause were kept in the clause space (as many as 5 or 6 versions of the same clause). When canonicalization was used in paramodulation runs, many of the inferences actually in the hyperparamodulation proof were blocked. For example, a very early step in that proof is to generate B(A(AA))=B by first getting Bx=A(Bx) from associativity and BA=B, and then using another generated equality, B(AA)=B, and finally canonicalizing. However, with paramodulation, the intermediate step Bx=A(Bx) itself is canonicalized and demodulated to Bx=Bx and deleted by subsumption. Paramodulation does find some of these blocked inferences by other deductive paths, but either cannot get them all or at least does not get them before time or memory runs out. Finally, we note that the hyperparamodulation proof contained 32 steps — 23 hyperparamodulants and 9 demodulations by generated demodulators. There were 490 new clauses added to memory when the proof was found, giving a penetrance of .065. This is still quite low, but significantly higher than normally found for paramodulation proofs.

Finally, we obtained proofs by hyperparamodulation even when we varied many of our parameters that control features like weighting, number of variables allowed in a clause, and others. In fact, the run times and other statistics were remarkably close for different runs of the same problem.

3.5. Heuristics of Choice

The extensive user control of the into and from roles that hyperparamodulation provides for generalized equality substitution captures well our position that programs should be flexible tools for theorem-proving research. However, we understand that this capability can be a disadvantage for a user not familiar with the peculiarities of equality in theorem proving. We therefore supply a technique by which users can choose both the nuclei and into terms as well as the satellites.

Recall that one motivation for the formulation of hyperparamodulation was that of borrowing the good properties of hyper-resolution. This fact leads one to surmise that, from among the input clauses, a reasonable choice for nuclei is the set of clauses which, though now written as equalities, correspond to the nuclei of hyper-resolution. For example, in the non-equality representation, associativity, commutativity, and distributivity are hyper-resolution nuclei. The equality counterparts to such clauses would then be chosen as nuclei for hyperparamodulation. For those hyper-resolution nuclei which are themselves equality substitution axioms (and hence are not present in the equality representation), we simply remark that paramodulation in part already automates their role -- but only in part, especially when refined to hyperparamodulation. Finally, our experience indicates that certain axioms generally should not be used as nuclei. Commutativity is the major one of these.

As for the terms which eventually get starred, we suggest those terms which are the correspondents to the negative literals of the hyper-resolution nuclei. We would, for example, star f(x,y), f(y,z), f(f(x,y),z), and f(x,f(y,z)) in the clause EQUAL(f(f(x,y),z), f(x,f(y,z))) representing associativity. The implicit underlying map from the non-equality representation to the equality representation maps both clauses for associativity to the single given clause. Thus, each of the four terms above corresponds to a negative literal. Closure is mapped to reflexivity, which itself has additional roles. This observation explains an earlier discussion of certain satellites not being the analogue of P1-resolvents. It also gives the theoretical justification for the programmed convention of automatically removing stars on subterms of a term just used in the hyperparamodulation. (Detailed consideration of this mapping and of various choices of nuclei and starred terms therein can provide information about refutation completeness and about compatibility with set of support.)

Note that such choices dictated by a consideration of hyper-resolution will not, however, cause the program to behave precisely as if hyper-resolution itself were the inference rule. Here the earlier discussion of representation and its effect on the value of demodulation becomes quite relevant. Nor, obviously, will a hyperparamodulation run behave like one using paramodulation.

Section 4. A Detailed Example

The following example is taken from Robbins algebra and is but one of the conjectures which arose during our study of that algebra. The study is still in progress. It was motivated by considering the following open question: Does Robbins imply Boolean, that is, if S is a set satisfying the Robbins axioms given below, is S a Boolean algebra? For finite S, the answer is known to be yes. For the following axioms, think of o as union and n as complement.

Robbins axioms

o(x,y) = o(y,x) commutativity

o(o(x,y),z) = o(x,o(y,z)) associativity

n(o(n(o(x,y)),n(o(x,n(y))))) = x key property, axiom 3.

We also used EQUAL(o(x,o(y,z)),o(y,o(x,z))), which is deducible from the first two axioms.

This clause is a valuable demodulator, for it contributes sharply to efficiency when using canonicalization.

This open question was ideal for us. First, solving any open question, especially one discussed and not solved by Tarski, would be a coup. Second, heavy use of a theorem-proving program leading to the answering of such a question would be a sharp rebuttal to the various cynics. And third, hyperparamodulation, already under study, appeared to provide an excellent implementation of the kind of reasoning we wished to employ in our attack on this problem. (The results of the Robbins study are being prepared for journal submission.) And so we asked, "How much behavior found in Boolean algebra is exhibited by the Robbins algebra?"

In particular, let a be some arbitrarily chosen fixed element. Assume that o(a,n(a)) = n(a). Can one then prove, for example, that n(n(a)) = a? This result is in fact provable, and a mathematician's proof might go as follows.

First, note that o(a,o(n(a),x)) = o(n(a),x). Second, substitution of a for x and also for y in the third and key axiom yields n(o(n(n(a)),n(o(a,a)))) = a. Third substitution of n(n(a)) for x and o(a,a) for y in axiom 3 yields n(o(a,n(o(a,o(a,n(n(a))))))) = n(n(a)). Fourth, with o(a,o(a,a)) for x and n(a) for y, axiom 3 gives n(o(n(n(a)), n(o(a,o(a,o(a,n(n(a)))))))) = o(a,o(a,a)). Fifth, a for x with o(n(a),n(n(a))) for y gives n(o(n(o(a,n(o(n(a),n(n(a)))))),n(o(n(a),n(n(a)))))) = a. Sixth, a for x and o(a,o(a,n(n(a)))) for y gives o(a,o(a,a)) = a. Seventh, o(a,o(a,o(a,a))) for x and a for y yields n(o(n(a),n(n(a)))) = o(a,a). And last, by applying the seventh and sixth to the fifth, one has n(n(a)) = a.

With this mathematical proof at hand, we cite but two instances of hyperparamodulation taken from a computer run attempting to prove the conjecture under discussion. In fact, all of the steps above correspond to hyperparamodulants; however, space considerations require this brevity. In our run the nuclei were associativity and axiom 3. All terms therein were starred except those which consisted of a single variable. To focus attention on n(a) and also to force an appropriate lexical ordering, the clause EQUAL(n(a),b) was input giving a name to n(a) with a low lexical value. The first application of hyperparamodulation to be cited is that which derived the equivalent of the second identity above but, of course, in clause form, namely EQUAL(n(o(n(n(a)),n(o(a,a)))),a). This was accomplished by paramodulating into axiom 3 first into n(y), and then into o(x,n(y)) positionally of the corresponding transformed intermediate nucleus. As discussed in Section 3.3, at this point a satellite and an intermediate nucleus was formed. The satellite, when demodulated, gave the desired result. The second instance of hyperparamodulation cited here used satellites which were the clause equivalent of the seventh and sixth steps above and in that order, and illustrates how the added term b comes into play. Again the nucleus of hyperparamodulation was axiom 3 and with the starred terms n(y) and o(x,n(y)). The resulting satellite, after much demodulation for both simplification and canonicalization, was EQUAL(n(b),a), which completed the proof since -EQUAL(n(b),a) was input.

Briefly, this style of proof consisted of repeated substitution into axiom 3 and associativity, mostly into axiom 3. So hyperparamodulation was ideally suited. We tried, as an example of a different kind of problem, to prove the problem often referred to as "the commutator problem" in automated theorem proving. It states that, in a group, if the cube of x is e (the identity) for all x, then ((x,y),y) = e for all x and y, where (x,y) is the commutator of x and y. While the cited example from Robbins algebra was successfully proved with hyperparamodulation but unproved with various attempts with paramodulation, the group theory example was proved with both. However, paramodulation yielded a proof in but one-sixth the time that hyperparamodulation took, 2 seconds versus 12. The Robbins proof was obtained in 22 seconds. The difference in the two proofs, the Robbins proof and the commutator proof, is that the latter involved significant interaction of generated lemmas by way of full paramodulation of satellites into other satellites.

Section 5. Conclusions

Hyperparamodulation seems most effective on word problem type theorems — that is, theorems whose proofs consist mostly of substitution into key axioms with only limited interaction between lemmas. The substitution is accomplished by hyperparamodulation and the limited interaction between lemmas is accomplished by dynamic addition of demodulators. Of course, one does not know ahead of time if a particular conjecture is of this type; however, we have found that theorems with small axiom sets and only 1 or 2 complex identities generally seem to work well with hyperparamodulation when the nuclei are restricted to those complex identities.

Hyperparamodulation allows one to affect the proof search rather differently than other rules now in use. Its dependency on user direction appears to us at present to be an overall advantage. We also feel its non-dependence on various parameter settings is a significant indication that the rule itself is effective.

References

1. McCharen, J., Overbeek, R., and Wos, L., "Problems and Experiments for and with Automated Theorem-Proving Programs," IEEE Transactions on Computers, Vol. C-25, No. 8(1976), pp. 773-782.

2. McCharen, J., Overbeek, R., and Wos, L., "Complexity and Related Enhancements for Automated Theorem-Proving Programs," Computers and Mathematics with Applications, Vol. 2(1976), pp. 1-16.

3. Overbeek, R., "An Implementation of Hyper-Resolution," Computers and Mathematics with Applications, Vol. 1(1975), pp. 201-214.

4. Veroff, R., Canonicalization and Demodulation, to appear as an Argonne National Laboratory technical report.

5. Winker, S., and Wos, L., "Automated Generation of Models and Counterexamples and its Application to Open Questions in Ternary Boolean Algebra," Proc. Eighth Int. Symposium on Multiple-Valued Logic, Rosemont, IL, 1978, pp. 251-256.

6. Winker, S., Wos, L., and Lusk, E., "Semigroups, Antiautomorphisms, and Involutions: A Computer Solution to an Open Question," to be submitted.

7. Wojciechowski, W., and Wojcik, A., "Multiple-Valued Logic Design by Theorem Proving," Proc. Ninth Int. Symposium on Multiple-Valued Logic, Bathe, England, 1979, pp. 196-199.

The AFFIRM Theorem Prover: Proof Forests and Management of Large Proofs

Roddy W. Erickson
USC Information Sciences Institute
4676 Admiralty Way, Marina del Rey, CA 90291 US.

and

David R. Musser
Computer Science Branch
General Electric Research and Development Center
Schenectady, NY 12345 US.

Abstract

The AFFIRM theorem prover is an interactive, natural-deduction system centered around abstract data types. Since long proofs are often required to verify algorithms, we describe a model (called the "proof forest") which helps the user to visualize and manage the potentially large number of theorems and subgoals that can arise.

1. Introduction

The AFFIRM verification system [2, 11, 13] permits the user to both state specifications and perform theorem-proving using the language of abstract data types [1, 3, 4, 5, 6, 11, 12]. Data types are defined axiomatically, and may be created or edited by the user as necessary. Typically, one states and proves general properties, and then calls upon these facts as lemmas when demonstrating more complex theorems (such as protocol invariants or program verification-conditions).

Some systems have concentrated on heuristics to guide the theorem prover. (Sometimes these are concerned with handling built-in types, like sets.) Our philosophy is to make the system highly interactive, depending on the user for strategy. AFFIRM automatically performs symbolic manipulation and simplification, records the steps and status of the developing proof, and checks that soundness constraints are met; the user decides when to apply lemmas, instantiate variables, expand definitions, or perform induction. If (as is often the case) a theorem is incorrectly stated, this approach seems to help clarify just where the problem lies. We hope to gradually add more mechanical assistance, while still minimizing the tendency for the system to take off on inappropriate paths.

This paper is intended to describe features (and design motivations) of the AFFIRM theorem prover related to its interactive, "natural deduction" approach. We shall concentrate on the notion of a "proof forest", which aids the user in managing large proofs. The proof forest is defined and discussed in Section 2; section 3 gives a brief overview of relevant theorem prover commands.

A second main aspect of our approach, the extensive use of rewrite rule techniques, has been described elsewhere [11], and will not be treated in detail here. To give the reader a better overview of

Supported in part by the Defense Advanced Research Projects Agency under contract #DAHC15-72-C-0308 the theorem prover, however, a brief description is given in Section 4 of the uses of rewrite rules and other "logic utilities", and in the Appendix a transcript is given of a small example application of the theorem prover. Section 5 concludes with some observations about our experience with the theorem prover and a glimpse at our plans for further development.

2. The Proof Forest

2.1. Purpose

AFFIRM is intended to aid the search for a proof, rather than just to check one which is laid out beforehand. Those of us who are experienced users have found this search for proofs to often be quite challenging. In a typical session (which might last several hours, or, through intermediate state saves, extend over several days), one may attack dozens of related theorems (main goals) and generate hundreds of subgoals, only a few of which could have been anticipated before starting the session. Often, it is necessary to attack a theorem several times before finally discovering the right approach. Given this high degree of complexity, it is important that AFFIRM interfere as little as possible with the user's natural train of thought. This objective of naturalness has led us to the following design principles:

- We follow the *natural-deduction* proof style: the user repeatedly transforms a goal into (one or more) simpler subgoals whose conjunction (together with previously entered axioms, definitions, and induction schemas) implies the goal. (Thus the user works from the root out to the leaves of a proof.)

- *Simplification.* At each proof step, AFFIRM automatically applies any suitable axioms, and normalizes the logical form of the result. These operations are usually invisible, and do <u>not</u> make up structure in the proof tree.

- The order of attack is as *unconstrained* as possible: the user may backtrack and move among theorems, subgoals, and lemmas at will. Subgoals are proven independently. The only constraint is one which forbids circular reasoning.

- The *structure* used to organize and present a proof caters to the use of lemmas and case analysis.

- *Persistence.* Facts demonstrated as part of one proof may be used elsewhere in that session.

- *Soundness.* AFFIRM monitors the logical dependencies among theorems, subgoals, and lemmas; it insists that all lemmas be accounted for before a theorem is pronounced "proven".

2.2. Structure

When a goal is broken down into one or more subgoals, a logical dependency is established. This dependency relation on the set of goals corresponds to a directed acyclic graph, which we use to represent the state of a proof. Nodes (which correspond to logical propositions) are annotated to indicate the AFFIRM command which transformed them. If a node has several children, each arc may be labelled to identify it by case. For example, an induction on sets might result in the following tree fragment:

```
              employ Induction(s)
                     /   \
        NewSet:     /     add:\
                   /           \
          (Immediate)           ...
```

"NewSet" and "add" are the constructors for the Set data type and are used as the labels of the cases of the induction argument (see the Appendix for further examples and explanation). The leaf label

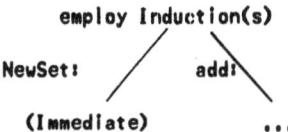

"(Immediate)" indicates that the proof of the NewSet case follows directly from the axioms.

If we suppose that no two cases of a proof are identical, then the proof state of a theorem may be visualized as a tree, rooted in the theorem. This is the model we present to the user. As the proof progresses, structure is added at the leaves (except for backtracking). When all leaves are the constant **true**, the proof is complete. (If two subgoals in the forest happen to be identical, they have a common proof.)

In presenting the proof of a proposition, it is useful to regard any auxiliary facts as atomic; while their status is relevant, details of their proofs are not. We therefore denote certain nodes as **theorems**: propositions of interest.[1] These include the user's original goals and verification conditions, as well as all subgoals entered by the user; the latter are generally of simple form and are called lemmas. (We expect that mechanically-produced subgoals will generally not be of particular interest.) Any logical dependencies upon theorems are recorded separately from our dependency graph. Each theorem therefore is the root of a (possibly degenerate) proof tree in this graph, which we term the **proof forest**. When summarizing a proof tree we note any lemmas used (along with their status) but leave their proofs separate.

2.3. Manipulation

The user has a cursor which may be moved at will about the nodes of the proof forest. The node (termed the **Current Proposition**) on which the cursor rests is the focus of the theorem prover's attention, and the target of all subgoaling commands. (If this node is already the parent of subgoals, because of an earlier proof attempt, those old arcs will be deleted when new ones are made.)

AFFIRM offers the user status information and guidance, lest the permissive control philosophy prove confusing. When all the leaves in a theorem's tree have been finished, AFFIRM notifies the user of any lemmas which remain to be shown. Notice is also given whenever a theorem enters the "proven" state. At any point, the user may ask for a reasonable "next" point of attack. AFFIRM favors (nontheorem) leaves in the current tree, after which it moves on to unproven lemmas and then unproven ancestors. A record is kept of the order in which theorems are visited, so AFFIRM can favor recent nodes in its suggestions.

Any node may be assigned a mnemonic name; this is useful, not only for lemmas, but also to identify key cases in proofs. Arbitrary annotations may also be attached to nodes.

At any time, the proof tree of one or more theorems may be displayed. This summarizes the status and lemma dependencies for each, and prints (in preorder sequence) the AFFIRM command used to justify each step, along with any annotations. For more detail, one can ask for the intermediate goal at each node. AFFIRM can also generate a table showing, for each theorem, its proof status, lemmas used, and proofs where it is applied.

---

[1] Thus we have both "proven theorems" and "unproven theorems". Sometimes, the unproven theorems are in fact found to be false and get discarded.

3. Directing the Theorem Prover

We give only a highly abbreviated description of a few commands. A more complete description can be found in [2, 13]. All expressions are strongly type-checked upon entry.

3.1. Entering theorems into the system

theorem T	Enters an proposition, which is marked as a theorem.
genvcs U	Creates verification conditions for a program unit; these are theorems.

Theorems can also be created by the **apply** and **try** commands. Whenever the user supplies a theorem, it may be given a name at the same time.

3.2. Splitting

apply L	Uses a theorem as a lemma: it is added as an hypothesis to the Current Proposition. (Variables of the applied theorem are renamed to avoid name conflicts.)
augment A	The Current Proposition is of the form "H imp C". We wish to use A as a hypothesis in the proof, and separately show that A follows from H. Two subgoals are produced: "(H∧A) imp C" and "H imp A"
suppose P	If the Current Proposition is G, produces two subgoals: "P imp G" and "(¬P) imp G"
employ S(v)	Performs structural induction on v, setting up the cases as subgoals. V is a universally-quantified variable in the Current Proposition, and S is a schema for induction or case analysis on v's abstract data type.

3.3. Logical manipulations

put x=t...	Instantiates existentially-quantified variables, subject to Skolem dependency constraints (see Section 4.5).
search	Uses the method of *chaining and narrowing* to seek instantiations which will prove the Current Proposition. (See Section 4.6)
replace	Performs substitutions using equality hypotheses in the Current Proposition.
invoke D	Expands definitions.
cases	Applies the case-distribution rule to remove embedded conditionals (see Section 4.3).

3.4. Movement

try m	Moves the cursor to a specified theorem or subgoal.
next	AFFIRM suggests the next node to attempt by moving its cursor.
up	Raises the cursor one level towards the root.
down	The cursor descends to a child of the current node. (Unless a particular child is specified, AFFIRM picks the first unproven one.)
resume t	t is a theorem; the cursor is moved to that node in t's tree which was most recently visited.

3.5. Marking

assume A	Marks theorem A as "assumed" -- it is treated as if proven (but its special status is noted). This is useful for lemmas proven in other sessions.

4. Logic Utilities

This section gives a brief overview of the basic logical manipulations performed by the AFFIRM theorem prover, as additional background for the example in the Appendix.

4.1. Rewrite rules

AFFIRM encourages the use of *equational specifications* for data abstractions, since the theorem prover is oriented toward performing deductions by making equational substitutions. The theorem prover is able to make such deductions automatically by treating equations, whenever possible, as *rewrite rules*. These are rules of the form

left → right

where *left* and *right* are expressions (possibly containing variables). The rules are used to rewrite expressions by replacing with *right* all subexpressions which are matched by *left*. Rewrite rules are applied to an expression until no further rewriting is possible; their order of application is unspecified.

An essential property of a set of rewrite rules is *finite termination*: no infinite sequence of rewrites is possible. Another extremely useful property is *unique termination*: any two terminating sequences of rewrites starting from the same expression will have the same final expression (no matter what choice is made of which subexpression to rewrite or which rule to apply first). A set of rules with the finite and unique termination properties is said to be *convergent*. If a set of axiomatic equations can be treated as rewrite rules, and these rules (or a finite set of rules derived from them) are convergent, then one can decide when an equation is provable from the axioms just by rewriting both sides to their final expressions and checking these for identity. Thus, the AFFIRM system attempts to form the equational parts of data type specifications into rewrite rules with this convergence property.

At present, the system does not attempt to prove that the finite termination property is maintained when a new rule is added to its data base, but rather assumes this property (although some simple tests are applied that may reveal its absence, in which case it refuses to add the rule).

The system checks for unique termination using an algorithm based on the Knuth-Bendix method [7, 8]. This is able to generate additional rules which may restore unique termination in case an added rule is found to violate this property [11]. (Rules which are redundant will be discarded.) If a new rule must be generated to preserve unique termination, and the rule's direction is open to question, the user will be asked to decide. If the convergence process finds a contradiction, it discards any rules it has added (including the rule the user was trying to add) and restores any rules it has discarded.

Notation about a data type may be introduced through the use of **definitions**. These are equational rules which only rewrite on command, and thus are not part of the Knuth-Bendix mechanism. They serve the purpose of deferring expansions which might make propositions too complex.

4.2. Propositional logic

In AFFIRM, all logical connectives are translated into an internal conditional expression form. The if-then-else form is automatically simplified in a manner similar to that of [10]; these simplification rules are sufficient to recognize any propositional tautology. For output to the user, the if-then-else form is in most cases translated back into a form using the traditional logical connectives.

4.3. Case distribution

Suppose f is a function symbol, and $x_1 \dots x_n$, a, b, c are expressions. The case-distribution rule

$f(x_1, \dots, \text{(if a then b else c)}, \dots, x_n) \rightarrow$

if a then $f(x_1, \dots, b, \dots, x_n)$ else $f(x_1, \dots, c, \dots, x_n)$

"raises" embedded if-then-else expressions across function symbols to the outer level of expressions.

Such embedded conditionals can arise from the automatic application of rewrite rules or from (user) controlled invocation of definitions. Should this rule be applicable, the user is informed. It does not take effect until requested, however, since further logical manipulations might simplify the embedded conditional expression (preventing "case explosions").

4.4. Quantification and Skolemization

First order quantification is permitted in formulas input to the theorem prover, using the keywords all and some for universal and existential quantification, respectively; e.g.

(all x (P(x) implies some y all z (Q(x,y,z) and some w R(w,z)))).

Internally, formulas are subjected to transformations that produce an equivalent form in which the quantifiers have been drawn out to the front, with all quantifiers appearing first, then some quantifiers, then a quantifier free body. E.g., the above formula becomes

(all x, z(y) some y(x), w(x,z) (P(x) implies (Q(x,y,z) and R(w,z))).

This is the "symmetric Skolemized form": Skolem functions are introduced for both universally and existentially quantified variables. (These Skolem functions are placed at the front, in the place of the binding variables, instead of being inserted into the body itself. When instantiation is later performed, then the Skolem functions enter into the body.) The main reason for choosing this approach to quantification is to be able to Skolemize definitions and replace defined symbols with their definitions without unSkolemizing and reSkolemizing.

4.5. Instantiation

With the **put** command, the user can direct the system to instantiate variables that appear in the some list of the Current Proposition. For the above example, one could say

put y=f and w=g

where f and g are formulas. The argument lists for y and w in the some list indicate the restrictions placed on f and g: in f the only free variable that can occur is x and in g only x and z can occur free. Note that this prohibits circular dependencies that would occur if, say, f contained z free (looking back at the original formula, we see that z is quantified inside the scope of some y and thus it is not meaningful to permit it in a value for y).

4.6. Chaining and narrowing

The theorem prover can also be directed to search for an instantiation. The **search** command implements a procedure called "chaining and narrowing" [9]. This can be viewed as a generalization of the propagation of assumptions and denials through the branches of conditional expressions. The generalization occurs in using the ***most general unifier*** of subexpressions in the condition and branch expressions. For example,

if P(x) then P(a) else <u>true</u>

can be "chained" to produce

if P(a) then <u>true</u> else <u>true</u>

The effect of the algorithm is the determination of whether a quantifier free first order formula (arising from Skolemization) in if-then-else form has a single ground instance that is a propositional tautology. (Thus <u>some</u> x(P(x) imp P(0)) is provable by this method, but <u>some</u> x(P(x) imp (P(0) and P(1))) is not since it requires two instantiations.)

5. Conclusions and Future Objectives

Probably the most significant observation after several months of experience with the AFFIRM theorem prover is the degree to which the user is working in exploratory and recovery mode. Generally, one is ultimately able to obtain short and well-structured sets of goals and proofs. But pitfalls lie along the way:

- unprovable or invalid lemmas are often stated, used, and subjected to considerable proof effort before discovering the error

- the user can get completely swamped in a deeply nested set of cases

- expressions may be misread or instantiations may be mistyped, producing further errors which may not show up for several steps.

Such fallibility seems (unfortunately) normal, although it is reducible by experience or another person participating in the proof development. We hope to gradually improve AFFIRM's ability to help prevent or recover from these. For example, when a lemma is applied, AFFIRM might suggest variable instantiations. If a lemma is unprovable as stated because it lacks an hypothesis, it might be possible to note that assumption, requiring that it be discharged wherever the lemma is used.

A major area of further development being planned is the display and query facilities which so strongly influence human productivity in using both the theorem prover and the AFFIRM system as a whole. Numerous command options need refinement and extension to enable smooth interaction. Of a more extensive nature is the need for additional support for proofs which persist over several sessions (as will all non-toy proofs). We could like a data base of theorems specific to a proof, along with how these depend on (possibly changing) sets of axioms. A more permanent, global data base (of which the current AFFIRM type library is a rudimentary facility) would also be helpful.

6. Acknowledgments

The members of the Program Verification project at ISI (S. Gerhart, D. Thompson, R. London, D. Baker, R. Bates, D. Taylor, and D. Wile) have made numerous suggestions concerning the design of the theorem prover (particularly its user-visible portions) as well as serving as guinea pigs during its development. D. Lankford made valuable contributions to the original design of the theorem prover, and the work of W. Bledsoe and his group inspired our use of the natural deduction approach.

References

1. Burstall, R. M. and Goguen, J. A., "Putting Theories Together to Make Specifications," in *Proceedings of the Fifth International Joint Conference on Artificial Intelligence*, pp. 1045-1058, IEEE, August 1977.

2. Gerhart, S. L. *et al.*, *An Overview of AFFIRM: A Specification and Verification System*, USC Information Sciences Institute, Technical Report RR-9-81, 1980. Also to appear in Proc. IFIP 80

3. Goguen, J. A., Thatcher, J. W., and Wagner, E. G., "Abstract Data Types as Initial Algebras and the Correctness of Data Representations," in Yeh, R. T. (ed.), *Current Trends in Programming Methodology, Volume IV*, Prentice-Hall, 1978.

4. Guttag, J. V., "Abstract Data Types and the Development of Data Structures," *CACM* 20, June 1977, 397-404.

5. Guttag, J. V., Horowitz, E., and Musser, D. R., "Abstract Data Types and Software Validation," *CACM* 21, December 1978, 1048-1064. (Also USC Information Sciences Institute RR-76/48, August 1976.)

6. Guttag, J. V., "Notes on Type Abstraction," *IEEE Transactions on Software Engineering* SE-6, (1), April 1979, 13-23.

7. Huet, G., *Confluent Reductions: Abstract Properties and Applications to Term Rewriting Systems*, IRIA - LABORIA, Technical Report LABORIA Report No. 250, 1978.

8. Knuth, D. E., and Bendix, P. B., "Simple Word Problems in Universal Algebras," in Leech, J. (ed.), *Computational Problems in Abstract Algebra*, pp. 263-297, Pergamon Press, New York, 1970.

9. Lankford, D. S. and Musser, D. R., On Semi-deciding First-Order Validity and Invalidity, 1978. unpublished manuscript

10. McCarthy, J., "A Basis for a Mathematical Theory of Computation," in Braffort and Hirschberg (eds.), *Computer Programming and Formal Systems*, pp. 33-70, North-Holland, 1963.

11. Musser, D. R., "Abstract Data Type Specification in the AFFIRM System," *IEEE Transactions on Software Engineering* SE-6, (1), April 1979, 24-32.

12. Spitzen, J., and Wegbreit, B., "The Verification and Synthesis of Data Structures," *Acta Informatica* 4, 1975, 127-144.

13. Thompson, D. H., ed., *AFFIRM Reference Manual*, USC Information Sciences Institute, 1979.

Appendix: A Sample Data Type and Proof

The data type 'SetOfInteger' has constructors 'NewSetOfInteger' and 'add'. Using those, we define the meaning of 'in' (set membership), 'rem' (removal of an element), and 'subset'. Equality between nonempty sets is given in terms of subsets. Equal and subset have 'definitions', which are only expanded by the 'invoke' command. The schema 'Induction' tells us that, in order to prove P(s) (for s a Sequence), it is sufficient to show "P(NewSet)" and "P(s) imp P(s add i)". Here is the entire type:

```
type SetOfInteger;

declare ii, i, i1, i2, x: Integer;
declare s, s1, s2, ss: SetOfInteger;

interfaces
        NewSetOfInteger, s add x, s rem i, s diff s1,
        s int s1, s union s1: SetOfInteger;

interfaces
        i in s, isNewSetOfInteger(s), s subset s1,
        s=s1, Induction(s), NormalForm(s): Boolean;

axioms  s=s == TRUE,
        NewSetOfInteger = s add i == FALSE,
        s add i = NewSetOfInteger == FALSE,

        NewSetOfInteger rem i == NewSetOfInteger,
        (s add x) rem i == if x=i then s rem i else (s rem i) add x,

        NewSetOfInteger diff s == NewSetOfInteger,
        (s add x) diff s1 == if x in s1 then s diff s1
                             else (s diff s1) add x,

        NewSetOfInteger int s1 == NewSetOfInteger,
        (s add x) int s1 == if x in s1 then (s int s1) add x
                            else s int s1,

        NewSetOfInteger union s1 == s1,
        (s add x) union s1 == (s union s1) add x,

        x in NewSetOfInteger   == FALSE,
        i in (s add x) == ((i=x) or (i in s)),

        isNewSetOfInteger(s)
          == (s = NewSetOfInteger);

define  s=s1 == ((s subset s1) and (s1 subset s)),

        s subset s1 == all x (x in s imp x in s1);

schemas Induction(s) == cases(Prop(NewSetOfInteger),
                              all ss, ii (IH(ss) imp Prop(ss add ii))),

        NormalForm(s) == cases(Prop(NewSetOfInteger),
                              all ss, ii (Prop(ss add ii)));


end {SetOfInteger} ;
```

The following is a transcript of the proof of a simple fact about sets: removing an element and then adding it results in an identical set. Commands to the theorem prover are in **boldface**; *comments in italics were added later.*

1 U: **try RemAdd, x in s imp (s rem x) add x = s;**
Making node RemAdd a theorem.
RemAdd untried
forall x, s : x in s imp (s rem x) add x = s

2 U: **invoke Equal;**
```
forall x, s :
  x in s imp    ((s rem x) add x) subset s
          and s subset ((s rem x) add x)
```

3 U: **split;**
splits 'p and q' into 'p', 'p imp q'
(first:)
forall x, s : x in s imp ((s rem x) add x) subset s

4 U: **invoke subset;**
```
forall x, s, x' :
  x in s and ((x'=x) or x' in (s rem x)) imp x' in s
```

5 U: **split;**
splits on the 'or'
(first:)
forall x, s, x' : x in s and (x'=x) imp x' in s

6 U: **replace;**
TRUE

7 U: **print proof;**
'?' shows unproven leaves. '->' denotes the Current Proposition. Numbers like 6: show the event number for the command, so it can be found in the transcript. Other numbers identify subgoals.

RemAdd is: x in s imp (s rem x) add x = s

```
proof tree:
2:|    RemAdd
           invoke  first  Equal
3:      2    split
4:       first:
            3    invoke  first  subset
5:          5    split
6:           first:
               6    replace
6:->           (proven!)
?            second:
               7
?        second:{RemAdd}
            4
```

8 U: **next;**
Going to leaf second:
We're now at node #7, above. The other half of the or.
```
forall x, s, x' :
  x in s and not (x'=x) and x' in (s rem x) imp x' in s
```

9 U: **apply RemInS, y in (s rem z) imp y in s;**
Please declare y (then type ok; to continue).
Command 9 is held in abeyance...

10 (1) U: **declare y,z:Integer;**
y, z: Integer

11 (1) U: **ok;**
...and continued
find y, s', z : y in (s' rem z) imp y in s'
The lemma, with s renamed

12 U: **searh;**
(searh => search)
1/4: (z = x) and (s' = s) and (y = x')
Proved by chaining and narrowing
using the substitution
(z = x) and (s' = s) and (y = x')

TRUE

13 U: **next;**
Going to leaf second:

```
forall x, s :
  x in s and ((s rem x) add x) subset s
      imp s subset ((s rem x) add x)
```

14 U: **invoke all subset;**
```
forall x, s, x" find x'(x, s, s) :
        x in s
   imp if x'=x
         then      x' in s and x" in s
                imp (x"=x) or x" in (s rem x)
         else      (x' in (s rem x) imp x' in s) and x" in s
                imp (x"=x) or x" in (s rem x)
```
Both clauses contain x" in s imp x" in (s rem x)

15 U: **try InSInRem, (y in s and y ne z) imp y in (s rem z);**
We decide to prove the lemma before applying it
Making node InSInRem a theorem.
InSInRem untried
forall y, s, z : y in s imp (y=z) or y in (s rem z)

16 U: **employ Induction(s);**
Case NewSetOfInteger: Prop(NewSetOfInteger) proven.
Case add: forall ss, ii : IH(ss) imp Prop(ss add ii)
remains to be shown.
(add:)

```
forall ss, ii, y, z :
        IH(ss) and ((y=ii) or y in ss)
   imp (y=z) or
        y in (if ii=z then ss rem z else (ss rem z) add ii)
```

(The 'cases' command is applicable)

17 U: **cases;**
```
forall ss, ii, y, z :
        IH(ss)
   imp if y=ii
         then not (y=z) and (ii=z) imp y in (ss rem z)
         else y in ss imp (y=z) or y in (ss rem z)
```

18 U: **replace ii;**

```
forall ss, ii, y, z :
      IH(ss) and not (y = ii) and y in ss
   imp (y = z) or y in (ss rem z)
```

19 U: **invoke IH;**

```
forall ss, ii, y, z
   find y'(ss, z, y, ii), z'(ss, z, y, ii) :
          y' in ss imp (y'=z') or y' in (ss rem z')
       and not (y = ii)
       and y in ss
   imp (y = z) or y in (ss rem z)
```

20 U: **search;**

```
1/3:  y' = y
    1/2:  z' = ii
    2/2:  z' = z
```

Proved by chaining and narrowing
using the substitution
(z' = z) and (y' = y)

TRUE
InSInRem proved

21 U: **print status;**

InSInRem is proved.
RemAdd is tried.
RemInS is untried.

22 U: **try RemAdd;**

RemAdd tried (may be resumed)
forall x, s : x in s imp (s rem x) add x = s

23 U: **resume;**

Takes us back to where we were

```
forall x, s, x" find x'(x, s, s) :
      x in s
   imp if x'=x
       then      x' in s and x" in s
              imp (x"=x) or x" in (s rem x)
       else        (x' in (s rem x) imp x' in s)
              and x" in s
              imp (x"=x) or x" in (s rem x)
```

24 U: **apply InSInRem;**

find y, s', z : y in s' imp (y=z) or y in (s' rem z)

25 U: **search;**

```
1/11:  (s' = s) and (y = x)
     1/3:  x' = x
     2/3:  (z = x) and (x' = x)
     3/3:  x' = x"
2/11:  (s' = s) and (y=x')
     1/3:  x' = x
     2/3:  x' = x"
         1/1:  z = x
```

Proved by chaining and narrowing
using the substitution
(z = x) and (x' = x") and (s' = s) and (y=x')

TRUE
RemAdd awaiting proof of lemmas RemInS.

26 U: **next;**

Going to lemma RemInS.

RemInS untried
forall y, s, z : y in (s rem z) imp y in s

27 U: **employ Induction(s);**

Case NewSetOfInteger: Prop(NewSetOfInteger) proven.
Case add: forall ss, ii : IH(ss) imp Prop(ss add ii)
remains to be shown.

```
 (add:)
forall ss, ii, y, z :
         IH(ss)
     and y in (if ii=z then ss rem z else (ss rem z) add ii)
   imp (y=ii) or y in ss
```

(The 'cases' command is applicable)

28 U: **invoke IH;**

```
forall ss, ii, y, z
   find y'(ss, ii, y, z), z'(ss, ii, y, z) :
            y' in (ss rem z') imp y' in ss
        and y in (if ii=z then ss rem z else (ss rem z) add ii)
   imp (y=ii) or y in ss
```

(The 'cases' command is applicable)

29 U: **search;**

1/1: y' = y
Unsuccessful.

30 U: **put x'=x and y'=y;**

User error: Variable x' doesn't appear in the find list.
Instantiation not done.
(... put aborted: effects undone)

31 U: **up;**
we're confused; try something else

```
forall ss, ii, y, z :
      IH(ss)
   and y in (if ii=z then ss rem z else (ss rem z) add ii)
  imp (y=ii) or y in ss
```

(The 'cases' command is applicable)

32 U: **cases;**

```
forall ss, ii, y, z :
     IH(ss)
  imp if ii=z then y in (ss rem z) imp (y=ii) or y in ss
        else not (y=ii) and y in (ss rem z) imp y in ss
```

"y in (ss rem z) imp y in ss" follows from IH

33 U: **invoke IH;**

```
forall ss, ii, y, z
   find y'(ss, y, z, ii), z'(ss, y, z, ii) :
      y' in (ss rem z') imp y' in ss
   imp if ii=z then y in (ss rem z) imp (y=ii) or y in ss
         else not (y=ii) and y in (ss rem z) imp y in ss
```

34 U: **search;**

1/2: (z' = z) and (y' = y)
Proved by chaining and narrowing
using the substitution
(z' = z) and (y' = y)

TRUE
RemInS proved
RemAdd proved

35 U: **print uses;**

Theorem	Uses:	Used by:
RemInS, proved	-	(RemAdd)
RemAdd, proved	(RemInS InSInRem)	-
InSInRem, proved	-	(RemAdd)

36 U: **print proof theorems;**

RemInS is: y in (s rem z) imp y in s

```
proof tree:
27:!   RemInS
          employ  Induction(s)
       NewSetOfInteger:
          Immediate
32:     add:
          21   cases
33:        23   invoke  first  IH
34:        24   put  (z' = z) and (y' = y)  {search}
34:->      (proven!)
```

RemAdd is: x in s imp (s rem x) add x = s
uses <RemInS!,InSInRem!>

```
proof tree:
2:!    RemAdd
           invoke  first  Equal
3:       2   split
4:      first:
           3    invoke  first  subset
5:         5    split
6:          first:
              6    replace
6:            (proven!)
11:          second:
              7    apply  RemInS
12:            10   put  (z = x) and (s' = s)
                         and (y = x')  {search}
12:            (proven!)
14:     second:{RemAdd}
          4    invoke  all  subset
24:        12   apply  InSInRem
25:        19   put     z = x and (x' = x")
                     and (s' = s) and (y=x')  {search}
25:       (proven!)
```

InSInRem is: y in s and not (y=z) imp y in (s rem z)

```
proof tree:
16:!   InSInRem
          employ  Induction(s)
       NewSetOfInteger:
          Immediate
17:     add:
          14   cases
18:        15   replace  ii
19:        16   invoke  first  IH
20:        17   put  (z' = z) and (y' = y)  {search}
20:       (proven!)
```

37 U: **quit;**

Type CONTINUE to return to AFFIRM.

DATA STRUCTURES AND CONTROL ARCHITECTURE FOR IMPLEMENTATION OF THEOREM-PROVING PROGRAMS

Ross A. Overbeek
and
Ewing L. Lusk

Northern Illinois University
DeKalb, IL 60115/USA

ABSTRACT

This paper presents the major design features of a new theorem-proving system currently being implemented. In it the authors describe the data structures of an existing program with which much experience has been obtained and discuss their significance for major theorem-proving algorithms such as subsumption, demodulation, resolution, and paramodulation. A new architecture for the large-scale design of theorem proving programs, which provides flexible tools for experimentation, is also presented.

I. Introduction.

Over the last nine years a group of researchers based at Argonne National Laboratory and Northern Illinois University have developed a large and powerful theorem proving system. Using this program, we were able to study the behavior of a number of algorithms on a wide variety of problem domains, and many valuable insights resulted. The results of this research are published elsewhere [1, 2, 3, 4, 5, 6, 8, 9, 10, 11, 12]

Recently the system has been used successfully as an actual research tool in other disciplines. In particular, it obtained solutions to three open questions in mathematics [8, 9,11] and is used for studies in the design of electronic circuits [13]. Such experience leads us to believe that theorem-proving programs will, in the near future, be extremely useful in attacking a limited class of problems.

This paper is a description of the major design features of a new theorem-proving system, which is currently in the implementation process. This project was begun, in parallel with continued work on our current system, for reasons given below. We hope to incorporate into it those features of the current system which have, over the years, proven most useful. We will describe here some of those features, particularly in the area of data structures and some critical algorithms, which we believe are essential to any powerful implementation. Thus, we hope that this paper will be of considerable utility to other researchers who are currently confronting implementation decisions.

First, we discuss some shortcomings of our current system, and some considerations for the future, which led us to undertake a redesign of our existing system. Next we will discuss the principle design features of the new system. The first of these is an external, portable representation of data which we offer as a standard for the exchange of information among widely differing theorem proving systems. The next topic we describe is the internal representation of data in the new system. Experience has taught us that the choice of internal data structures is perhaps the single most critical design decision to be made in an implementation. Then we discuss the utilization of our choice of internal data structures in major theorem proving algorithms such as subsumption, demodulation, resolution, and paramodulation. Finally we discuss control structures for theorem-proving strategies, which is the area in which the new implementation departs most radically from the existing one.

2. The Need for a New Theorem Prover.

The current theorem prover was originally designed as a tool for the investigation of a limited class of theorem proving algorithms, not as a research support tool for general use. As a result, it has certain shortcomings which will impair its usefulness, both to those who wish to continue a wide class of theorem proving experiments and to others who want it as a tool for study in application areas. The main problems are the following:

a) The current program's portability is extremely limited. Powerful machines capable of supporting our system will continue to exist in the foreseeable future, but the increasing availability of relatively inexpensive, powerful research machines provides strong motivation for creating a portable system.

b) The current system is not as easily modifiable as is now desired. The experience of the last five years has taught us that unforeseen ideas will continue to be implemented as progress in the field is made. Only an extremely limited group of people are familiar enough with the overall design (due to architectural complexities, inadequate documentation, and the use of assembly language for lower-level routines) to implement enhancements easily. Although limited distribution of our system has occurred, it will be of greater use to experimenters if its architecture is straightforward yet flexible. This issue will be principally addressed below in the section on control structures.

c) Perhaps the major drawback of the current system is its inability to utilize effectively multiprocessor systems. Systems supporting a relatively large number of processors with massive shared memory will soon be available. High-level languages which support the required multiprocessing primitives and the features required for efficient implementation are now beginning to appear. In this paper we present the architecture of the new theorem prover now being implemented, which will allow it to take full advantage of the hardware of the near future.

This is not to apologize for the design of the existing system. Requirements have changed since its creation. If it had not been designed the way it was, it would not have been as successful as it has been.

In summary, we have undertaken the project described here because we believe that it is time to create a multiprocessed system based on generalizations of the design principles that have proven successful in the current program. Every attempt should be made to make it portable, easily modifiable as a research tool, and to make it support multiple, convenient user interfaces.

3. Representation of Data.

Choice of the underlying formalism for a system of the sort we propose is critical. Some of the most commonly advocated alternatives are these:

a) The clause format is widely used. It is the representation used in our current system.

b) The use of unnormalized statements in the first order predicate calculus characterizes a number of highly successful "natural deduction" systems.

c) Exploration of the use of higher-order logics has increased recently.

d) A variety of data structures for statements about models of reality have been proposed in the artificial intelligence literature.

To a large extent it has been impossible to analyze the relative merits of such alternatives because comparable implementations supporting them do not exist.

There will be three levels of representation for data. The highest level should represent a convenient user format. In the long run there will be many such user formats. Examples of these might be clause form, first order predicate calculus, higher order logic expressions, or typed lambda calculus [7]. Each of these user formats will be translatable into a common language which we here call external format. This will be translated into a memory resident format, the internal format, which will be the representation manipulated by the fundamental theorem-proving algorithms.

3.1 The External Format.

Several considerations dictate the choice given here for the external format.

a) It must allow representation of a wide class of user-oriented languages. One of the most general such languages is the typed lambda calculus, and our external format is loosely based on it. It thus allows representation of sets of clauses, first order predicate calculus, and higher order logics.

b) It must support a variety of special-purpose features which cannot be predicted, since they will arise in the course of future experimentation. We have attempted to provide this flexibility by allowing expressions to have arbitrary, user-defined "attributes," whose values are binary trees with integer values in the leaves.

c) It must be portable in the sense that it is not bound to any processor or device. In particular, the external format, unlike the internal format, contains no pointers. This allows data to be moved from one processor or device to another in the course of theorem-proving operations, as well as providing a standard language for the interchange of data among radically different theorem-proving systems.

Since the syntax we propose is short, we give it here in Backus-Naur form.

```
<external term> ::= <term>;
<term> ::= <name> | <term>(<argument list>)
         | <qualifier><variable list>(<term>)
         | <term>:<extension>
<name> ::= <variable name> | <function name>
<argument list> ::= <term> | <term>,<argument list>
<qualifier> ::= (<name>)
<variable list> ::= <variable name> | <variable name>,<variable list>
<extension> ::= <attrs> | <type> | <attrs>:<type>
<type> ::= T(<name>)
<attrs> ::= <attribute> | <attribute>:<attrs>
<attribute> ::= A(<name>/<list>)
<list> ::= <number> | (<list>,<list>)
```

A <name> is a character string consisting of numbers, letters, and underscore characters and beginning with a letter. The variable names are those which begin with letters in the range S-Z; the others are function names. A <number> is a nonnegative integer. Intuitively, a <qualifier> can be used to represent quantifiers and in formulating lambda-expressions, among other things.

This syntax is simple and allows encoding of almost all types of statements currently used as input to theorem-proving programs. It is not particularly convenient from a user's point of view, but various user-oriented formats can be represented. Here are three examples.

a) Clause format. This is approximately our current user interface.

CL ¬P(X,F(A)) Q(B);

translates to

(ALL)X(OR(NOT(P(X,F(A))),Q(B)));

Here we assume for simplicity of the example that there are no attributes and types are allowed to default.

b) Unnormalized first order predicate calculus.

(Ax)(P(X,F(A)) --> Q(B))

translates to

(ALL)X(IMPLIES(P(X,F(A)),Q(B)))

Note that the quantifier "for all" appears in the syntax as a qualifier.

c) Lambda calculus

P(X(G(X)),A)

translates to

P((LAMBDA)X(G(X)),A);

Note that the fits into the syntax as a qualifier.

Those features of our current design which do not generalize easily will be implemented to operate only on statements which are clauses. However, those features which do generalize will be included in such a way that no unnecessary restrictions are imposed.

This should be a first step in creating an environment in which realistic comparisons of a wide variety of theorem-proving schemes can be attempted.

3.2 The Internal Format.

The success of the current program has been due in large part to the structure used to represent a clause in memory. The key feature of that structure is that there is only one copy of any clause, literal, or term in the permanent data structures. Thus if a literal occurs in five clauses there is only one copy of the literal; each of the containing clauses has a pointer to the literal, and the literal has pointers to each of the clauses. As will be seen below, this has significant implications for several major algorithms.

These concepts generalize in a natural way to the internal representation of terms in the new formalism. Clauses and literals will be special cases of terms. Since a detailed description of this internal format will clarify a number of concepts used in later sections, we describe it here, and give a PASCAL implementation in the appendix.

A term node will always contain at least the following fields:

a type: The inclusion of types in the internal format allows implementation of typed calculi an id: This unique id will be used for output functions and during the process of integrating new terms into the permanent structures

attribute header: There will be a set of attributes associated with each term. Each attribute has an identifying code and a value. The value is a binary tree with integer values in the leaves.

occurs-in header: this doubly-linked list includes all argument nodes used to reference the given term as a subterm. (An argument node connects a term to a subterm, and vice versa). Thus, by following this list, all occurrences of the term can be located (There is only one copy. By following this list, the containing terms can be determined).

There are three distinct types of term nodes:

a) Variable nodes contain an integer identifying the variable.

b) Function symbol nodes contain the name of the function.

c) Application nodes contain a header to a linked list of other term nodes. The first of these represents a function; the rest represent the arguments.

4. Access to Terms in Internal Format.

The existing version of the above internal format, together with certain auxiliary access structures for accessing terms, has significantly enhanced performance of the current program. In this section we will generalize some of our existing techniques and show how they are used in the new system.

First we discuss the process of integrating a term into the permanent data structures. Then we present a technique for rapidly isolating a desired set of terms. Finally we give in detail a complete subsumption checking algorithm which utilizes these features, and briefly discuss their implications for the algorithms which compute resolvents and paramodulants.

4.1 Integrating a Term into the Permanent Data Structures.

During the course of a theorem-proving run, new terms are constantly being produced. A newly generated term, before it has been determined to be worthy of retention, will be called non-integrated. It may contain new copies of terms which already exist, and some new terms, as well as multiple copies of subterms which occur in it more than once. If it is decided to keep the new term, nodes in it which already exist in the permanent data structures must be replaced by the permanent copies by adjusting pointers, and permanent copies of new terms must be created. Once this is done, the term will be called _integrated_.

The key to accomplishing this efficiently is a hashed structure which provides fast access to each node in the permanent data structures. Note that each node in the permanent data structures is assigned a permanent identifier when it is created. This identifier is an arbitrarily assigned unique integer, and is not to be confused with the keys used for the hashing mechanism. These keys are constructed as follows:

a) The key of a function node is
   (function symbol, type)
b) The key of a variable node is
   (variable number, type)
c) The key of an application node is
   (identifier of node representing the function,
   identifiers of the arguments)

A term being integrated is processed as a tree, working up from the leaves, each of which is a function symbol or variable node. A leaf is examined to see if it is already in the permanent structures by looking it up in the hash table, which contains a pointer to the permanent copy. If the leaf occurs in the permanent structures, the temporary copy can be freed and pointers to it altered to reference the permanent copy. If the leaf does not occur, it can be assigned an identifier and added to the permanent structures. This process can then be repeated on nodes at the next level up in the tree, and so on.

4.2 _Isolation of a Set of Desired Terms_

In the existing program it has been found extremely useful to be able to locate very rapidly all literals of a given sign and predicate fulfilling the added requirement that a specified argument begin with some specified function symbol. The mechanism implemented to support such a feature has been previously described [6]. A similar feature was implemented to increase the efficiency of demodulation routines [3]. Here we present a generalization and unification of these techniques which is being incorporated into our new theorem-proving system.

The general feature desired is the ability to locate a set of terms that fulfill a specified condition. This problem is completely identical to one that arises in any data base management system supporting a sophisticated query language.

Conditions are of two kinds, simple and complex. A simple condition is of the form <property> <relation> <value>. Complex conditions are Boolean combinations of simple conditions. In order to give this feature the widest possible scope, we allow great flexibility in the kinds of properties allowed. A simple example of a property would be an attribute name. A more complex property might perhaps be formulated as a vector. In order to illustrate the types of properties which have proven useful to single out, and to demonstrate how such a feature is used in fundamental algorithms, we present here a fairly detailed example.

Suppose that in the domain of first-order logic we wish to find all the terms (in the permanent data structures) which unify with

F(A,F(B,X)),

where A and B are constants and X is a variable.

If a term is viewed as a tree, then each occurrence of a function symbol or variable symbol is a leaf on the tree, and its position can be denoted by a vector. In F(A,F(B,X)), for example, we have the following position vectors:

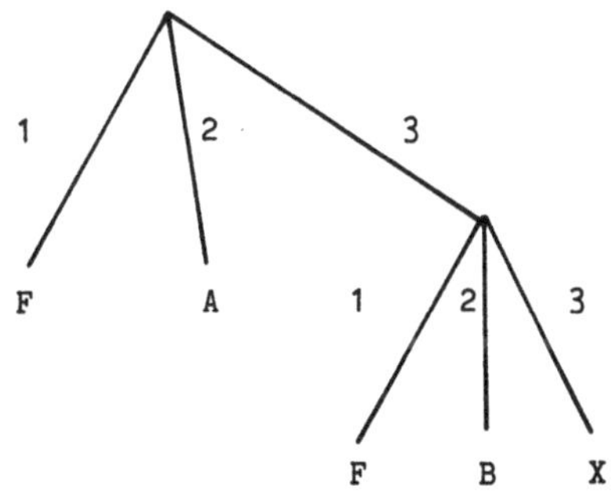

leaf	position
F	(1)
A	(2)
F	(3,1)
B	(3,2)
X	(3,3)

Each position vector then names a property. The values such a property can take on are of two kinds:

a) The The function symbol occuring in the given position, if that position is not occupied by a variable

b) A value (for example, "*") meaning that the position contains a variable.

Thus the complex condition, satisfied by the above term, of having as F as is major function symbol and a variable in the second argument of its second argument, could be written as

((1) = F) and ((3,3) = *)

Now in order to locate terms which will unify with

F(A,F(B,X))

one can begin by locating terms which meet the following complex condition:

((1) = F) and
(((2) = A) or ((2) = *)) and
(((3) = *) or (((3,1) = F) and (((3,2) = B) or ((3,2) = *))))

A unification check must still be performed on such terms, since, for example, F(X,X) satisfies the above condition but does not unify with F(A,F(B,X)). Still the number of terms on which the unification check must be performed has been drastically reduced.

Implementation of this capability to locate terms meeting complex conditions involves creating structures (such as the FPA lists described in [6]) which allow rapid access to sets of pointers to terms satisfying basic conditions. Then computation of a set of pointers to terms satisfying complex conditions can occur rapidly by taking unions and intersections of these sets.

One may object that the use of such techniques to isolate sets of desired terms is unnecessarily complex. However, these methods become essential when processing sets of terms which may become arbitrarily large. These approaches will be absolutely required to implement reasonably efficient subsumption checks, computation of resolvents, simplification techniques, etc. On large problems, efficient implementations are necessary; standard theorem-proving methods are useless without them. In order to illustrate the effect of the facilities described in this section on such algorithms, we next present a version of the subsumption algorithm.

4.3 Example Algorithms

In this section we discuss the effect of our choice of data structures on theorem-proving algorithms. Some features of these data structures which have proven particularly useful are:

a) The ability to access easily all the terms which contain a given term. (Discussed in section 3.2)

b) The ability to access a set of terms which have a high probability of unifying with a given term. (Discussed in section 4.2)

The main vehicle for illustrating the impact of these features will be a detailed presentation of the algorithm currently used in our resolution-based system for determining whether or not a given clause (perhaps a newly generated one) is subsumed by an existing clause in the permanent data structures. We have chosen subsumption because it illustrates the above features and is somewhat non-intuitive. It is also critical to the ability of our current theorem prover to handle relatively large sets of clauses. The principal difference between this and more straightforward subsumption algorithms is that clauses are not checked pairwise. Rather the selection of candidate literals precedes an attempt to locate a subsuming clause.

Input to the procedure FORWARD-SUBSUMPTION consists of a clause C containing N literals. The output is a return code RC representing the value "subsumed" or "not subsumed."

```
FORWARD-SUBSUMPTION:  PROC (C,RC)
   Set RC to "not subsumed"
   Set I to 1
   DO WHILE (I ≤ N & RC = "not subsumed")
      Acquire (using the techniques of section 4.2) the set S of
        literals which probably have the Ith literal of C as an
        instance.  Suppose that S contains M literals.
      Set J to 1
      DO WHILE (J ≤ M & RC = "not subsumed")
         If  (the Ith literal of C is an instance
            of the Jth literal in S)
               CALL ONE-LITERAL (C, RC, Jth literal in S,
                 substitution*)
         ENDIF
         Set J to J+1
      ENDDO
      Set I to I+1
   ENDDO
ENDPROC
```

*The exact format of a substitution here is irrelevant. The substitution being referred to is the substitution of terms for the variables in the Jth literal of S which make it identical to the Ith literal of C.

ONE-LITERAL is passed a literal P and a substitution. Its function is to attempt to find a clause D containing the literal and to extend the substitution, mapping the remaining literals of D into the literals of C.

```
ONE-LITERAL:  PROC (C, RC, P, substitution)
     Form the set F of clauses that contain P.  Suppose that
       there are L clauses in F.
     Set K to 1
     DO WHILE (K < L & RC = "not subsumed")
        Form the set D* of all literals which occur in the Kth
          clause D of F, except P.
        CALL SUB-TEST (C, RC, D*, substitution)
        Set K to K+1
     ENDDO
ENDPROC
```

SUB-TEST is a general-purpose routine which can also be used in back- ward subsumption (determining whether a given clause subsumes any existing clauses in the permanent data structures). It is passed two sets of literals, which we will call the <u>domain set and range set</u>, and a substitution for variables occuring in the domain-set. Its function is to find an extension of the substitution which supports a mapping of the domain-set DS into the range set RS. The image of a literal T under the mapping must be a literal of the form $T\phi$, where ϕ is the extended substitution.

```
SUB-TEST: PROC (RS, RC, DS, substitution)
   Set I1 to the number of literals in RS
   Set D1 to the number of literals in DS
   SET J1 to 1
   DO WHILE (J1 ≤ I1 & RC = "not subsumed")
      Determine if the substitution can be
        extended to allow a mapping of the first literal of DS
        into the J1 nth literal of RS.
      IF such an extension can be made
         If  (D1 = 1)
            Set RC to "subsumed"
         ELSE
            CALL SUB-TEST (RS, RC, DS-(first literals in DS),
              the extended substitution)
         ENDIF
      ENDIF
      Set J1 to J1 + 1
   ENDDO
ENDPROC
```

This is a somewhat simplified version of the subsumption algorithm which exists in the current program. Backward subsumption is similar.

Pseudo-subsumption, which takes into account the clashability of literals in clauses, is substantially more complex.

We will not present here the algorithms for calculating resolvents or paramodulants. However, the features of our data structures discussed above are heavily used there as well. In particular,

a) As is explained in section 4.1, unification can effectively be speeded up by using it on a pre-screened set of literals.

b) Once Once two literals of opposite sign have been unified, an entire set of resolvents can rapidly be constructed by considering the clauses which contain the matched literals.

c) Similarly, Similarly, once an argument in an equality literal has been unified with a term T, an entire set of paramodulants can be formed - one for each occurrence of T.

Restrictions are imposed on the computation of resolvents and paramodulants. However, the ability to rapidly isolate relevant literals and terms and then to infer a set of new clauses is significant. Too often theorem proving programs have been designed with small sets of statements in mind. Successful programs in the future will have to have the capability of maintaining very large sets of active terms without suffering serious degradation in performance. Versions of the techniques discussed in this section have proven very useful in the current system and generalizations of them will form the backbone of the new system.

5. The Multiprocessing Architecture.

In the preceding section we described features of the new theorem-proving system which will be generalized versions of proven techniques in the existing system. In this section we describe the multiprocessing architecture of the new system, which is very different from that of the current system and is intended to provide two significant advantages.

a) The theorem prover will be able to take advantage of the computing power made available by networks of processors, as the required hardware and software become available over the next few years.

b) The architecture will encourage and facilitate experimentation through substitution and interconnection of small, interchangeable process modules. This will enable a user to configure a theorem prover for special purposes without having familiarity with the details of the entire system.

5.1 General Multiprocessing Concepts

In order to more clearly present the architecture of the new theorem proving system, we discuss here some general concepts.

a) In a multiprocessing algorithm distinct components are thought of as operating simultaneously. It is therefore natural to visualize each process as a separate machine. This does not require, however, that each process actually run on a separate machine; it is quite common to execute multiple processes on a single computer by executing a number of instructions from each process in turn. A multiprocessing algorithm can thus be implemented on either multiprocessor or uniprocessor systems.

b) Each process should be visualized as possesing its own separate memory; there are no shared variables. If two processes require access to the same data structure, they will do so by submitting requests to a third process which owns the data structure.

c) A program is used to define the behavior of a process. It is reasonble for two distinct processes to be defined by a single program. A language for describing these programs requires, in addition to the usual instructions for manipulating data and indicating flow of control, commands for interprocess communication.

5.2 Structure of a Process

Our goal is to provide a theorem-prover which can be easily modified and reconfigured with a minimum of programming. In order to do this we will give each process the structure of a machine with several input and output ports. For example,

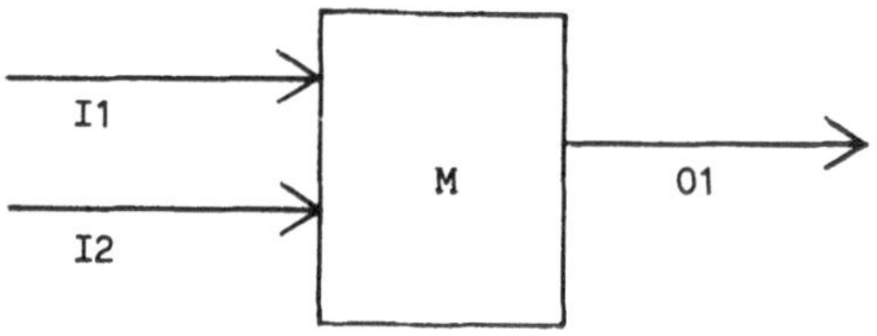

depicts a process M with two input ports and one output through which data (in the form of messages or terms, for example) may flow.

The primitives for interprocess communication which must be supported by the language in which the program defining such a process is written are

a) SEND data-item TO output port #
b) RECEIVE-SPECIFIC buffer FROM input port #.
c) RECEIVE buffer (from any input port)

The execution of either RECEIVE command causes the process to wait for a data item to appear at an input port.

5.3 Networks of Processes

Now we are in a position to describe how a theorem proving program can be configured, once individual processes have been defined. This is done in a way analagous to the "patchboard" method of connecting the inputs and outputs of electronic components. What is needed is a language for describing patterns such as the following:

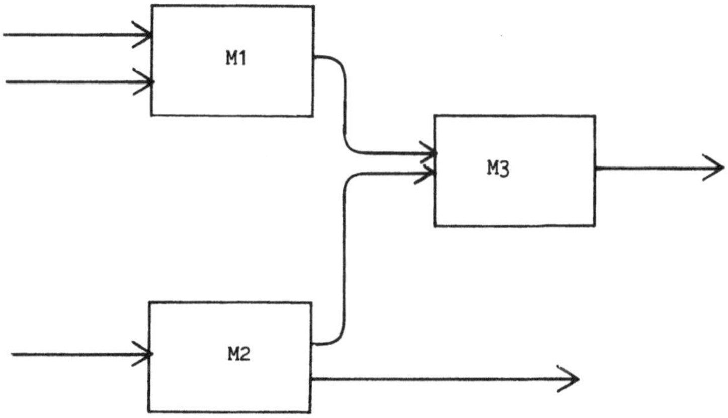

This is easily done and we will not describe such a language here.

It should be emphasized, however, that such an approach provides heretofore unavailable levels of flexibility in the configuration of theorem proving programs, once a standard library of well-defined processes has been implemented. Experimentation over the last nine years has shown us that the exact requirements of a user cannot be predicted. It must be possible to implement, with a minimum amount of work, an algorithm of the following kind:

> "Once a clause has been generated, eliminate all literals having a predicate beginning with 'AR'. If this cannot be done, throw out the clause; otherwise, integrate it into the permanent data structures."

Flexibility of this kind will be provided by the language for the interconnection of processes.

The library of processes will also contain a number of standard networks, accessible by name, which can themselves be connected into larger networks. This simplifies the configuration of a complete theorem-proving system, which may contain a large number of individual processes.

6. Conclusion

In this paper we have reported on the basic design of a new theorem-proving system which is now in the implementation process. Our intent is to provide a very general tool for experimentation with theorem-proving algorithms. Neither its external nor internal data format severely restricts it to a particular outlook. We have presented generalizations of some of the fundamental concepts currently used in our current system which we believe will be important building blocks of the new system. Finally, we have described its multiprocessing architecture, which will allow it to exploit new hardware advances while serving as the foundation of its generality and flexibility.

REFERENCES

1. E. Lusk and R. Overbeek, Experiments with Resolution-Based Theorem-Proving Algorithms, Comp. Math. with Appls, to appear.
2. E. Lusk, and R. Overbeek, Experiments with Resolution-Based Theorem-Proving Algorithms (extend abstract), Third Workshop on Automated Deduction, Boston, 1977.
3. J. D. McCharen, R. A. Overbeek, and L. Wos, Complexity and related enhancements for automated theorem proving programs, Comp. Maths. with Appls., 2, 1-16 (1976).
4. J. D. McCharen, R. A. Overbeek, and L. Wos, Problems and Experiments for and with automated theorem-proving programs, IEEE Trans. on Computers, C-25, No. 8, 773-782, (1976).
5. R. A. Overbeek, A New class of automated theorem-proving algorithms, JACM, 21, 191-200 (1974).
6. R. A. Overbeek, An implementation of hyper-resolution, Comp. Maths. with Appls., 1, 201-214 (1975).
7. J. A. Robinson, Mechanizing higher-order logic, Mach. Intelligence 4, Amer. Elsevier Pub. Co., Inc., 151-170 (1969).
8. S. Winker and L. Wos, Automated Generation of Models and Computer Examples and its Application to Open Questions in Ternary Boolean Algebra, Proc. Eighth Int. Symposium on Multiple-Valued Logic, pp. 251-256. Rosemont, Illinois (1978); IEEE (1978)
9. S. Winker, L. Wos, and E. Lusk, Semigroups, involutions, and antiantomorphisms: a computer solution to an open question, in preparation.
10. S. Winker, Generation and verification of finite models and counterexamples using an automated theorem prover, answering two open questions, Proceedings of the Fourth Workshop on Automated Deduction, Austin, 1979.
11. L. Wos, S. Winker, and L. Henschen, Hyperparamodulation: a refinement of paramodulation, in preparation.
12. W. Wojcieckowski and A. Wojcik, Multiple-valued logic design by theorem proving, Proceedings of the Ninth International Symposium on Multiple-Valued Logic, Bathe, England, May 1979.

APPENDIX

Internal Format in PASCAL

```
{namear - array containing a name}
{
     A namear is just an array of characters containing a name.  The name
     will be the initial characters in the array and will be terminated
     by a blank .
}
     namear = array[1..maxlname] of char;



{typetable - the table defining types}
{
     The typetable defines the type trees.  Each type will be of the form

            (t2,t2,t3,...) -> tn

     where n can be 1 (for a basic type).  The argument types are given in the
     list chained off the table entry.  If the typecode in the table entry is
     equal to the subscript, it is a primitive type.  It is assumed that

        type code 1 is used for the primitive type "truthvalue" (TF)

        type code 2 is used for the unary functions from truthvalues into
          truthvalues (UBOOL = (TF) -> TF)

        type code 3 is used for binary functions of truthvalues into truthvalues
          (BBOOL = (TF,TF) -> TF)

        type code 4 is used for the first individual type (IND)

        type code 5 is used for (IND,TF) -> TF  (used for quantifiers)

        type code 6 is used for (IND,IND) -> TF (used for equality)
}
      typeent   = record
                    typecode: integer;  {the tn value}
                    nexttype: ^typeent  {pointer to list of t1,t2,...}
                  end;
      typetable = array[1..maxtype] of typeent;


{attrnode - an attribute of a term or occurrence of a term}
{
      An attribute node can be chained off a term node
      It contains a link to another attrnode (in case
      there are more than 1 attribute for the given term node)
      an attrtype (which is a subscript into the attribute symbol table)
      and a binary tree of integers.
}
      attrnode = record
                   nextattr: ^attrnode;  {link to next attribute}
                   attrtype: integer;    {subscript into the AST}
                   attrvals: ^listnode    {pointer to value tree}
                 end;
      attrptr = ^attrnode;
```

```
{listntype - types of list nodes}
{
        A listnode is either a leaf or a nonleaf
}
        listntype = (leaf,nonleaf)


{listnode - node in binary tree of integers}

        listnode = record
                     case ltype:listtntype of
                       leaf: (leafval:integer);
                       nonleaf: (lchild:  listnode
                                 rchild:  listnode);
                     end;


{terms - types of term nodes in internal format}
{
      A term node can represent a variable, a function, or the
      application of a function to a sequence of arguments.
}
      terms = (variable, funct, application);



{term - internal format of term}

{
      In the internal format all logical operators, quantifiers, and
      lambda expressions are treated as functions.
      A term in the internal format will be referred to as "integrated"
      iff it has been integrated into the central data structure.  An
      integrated term will be in a structure in which there is only one
      copy of any term.  Hence, the OCCURSIN lists can meaningfully be filled
      in to allow access to all terms containing a given term.  In a
      nonintegrated term multiple copies of the same term can occur.
}

     term = record
              avllink:   ^term;      {used only for unallocated terms}
              termtype: integer;    {subscript into the typetable}
              termid:   integer;    {unique identifier}
              attrs:     ^attrnode; {header to attribute list}
              foccursin:^argnode;   {header to containment list}
              loccursin:^argnode;   {pointer to last containment node}

              case ttype: terms of

                variable: (varnum: integer);

                funct: (namef: namear);

                application (farg: ^argnode;
                             larg: ^argnode)  {header to argument list}

                  {the args pointer points at a list of argnodes.  the first
                   connects the application to the function, and the rest
                   connect it to the arguments
                  }
             end;

      termptr = ^term;
```

```
{argnode - relationship node connecting a containing and contained term}
{
      An argnode should be viewed as a relationship node implementing
      a bidirectional relationship between an application term and an
      argument (or the function being applied).
}
      argnode = record
                  narg:       ^argnode;  {for access to next arg from application}
                  parg:       ^argnode;  {for access to previous arg node}
                  noccursin:  ^argnode;  {for access of next containing term from
                                          the contained argument}
                  poccursin:  ^argnode;  {for access of previous containing term
                                          from the contained argument}
                  argof: ^term;         {pointer to application node}
                  argval: ^term;        {pointer to argument}
                end;

     argptr = ^argnode;
```

A NOTE ON RESOLUTION: HOW TO GET RID OF FACTORING WITHOUT LOOSING COMPLETENESS

Helga Noll

Keywords: theorem proving, resolution, factoring, first order logic

Abstract

It is often useful to simplify the resolution inference system by elimination of factoring. Factoring, however, cannot be ignored entirely without loosing completeness. This paper studies to what extent factoring is necessary to preserve it. We can show that it is sufficient to factorize only one of the two parent clauses of a resolvent. We apply this basic result to a class of well known refinements and describe for each rule which clause (the so called "selected parent") has at most to be factored.

We try to achieve the results in a transparent manner using only elementary notions of resolution theory, so that this note should be readable without detailed knowledge of resolution strategies.

1. Introduction

The most widely studied and best understood general method for automated theorem proving is due to the resolution principle introduced by J.A. Robinson [Rob65]. This resolution principle is a theory consisting of a system of first-order logic with just one inference rule, the resolution rule. Since 1965 a lot of research has been done in developing resolution proof procedures. In the beginning strong emphasis has particularly been laid upon those strategies which enforce a restriction of the search space and increase the efficiency of the procedures. The aim was to support the efforts in constructing "intelligent" computer programs. Later the interest in resolution theory shifted to the aspect that deductions are computations. The applicability of this fact was first demonstrated in [Gre69]. Especially it was found in [Kow74] that resolution proof procedures can be adapted to act like an interpreter for a programming language.

Resolution proof procedures are based on the resolution rule. This resolution rule can be regarded either as a single rule or as two rules: a factoring rule for generating instances of clauses and a binary resolution rule for resolving factors. The last representation is common now and can be found in textbooks, see, e.g. [CL73]. If the resolution rule is represented as a single rule - for reasons of convenience we will employ this approach - then factoring is contained implicitly.

Factoring causes not only redundancies in the search space, but also complications in dealing with resolution theory, so that restrictions are desirable. Refinements of factoring are well known, e.g. the m-factoring rule [Kow70], the SL-resolution [KK71], the $\hat{K}$-refinement [Hay75], and the α-factoring [Nol76]. The reduction of factoring introduced by these rules are of different type depending on the intended applications.

It is often useful to simplify the resolution inference system drastically by elimination of factoring. For example, the above mentioned procedural interpretation of resolution logic uses binary resolution only. It is well known that factoring cannot be ignored entirely without loosing completeness. Therefore we study in this paper to what extend factoring is necessary to preserve it. To analyse the situation, we introduce as a tool a refinement of the resolution rule, the "resolution rule with half-factoring $\tilde{R}$" defined by a restriction of the use of factoring. We will show that $\tilde{R}$ is complete and that the combination of $\tilde{R}$ with some known simple refinements of the resolution rule yields statements about to what extent factoring is necessary for completeness.

The results obtained are of following type (see corollary 2):

- If the set of clauses to be proved contains its factors and if we use linear resolution then at most the far parents have to be factored.
- If the resolution rule with respect to a model is used to prove a set of clauses, then at most the semantically restricted clauses have to be factored; especially for the p_1-resolution this means that at most the positive clauses have to be factored.

Corollaries 3 and 4 of this paper are just theorems 1 and 2 in [Hen74]: Input resolution and unit resolution without factoring are complete inference rules for sets of Horn clauses.

The construction of the resolution rule with half-factoring and the combination of this rule with refinements is quite straightforward, and we believe that it can be very easily applied as well to refinements other than those used in our examples.

In the following we will assume that the reader is familiar with the basic concepts of resolution theory [Rob65], [CL73]. A full and elementary description of the terminology and preliminaries used here is given in [BN77].

2. Elimination of Factoring

Let $C \in R(C_1,C_2)$ *) be a resolvent of the parent clauses C_1 and C_2, $N_i \subset C_i$ $(i=1,2)$ the sets of literals resolved upon and - without loss of generality - let C_2 be factored, i.e., $N_2 = \{L_o,\ldots, L_{m-1}\}$ with $m \geq 2$. It is quite straightforward to get rid of factoring in C_2: instead of resolving upon the literals in N_2 simultaneously we take one after the other. In other words, we transform the resolvent $C \in R(C_1,C_2)$ to a derivation D^C:

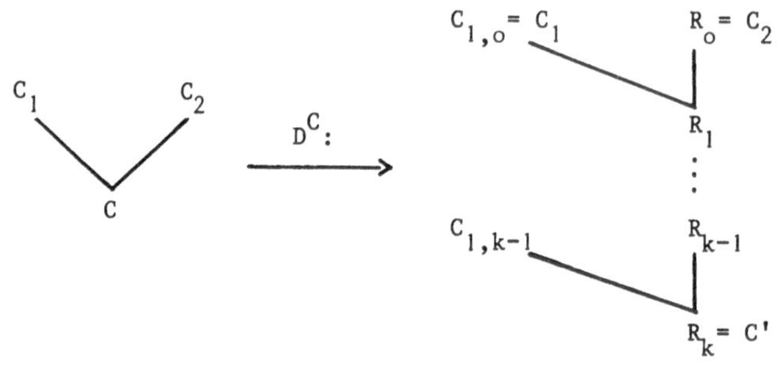

Here, the clauses $C_{1,o},\ldots,C_{1,k-1}$ are alphabetic variants of C_1, which are named in such a way that D^C is standardized and $R_n \in R(R_{n-1},C_{1,n-1})$ $(n=1,\ldots,k$ and $1 \leq k \leq m)$ are the resolutions that we execute with just one literal resolved upon from $R_o = C_2,\ldots, R_{k-1}$, respectively. The following example illustrates the procedure (the literals resolved upon are underlined):

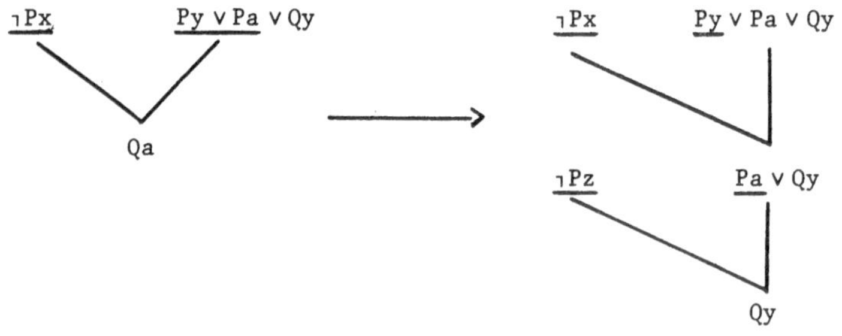

The construction produces a derivation from an arbitrary resolvent such that every resolvent in the derivation has at most one factored parent. This property defines a refinement of the resolution rule R, that we call *resolution rule with half-factoring* $\tilde{R}$. In more formal notation we have: $C \in \tilde{R}(C_1,C_2)$ iff $C \in R(C_1,C_2)$ and C_i is not factored for one $i \in \{1,2\}$. We are now able to state the following

---

*) If ρ is the resolution rule R or a refinement of R we write $\rho(C_1,C_2)$ for the set of conclusions of C_1,C_2 with respect to ρ.

Basic Lemma. If C is a resolvent of the clauses C_1 and C_2, then there exists a derivation D^C of a clause C' from the set of clauses $\{C_1,C_2\}$ with respect to the resolution rule with half-factoring $\tilde{R}$ and C is an instance of C'.

Proof: The proof although based only on elementary set theoretic manipulations is cumbersome to carry out; it has been placed in Appendix A.

As a generalization of the above basic lemma we obtain

Theorem 1. For every derivation D of a clause C from a set of clauses S with respect to the resolution rule R there exists a derivation D' of a clause C' from S with respect to the resolution rule with half-factoring $\tilde{R}$, and C is an instance of C'.

Proof: We assume that derivations can be represented as binary, rooted trees in the usual way. The proof is by induction on the level $\ell(D)$ of D with the basic step: D'= D if $\ell(D)$ = 0. We give a diagrammatic representation to sketch the induction step:

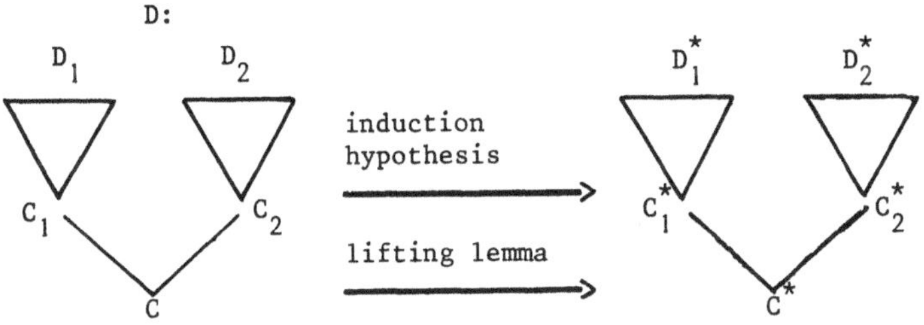

- D_i^* are derivations of C_i^* from S with respect to $\tilde{R}$ and $C_i^*\beta_i^* = C_i$ (induction hypothesis).
- $C^*\alpha = C$ (lifting lemma).

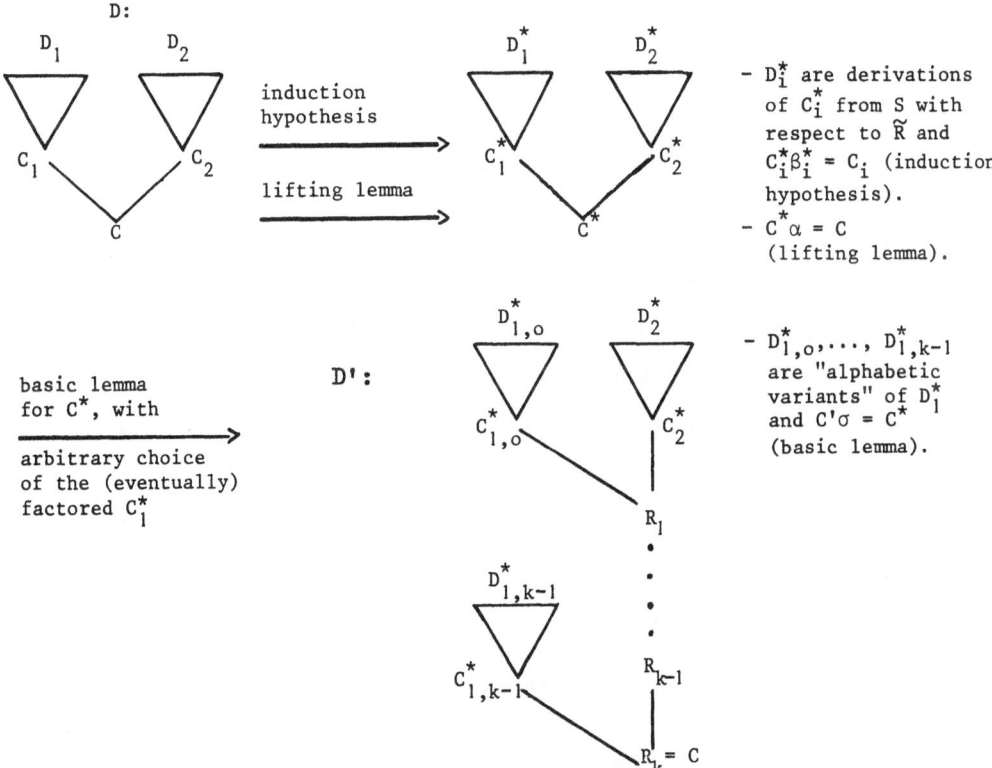

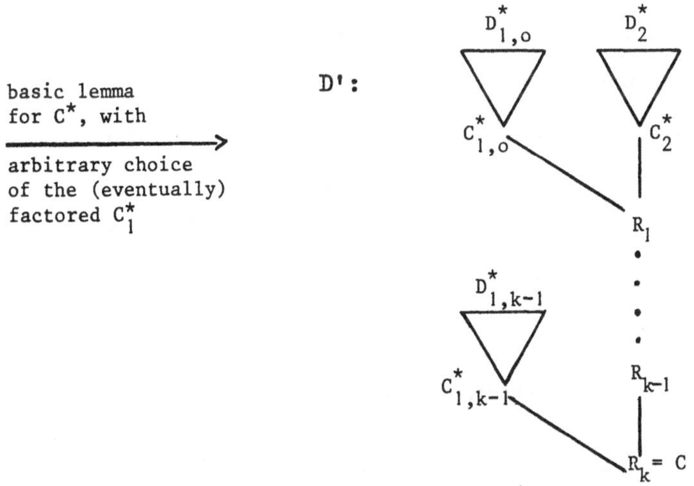

- $D_{1,o}^*, \ldots, D_{1,k-1}^*$ are "alphabetic variants" of D_1^* and $C'\sigma = C^*$ (basic lemma).

Taking into account the completeness of the resolution rule R we obtain from theorem 1:

> Corollary 1. (Completeness of $\tilde{R}$) If S is an unsatisfiable set of clauses, then there exists a refutation of S with respect to the resolution rule with half-factoring $\tilde{R}$.

Remark: Theorem 1 was suggested by chapter 4.6 [Kow70].

3. Selected-Parent-Refinements and Elimination of Factoring

We will introduce some well known refinements in order to get a sharpening in our statements about elimination of factoring.

In general, a refinement R_v of the resolution rule R is defined by means of a property P_v of the resolvent. Very often, however, this property is the property of just one parent of the resolvent, i.e., we have: $C \in R_v(C_1,C_2)$ iff $C \in R(C_1,C_2)$ and $P_v(C_i)$ for one $i \in \{1,2\}$. We call such a refinement a *selected-parent-refinement* and the parent C_i with $P_v(C_i)$ the *selected parent*. We will now prove results concerning the following well known selected-parent-refinements: the resolution with respect to a model R_M, the p_1-resolution R_p, the linear resolution (or ancestry filter) R_L, the input resolution R_I, the unit resolution R_U, and the positive unit resolution, R_{Up}. The selected parents C_i have the following properties:

- R_M : C_i is not satisfied in M.
- R_p : C_i is positive, i.e., all literals in C_i are positive.
- R_L : C_i is an input clause (the input parent of C) or C_i is an ancestor of C_j, $j \neq i$ (the far parent of C).
- R_I : C_i is an input clause.
- R_U : C_i is an unit, i.e., C_i contains one literal.
- R_{Up}: C_i is a positive unit, i.e., C_i contains one positive literal.

Remark: 1. In [Luk70] R_M, R_L have been denoted by R_1, R_3, respectively, and R_p is the p_1-resolution [Rob65a]. The latter one is a R_M-rule with a special M and R_I is a refinement of R_L. R_M, R_p, R_L are complete [Luk70], whereas R_I, R_U, R_{Up} are in general not complete.

2. Please notice that the term "ancestor" in property P_L must be regarded with respect to our representation of derivations as binary trees.

The construction in the basic lemma suggests to combine the resolution rule with half-factoring $\tilde{R}$ and a selected-parent-refinement R_v in such a way that at most the selected parent is factored. This combination forms a new refinement that we call *selected-parent-refinement with half-factoring* $\tilde{R}_v$. We have: $C \in \tilde{R}_v(C_1,C_2)$ iff $C \in R_v(C_1,C_2)$ and at most the selected parent C_i with $P_v(C_i)$ is factored. For the selected-parent-refinements stated above we can describe the restriction on factoring as follows: at most the

$\tilde{R}_M$: semantically restricted parent

$\tilde{R}_p$: positive parent

$\tilde{R}_L$: input parent or else far parent

$\tilde{R}_I$: input parent

is factored and with $\tilde{R}_U$, $\tilde{R}_{Up}$ no factoring is applied at all.

Let D be a derivation with respect to R_v. We will form a derivation D' with the use of theorem 1. Applying the basic lemma we always choose the selected parent to be the (eventually) factored parent $C_{1,o},\ldots, C_{1,k-1}$, respectively. Thereafter D' clearly will be a derivation with respect to $\tilde{R}_v$, provided that R_v is liftable in the following sense:

1. R_v is liftable in the usual sense, i.e., the refinement condition P_v is preserved under the lifting process. In the case of resolvents this implies: if $C^* \in R(C_1^*,C_2^*)$ lifts $C \in R_v(C_1,C_2)$ then we have $C^* \in R_v(C_1^*,C_2^*)$, too.

2. The property P_v is preserved under the transformation of the tree, defined by the above construction of D' from D.

Hence we obtain from theorem 1:

> <u>Theorem 2.</u> If R_v is a selected-parent-refinement that is liftable, then we have: For every derivation D of a clause C from a set of clauses S with respect to the selected-parent-refinement R_v there exists a derivation D' of a clause C' from S with respect to selected-parent-refinement with half-factoring $\tilde{R}_v$, and C is an instance of C'. Especially $\tilde{R}_v$ is complete if R_v is complete.

For the selected-parent-refinements stated above the following is true: R_M,R_p,R_L,R_I are liftable but R_U,R_{Up} are in general not. For R_M this statement is valid because the substitution rule is correct and for R_L it is valid because two "alphabetic variants" of subderivations occurring in D are transformed in the same way to "alphabetic variants" of subderivations occurring in D'. It follows that theorem 2 is true for R_M,R_p,R_L,R_I and - together with the fact that R_M,R_p,R_L are complete - we obtain:

Corollary 2. (Completeness of $\tilde{R}_M$, $\tilde{R}_p$, $\tilde{R}_L$) If S is an unsatisfiable set of clauses then there exists a refutation of S with respect to R_M, R_p, and R_L, respectively, and at most the selected parents in this refutations are factored, i.e., the semantically restricted parents, the positive parents, the input parents or far parents, respectively.

We have already mentioned that the unit resolutions R_U and R_{Up} are not liftable and not complete. But we can save these properties by using an additional assumption. Suppose S is a set of Horn clauses, i.e., every clause in S contains at most one positive literal. Then we have:

Corollary 3. (Theorem 1 in [Hen74]) If S is an unsatisfiable set of Horn clauses, then there exists a refutation of S with respect to the unit resolution R_{Up} and no factoring is applied.

Proof: If the input clauses of a derivation are Horn clauses, then all clauses in that derivation are Horn clauses. Clearly for Horn clauses R_{Up} is liftable, so that theorem 2 is true for R_{Up} under this assumption. The completeness of R_{Up} for sets of Horn clauses follows from the completeness of the p_1-resolution R_p. Employing theorem 2 we get the completeness of $\tilde{R}_{Up}$.

An analogous statement can be made about the input resolution R_I:

Corollary 4. (Theorem 2 in [Hen74]) If S is an unsatisfiable set of Horn clauses, then there exists a refutation of S with respect to the input resolution R_I and no factoring is applied.

Proof: Since R_I is liftable, theorem 2 is true for R_I, i.e., we have already shown: If there is a refutation of S with respect to R_I, then there exists a refutation of S with respect to R_I such that at most the input parents are factored. It remains to show the lemma [Hen74]: If S is an unsatisfiable set of Horn clauses, then there exists a refutation of S with respect to R_I and the input parents are not factored.

Proof: Apply the set-of-support-rule choosing as set of support the subset K of S consisting of the negative clauses (i.e., those clauses that contain only negative literals). Then all supported clauses are negative so that the positive literal in any resolution must come from a clause in S-K. These clauses are clearly not factored, because S is a set of Horn clauses.

Finally we want to consider $\tilde{R}$ as a selected-parent-refinement. Then $\approx$R is defined, which is just the binary resolution, i.e., resolution without any factoring. We cannot apply theorem 2 to $\tilde{R}$ because $\tilde{R}$ is not liftable. The resolvent $\Box \in R(\underline{Px \vee Py}, \underline{\neg Pu \vee \neg Pv})$ may serve as an example to demonstrate this fact (see example (a) in appendix A). At the same time it is an example of the (well known) fact, that the binary resolution is not complete.

4. Concluding Remarks

The corollaries of this paper are achieved in a comparatively transparent and easy manner, using only elementary notions of resolution theory. Theorem 2 states the common basis of the results and is received from theorem 1. Simplicity in deriving these theorems is obtained by shifting the complexity of the whole proof into the Basic Lemma. This complexity seems to be intrinsic in problems dealing with unification or - more precisely - with substitutions and instantiations in general. We have used the method of isolating complicated parts of a proof in a basic proposition which is independent of an application in a specific refinement. We think it is a simple and efficient, but not generally utilized method. It avoids to spread the proof complexity into all proofs of the specific results, proving the basic facts each time again, as it is done in some papers of that subject. Furthermore, such a method helps to remove not mastered complexity and illuminates the problem structure, thus allowing to write more readable papers.

As an additional example it should be mentioned that the "instantiation" of derivations is a problem of a complexity similar to that treated in this paper. This problem is strictly related to factoring and lifting, and its intricacy has been involved in several early theorem proving papers without being isolated from specific applications. Moreover, this example may show that we do not emphasize that elimination of factoring is always favourable. For example, it is possible to avoid the complications in the process of "instantiation" of derivations by permitting only such derivations which have an isomorphic counterpart by ground resolution (or - equivalently - which are lifting a ground derivation). This restriction can be obtained by a refinement such as the $\hat{K}$-refinement [Hay75] or the α-factoring [Nol76] that take care of factoring of a "sufficient" number of literals.

The elimination of the factoring rule is formulated in a refinement. Usually, refinements are intended to improve the efficiency of proof procedures based on that rule. But we have treated refinements just as subjects of research in resolution theory - as a tool useful for simplifying the inference system in accordance with the special application in mind. Investigations concerning the efficiency of resulting proof procedures are beyond the scope of this note.

Appendix A Proof of the Basic Lemma

Let $C \in R(C_1,C_2)$ be a resolvent as in the beginning of section 2, and m,k,D^C as described there. We have

$$C = (C_1 - N_1)\Theta \cup (C_2 - N_2)\Theta$$

where Θ is a most general unifier (m.g.u.) of $N_1 \cup \sim N_2$, $\sim N_2$ the set of complements of the literals in N_2. First we will represent the

Construction of the derivation D^C:

(1) A derivation is standardized, if two input clauses at a time contain no common variables. Let $C_{1,i} = d_{1,i} \dot{\cup} N_{1,i}$ $(i \geq 0)$ be alphabetic variants of C_1, where the variables are named in such a way that D^C will be standardized and for a suitable renaming μ holds $C_{1,i}\mu = C_{1,o} = C_1$, $N_{1,i}\mu = N_{1,o} = N_1$ and $(C_2 - N_2)\mu = C_2 - N_2$.

(2) From the decomposition lemma (see Appendix B) we obtain m.g.u.'s for the intermediate resolution steps of the derivation D^C:

let be $\quad X = \bigcup_{i=1}^{m} (N_{1,i-1} \cup \sim N_2)$,

and $\quad X_i = N_{1,i-1} \cup \sim L_{i-1}$ *)

where $\quad L_o,\dots, L_{n-1} \in N_2 \qquad (1 \leq i \leq n,\ 1 \leq n \leq m)$.

The X_i are the literals resolved upon in the i.th step. X is unifiable with m.g.u. $\mu\Theta$ and hence there exists for $(1 \leq i \leq n)$ m.g.u.'s Θi of $X_i\Theta_o\dots\Theta_{i-1}$ and a substitution λ_n with

$$\Theta_1\dots\Theta_n\lambda_n = \mu\Theta .$$

Because of standardization we get:

$$X_i\Theta_o\dots\Theta_{i-1} = N_{1,i-1} \cup \sim L_{i-1}\ \Theta_o\dots\Theta_{i-1} .$$

(3) Consider the following construction:

step 1 Set $R_o = C_2$, $E_o = N_2$, n=0 and take $L_o \in N_2$ arbitrary.

step 2 Set $n = n+1$ and the resolvent in the n.th step

$$R_n = (R_{n-1} - L_{n-1}\ \Theta_o\dots\Theta_{n-1})\Theta_n \cup (C_{1,n-1} - N_{1,n-1})\Theta_n,$$

where Θ_n is a m.g.u. of $N_{1,n-1} \cup \sim L_{n-1}\ \Theta_o\dots\Theta_{n-1}$ whose existence is ensured by (2) above.

Set the set of literals remaining for resolving after n steps

$$E_n = (E_{n-1} - L_{n-1}\ \Theta_o\dots\Theta_{n-1})\Theta_n.$$

*) Considering sets of literals $\{l_1,\dots,l_s\}$ we will omit the brackets and write $l_1,\dots,l_s$.

If $E_n = \emptyset$ then go to step 3; otherwise take

$$L_n\,\Theta_1 \ldots \Theta_n \in E_n, \text{ where } L_n \in (N_2 - L_o, \ldots, L_{n-1}).$$

If $L_n\Theta_1 \ldots \Theta_n \notin (C_2 - N_2)\Theta_1 \ldots \Theta_n$ then go to step 2; otherwise go to step 3.

(Remark: $E_n \subset (N_2 - L_o, \ldots, L_{n-1})\Theta_1 \ldots \Theta_n$ and

$$L_{n-1}\Theta_1 \ldots \Theta_{n-1} \in E_{n-1} \subset R_{n-1}).$$

<u>step</u> 3 stop; the number of resolutions in D^C is $k = n$ and we have $E_k = \emptyset$ or $E_k \subset (C_2 - N_2)\Theta_1 \ldots \Theta_k$.

We want to illustrate the construction with three examples, the last two with the case $k < m$:

<u>(a) m=k=2</u>

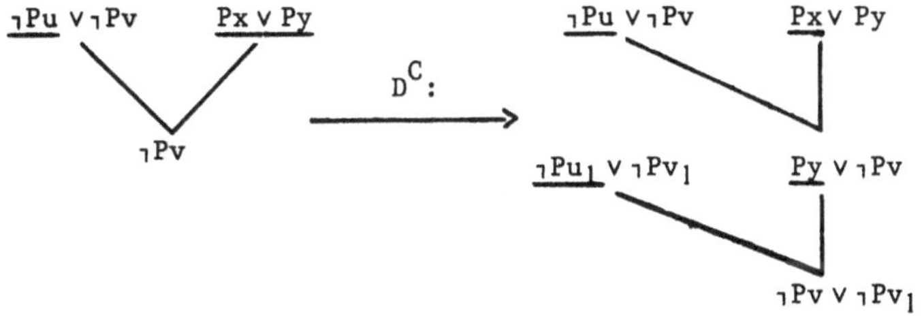

<u>(b) m=3, k=2</u>

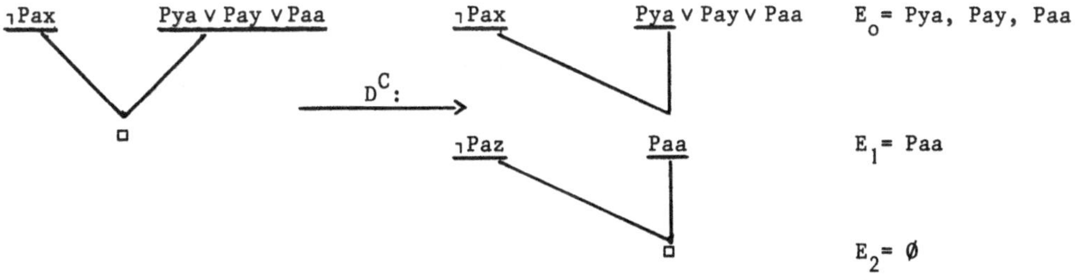

<u>(c) m=2, k=1</u>

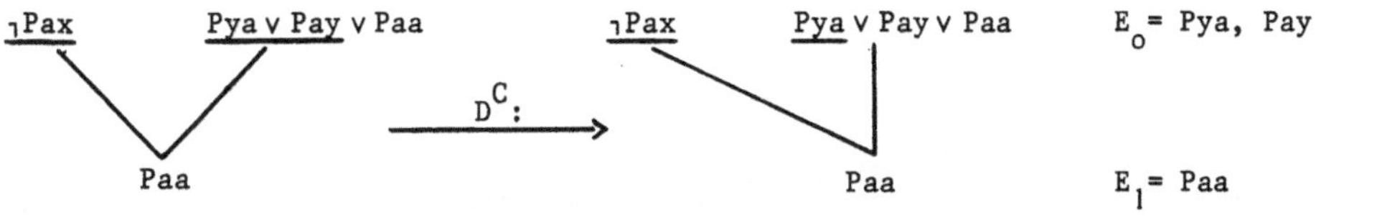

Remarks: 1. Our example in section 2 and example (a) show that C is not the root R_k of D^C but an instance of R_k. In the first example the reason for this fact is expressed in the decomposition lemma. In example (a) it is the case because R_k contains additional literals that are coming from the variants of C_1. Moreover, this example may be used to show that theorem 2 cannot be applied to $\tilde{R}$, as mentioned at the end of section 2.*)

2. Examples (b) and (c) illustrate two different reasons for $k \le m$. Firstly, literals from N_2 may collapse in the process of stepwise instantiation and are therefore lost for resolving (see example (b)). Secondly, there may exist literals in N_2 which match with literals from $(C_2 - N_2)$ in the process of stepwise instantiation (see example (c)). In our construction we do not use such literals for resolving (more precisely: we do not use literals $L_n \in (N_2 - L_o, \ldots, L_{n-1})$ with $L_n\Theta_1 \ldots \Theta_n \in (C_2 - N_2)\Theta_1 \ldots \Theta_n$, see step 2 in construction (3)), otherwise we only could show that C is subsumed by R_k.

3. D^C is not unique, for example take Pay $\in N_2$ as the first literal resolved upon in the examples (b) and (c).

This closes the construction of D^C. We prove the basic lemma by showing that C is an instance of the root R_k of the derivation. This is an immediate consequence of the following

Main Lemma $\quad R_n\lambda_n = C \cup E_n\lambda_n \quad (1 \le n \le k)$ and

$E_k\lambda_k \subset C$, where λ_n is a substitution constructed in (2)

Corollary $\quad R_k\lambda_k = C$

For the proof of the main lemma we state three lemmata $(1 \le n \le k)$:

Lemma 1 (representation)

(a) $\quad E_n = (\ldots ((N_2 - L_o)\Theta_1 - L_1\Theta_1)\Theta_2 \ldots - L_{n-1}\Theta_o \ldots \Theta_{n-1})\Theta_n$

Let $R_n = R_n^1 \cup R_n^2$, where R_n^1, R_n^2 are the literals in R_n that are coming from the variants of C_1 or from C_2. Then we have:

(b) $\quad R_n^2 = (\ldots ((C_2 - L_o)\Theta_1 - L_1\Theta_2)\Theta_2 \ldots - L_{n-1}\Theta_o \ldots \Theta_{n-1})\Theta_n$

*) Starting with □ ∈ R(Px v Py, ¬Pu v ¬Pv) we first apply the basic lemma and after that theorem 2, using the example (a) above. The application of the lifting lemma in our construction produces from □ ∈ $\tilde{R}$(Px_1 v Py_1, ¬Pv) a resolvent □ ∈ R(Px_1 v Py_1, ¬Pv v ¬Pv_1), that is no longer a resolvent with half-factoring.

(c) $R_n^{1} = (C_{1,n-1} - N_{1,n-1})\Theta_n \cup R_n^{11}$ with $R_1^{11} = \emptyset$ and

$$R_n^{11} \subset \bigcup_{i=o}^{n-2} (C_{1,i} - N_{1,i})\Theta_{i+1}\ldots\Theta_n \qquad (n \geq 2)$$

Proof: The proof is by induction on n with the definitions for E_n, R_n from the above construction (3). (a), (b) are obvious. To prove (c) execute the basis step and consider

$$R_n^{11} = (R_{n-1}^{1} - L_{n-1}\Theta_o\ldots\Theta_{n-1})\Theta_n \subset R_{n-1}^{1}\Theta_n ,$$

which is obtained from the definition of R_n and $R_{n-1} = R_{n-1}^{1} \cup R_{n-1}^{2}$. With the induction hypothesis follows (c).

Lemma 2 $\quad R_n^{2} = F_n \cup E_n$ with $F_1 = (C_2 - N_2)\Theta_1$ and

$$F_n = (\ldots ((C_2 - N_2)\Theta_1 - L_1\Theta_1)\Theta_2 \ldots - L_{n-1}\Theta_o\ldots\Theta_{n-1})\Theta_n \qquad (n \geq 2)$$

Proof: With $C_2 - L_o = (C_2 - N_2) \cup (N_2 - L_o)$ from lemma 1, (a) and (b).

Lemma 3 (instantiation)

(a) $R_n^{1}\lambda_n = (C_1 - N_1)\Theta$

(b) $F_n\lambda_n = (C_2 - N_2)\Theta$

(c) $E_k\lambda_k \subset (C_2 - N_2)\Theta$

Proof: From our construction we obtain:

1. $(C_{1,n-1} - N_{1,n-1})\Theta_n = (C_{1,n-1} - N_{1,n-1})\Theta_1\ldots\Theta_n$ [standardization of D^C].
2. $(C_{1,n-1} - N_{1,n-1})\mu = C_1 - N_1$ [construction (1)]
   $(C_2 - N_2)\mu = C_2 - N_2$
3. $\Theta_1\ldots\Theta_n\lambda_n = \mu\Theta$ [construction (2)]
4. $L_i\Theta_1\ldots\Theta_i \notin (C_2 - N_2)\Theta_1\ldots\Theta_i$ $(1 \leq i \leq k-1)$ [construction (3)]
5. $E_k \subset (C_2 - N_2)\Theta_1\ldots\Theta_k$ or $E_k = \emptyset$ [construction (3)]

(a) follows from lemma 1, (c) with 1., 2. and 3.. From the definition in lemma 2 and 4. we obtain $F_n = (C_2 - N_2)\Theta_1\ldots\Theta_n$ and with 2., 3. follows (b). The last point (c) we get from 5 with 3., 2..

Now the proof of the main lemma is easy:

$$
\begin{aligned}
R_n\lambda_n &= R^1_n\lambda_n \cup R^2_n\lambda_n \\
&= R^1_n\lambda_n \cup F_n\lambda_n \cup E_n\lambda_n && \text{[lemma 2]} \\
&= (C_1 - N_1)\Theta \cup (C_2 - N_2)\Theta \cup E_n\lambda_n && \text{[lemma 3, (a), (b)]} \\
&= C \cup E_n\lambda_n
\end{aligned}
$$

Lemma 3, (c) completes the proof.

Appendix B In our proof of the basic lemma we have used the following version of the decomposition lemma [Kow70]:

Decomposition Lemma Let X be a set of literals, that is unifiable and $X_i \subset X$ $(1 \le i \le n)$. Then there exist m.g.u.'s

Θ_i of $X_i\Theta_o \ldots \Theta_{i-1}$ (Θ_o the identity)

ρ_n of $X\Theta_1 \ldots \Theta_n$

and - for arbitrary choice of this m.g.u.'s - $\Theta_1 \ldots \Theta_n \rho_n$ is a m.g.u. of X.

Supplement If Θ is a m.g.u. of X then $\Theta = \Theta_1 \ldots \Theta_n \lambda_n$ for a substitution λ_n.

Proof: The proof is omitted, see [Kow70] or [BN77].

Acknowledgment

I wish to thank Larry J. Henschen for bringing this problem to my attention. Thanks are also due for helpful discussions with my former colleague Eberhard Bergmann.

References

[BN77] E. Bergmann und H. Noll: Mathematische Logik mit Informatik-Anwendungen. Berlin, Heidelberg, New York: Springer 1977

[CL73] C.L. Chang and R.C.T. Lee: Symbolic Logic and Mechanical Theorem Proving. New York, London: Academic Press 1973

[Gre69] C. Green: Applications of Theorem Proving to Problem Solving. Proceedings International Joint Conference.Artificial Intelligence, 219-239 (1969)

[Hay75] P.J. Hayes: An Abstract Lifting Theorem. University of Essex. Memo CSM-11, 1975

[Hen74] L.J. Henschen: Unit Refutations and Horn Sets. J. ACM 21, 590-605 (1974)

[Kow70] R. Kowalski: Studies in Completeness and Efficiency of Theorem Proving by Resolution. Ph.D. Thesis, University of Edingburgh, 1970

[Kow74] R. Kowalski: Predicate Logic as Programming Language. Proceedings of the IFIP-Congress, Stockholm, 569-574 (1974)

[KK71] R. Kowalski and D.G. Kühner: Linear Resolution with Selection Function. Artificial Intelligence 2, 221-260 (1971)

[Luk70] D. Luckham: Refinement Theorems in Resolution Theory. Proceedings IRIA Symposium on Automatic Demonstration. Berlin, Heidelberg, New York: Springer-Verlag, 163-190 (1970)

[Nol76] H. Noll: Eine konstruktive Formulierung des Lifting-Theorems unter expliziter Angabe der liftenden Substitutionen und eine Anwendung auf Faktorisierungs-Einschränkungen. Dissertation, Technische Universität Berlin, Fachbereich 20 - Informatik, 1976
This thesis contains a weaker version of the results presented in this paper.

[Rob65] J.A. Robinson: A Machine-Oriented Logic Based on the Resolution Principle. J. ACM 12, 23-41 (1965)

Abstraction Mappings in Mechanical Theorem Proving

D. A. Plaisted
Department of Computer Science
University of Illinois at Urbana-Champaign
Urbana, Illinois 61801

1. INTRODUCTION

A good problem solving technique is to convert a problem to a simpler problem, solve the simpler problem, and use the solutions of the simpler problem to guide the search for a solution to the original problem. This approach and the related use of analogy have been investigated to some extent [2, 4, 8]. We apply this technique to resolution theorem proving in the first-order predicate calculus. We define a class of "abstraction mappings" which map a set of clauses to a simpler set of clauses. We then present some strategies which use these mappings to guide the search for a proof from the original set of clauses. These strategies can be used to search for proofs of clauses other than NIL (the empty clause). The ideas presented are sufficiently general to apply to other inference rules and to higher order logics with minor modifications.

The basic idea of the strategies based on abstraction is to construct an outline of a proof and fill in the details later. This yields a global strategy. That is, each step of the search is controlled in a meaningful, non-trivial way by the structure of the problem as a whole. We doubt that local, problem-independent strategies will ever be sufficient by themselves to construct useful theorem proving programs.

We discuss the use of several levels of abstraction, the search at each level being guided by the next higher level. Also, we discuss the compatibility of abstraction with conventional strategies such as locking resolution [1] and P_1-deduction [7]. One abstraction strategy is presented which is compatible with equational theories. An advantage of abstraction is that it permits some of the benefits of set-of-support strategies to be realized in forward reasoning strategies such as P_1-deduction and locking resolution.

2. ORDINARY ABSTRACTIONS

Standard resolution theorem proving terminology will be assumed. In particular, we say a clause C1 <u>subsumes</u> a clause C2 if there is a substitution θ such that C1θ is a subset of C2. Also, clauses C and D are <u>variants</u> if they are instances of each other. That is, C and D are the same except for a renaming of variables.

<u>Definition</u>. An (ordinary) <u>abstraction</u> is an association of a set f(C) of clauses with each clause C such that f has the following properties:

1. If clause C3 is a resolvent of C1 and C2 and D3 ε f(C3) then there

exist D1 $\in$ f(C1) and D2 $\in$ f(C2) such that some resolvent of D1 and D2 subsumes D3.

2. f(NIL) = {NIL}. (NIL is the empty clause.)

3. If C1 subsumes C2, then for every abstraction D2 of C2 there is an abstraction D1 of C1 such that D1 subsumes D2.

If f is a mapping with these properties then we call f an _abstraction mapping_ of clauses. Also, if D $\in$ f(C) we call D an abstraction of C. Abstractions usually also satisfy the property that f(C) is a tautology if C is.

Definition. A _weak abstraction_ is an association of a set f(C) of clauses with each clause C such that f has the following properties:

1. If clause C3 is a resolvent of C1 and C2, and D3 $\in$ f(C3), then there exist D1 $\in$ f(C1) and D2 $\in$ f(C2) such that either D1 subsumes D3 or D2 subsumes D3 or some resolvent of D1 and D2 subsumes D3.

2,3. As in the definition of abstraction.

If f is such a function, we call f a _weak abstraction mapping_ of clauses. If clause D is in f(C), we call D a weak abstraction of C.

The following result gives us a fairly general method of constructing abstractions.

Theorem 2.1. Suppose F is a set of mappings from literals to literals. Suppose that for all $\phi \in$ F, for all literals L, $\phi(\overline{L}) = \overline{\phi(L)}$. If C is a clause, let ϕ(C) be $\{\phi(L):L \in C\}$, as usual. Suppose that if clause D is an instance of clause C, then for all $\phi 2 \in$ F there exists $\phi 1 \in$ F such that $\phi 2$(D) is an instance of $\phi 1$(C). Define f by f(C) = $\{\phi(C):\phi \in F\}$. Then f is an abstraction mapping.

Proof. Properties 2 and 3 of abstractions are easy to verify. We show that f satisfies property 1.

Suppose C3 is a resolvent of C1 and C2. Then there exist sets A1, A2 of literals such that A1 $\subset$ C1 and A2 $\subset$ C2 and there exist substitutions $\alpha 1$ and $\alpha 2$ such that C3 = (C1 - A1)$\alpha 1 \cup$ (C2 - A2)$\alpha 2$. Also, for some literal L, A1$\alpha 1$ = {L} and A2$\alpha 2$ = $\{\overline{L}\}$. We desire to show that for all $\phi 3 \in$ F, there exist $\phi 1 \in$ F and $\phi 2 \in$ F such that $\phi 3$(C3) is subsumed by a resolvent of $\phi 1$(C1) and $\phi 2$(C2).

Let $\phi 1$ and $\phi 2$ be such that $\phi 3$(C1$\alpha 1$) is an instance of $\phi 1$(C1) and $\phi 3$(C2$\alpha 2$) is an instance of $\phi 2$(C2). Such $\phi 1$ and $\phi 2$ must exist, by hypotheses concerning F.

Now, $\phi 3$(C3) = $\phi 3$((C1 - A1)$\alpha 1$) $\cup$ $\phi 3$((C2 - A2)$\alpha 2$). Also, $\phi 3$(A1$\alpha 1$) = $\{\phi 3(L)\}$ and $\phi 3$(A2$\alpha 2$) = $\{\phi 3(\overline{L})\} = \{\overline{\phi 3(L)}\}$. Hence $\phi 3$(C3) has a subset (possibly a proper subset) which is a resolvent of $\phi 3$(C1$\alpha 1$) and $\phi 3$(C2$\alpha 2$). Therefore, since $\phi 3$(C1$\alpha 1$) is an instance of $\phi 1$(C1) and $\phi 3$(C2$\alpha 2$) is an instance of $\phi 2$(C2), $\phi 3$(C3) is subsumed by some resolvent of $\phi 1$(C1) and $\phi 2$(C2). If f is an abstraction mapping as in the above theorem, then we say f is _defined in terms of literal mappings_. Not all abstractions are defined in terms of literal mappings.

2.1 *EXAMPLES OF ABSTRACTIONS*

Using these theorems, we can construct many abstractions. We now give some examples of abstractions, all of which can be obtained from the above theorem.

Examples of Syntactic Abstractions

1. The ground abstraction. If C is a clause, then f(C) = {C':C' is a ground instance of C}. Note that f(C) will usually be an infinite set of clauses.
2. The propositional abstraction. If C is the clause $\{L_1, L_2, \ldots, L_k\}$ then f(C) is {C'} where C' is the clause $\{L_1', L_2', \ldots, L_k'\}$ and L_i' is defined as follows, for $1 \leq i \leq k$:

   If L_i is of the form $P(t_1, \ldots, t_n)$ then L_i' is P. If L_i is of the form $\neg P(t_1, \ldots, t_n)$ then L_i' is $\neg P$.

   Thus f(C) is a clause in the propositional calculus.
3. Renaming predicate and function symbols. For clause C, f(C) = {C'} where C' is the clause in which all function and predicate symbols of C have been renamed in some systematic way. The renaming need not be one-to-one; two distinct predicate or function symbols may be renamed to the same symbol.
4. Changing signs of literals. Let Q be a set of predicate symbols. If C is the clause $\{L_1, \ldots, L_k\}$ then f(C) is {C'} where C' is the clause $\{L_1', \ldots, L_k'\}$ and L_i' is defined as follows, for $1 \leq i \leq k$:

   If L_i is of the form $P(t_1, \ldots, t_n)$ and $P \in Q$ then L_i' is $\neg P(t_1, \ldots, t_n)$.
   If L_i is of the form $\neg P(t_1, \ldots, t_n)$ and $P \in Q$ then L_i' is $P(t_1, \ldots, t_n)$.
   Otherwise, L_i' is L_i.
5. Permuting arguments. For clause C, f(C) = {C'} where C' is C with the order of the arguments of certain function or predicate symbols changed in some systematic way.
6. Deleting arguments. For clause C, f(C) = {C'} where C' is C with certain arguments of certain function or predicate symbols deleted. For example, $g(t_1, \ldots, t_n)$ may be replaced by $g(t_2, \ldots, t_n)$ everywhere. Note that the propositional abstraction is a special case of this (all arguments of all predicate symbols are deleted).

Example of a Semantic Abstraction

With each clause C, we associate a set f(C) of clauses as follows:

Let I be an interpretation of the set of clauses over some set of function and predicate symbols. Let $\mathcal{D}$ be the domain of the interpretation I. The interpretation I can treat equality as any other predicate symbol. That is, we may have $a_1 = a_2$ true in I even if a_1 and a_2 are distinct elements of $\mathcal{D}$.

With each ground literal of form $P(t_1, \ldots, t_n)$ we associate the literal $P(a_1, \ldots, a_n)$ where $a_i \in \mathcal{D}$ and a_i is the value of t_i in the interpretation I, for $1 \leq i \leq n$. With the literal $\overline{P}(t_1, \ldots, t_n)$ we associate $\overline{P}(a_1, \ldots, a_n)$.

With each ground clause $C = \{L_1, \ldots, L_k\}$ we associate $C' = \{L_1', \ldots, L_k'\}$ where L_i' is associated with L_i as indicated above. If C1 is an arbitrary clause then f(C1) = {D: D is associated with C for some ground instance C of C1}. We call f the <u>I-abstraction</u> or the <u>abstraction obtained from</u> I.

Example: If I is the usual interpretation of arithmetic then with the clause $\{\neg(x\leq y), \neg(y\leq z), x \leq z\}$ we associate the clauses $\{\neg(1\leq 2), \neg(2\leq 3), 1 \leq 3\}$, $\{\neg(1\leq 5), \neg(5\leq 2), 1 \leq 2\}$, $\{\neg(8\leq 7), \neg(7\leq 6), 8 \leq 6\}$ et cetera. The clauses in f(C) will be true in I if C is, but need not be in general.

Note that f(C) may contain infinitely many clauses if f is an I-abstraction as above and $\mathcal{D}$ is infinite. However, if $\mathcal{D}$ is finite, then f(C) will be finite for every clause C. Such abstractions appear to be particularly useful. In fact, such abstractions can be generated automatically by choosing $\mathcal{D}$ to be $\{1, 2, 3, \ldots, n\}$ for some small n and assigning to each function symbol in S an arbitrary function from $\mathcal{D}$ to $\mathcal{D}$. If abstraction f is defined in this way, then it is straightforward to compute f(C) given clause C.

Example of a Weak Abstraction

Suppose P is a predicate symbol. With the clause C we associate C' where C' is C with all literals containing P deleted.

2.2 *ALGEBRAIC PROPERTIES OF ABSTRACTIONS*

<u>Definition</u>. Suppose f_1 and f_2 are abstractions. The <u>composition</u> of f_1 and f_2, denoted f_2f_1, is defined by

$$f_2f_1(C) = \cup \{f_2(D) : D \in f_1(C)\}.$$

If f_1 or f_2 are weak abstractions, their composition is defined similarly.

<u>Definition</u>. The <u>identity</u> abstraction is the mapping f such that f(C) = {C} for all clauses C.

<u>Definition</u>. We say abstractions f_1 and f_2 are <u>inverses</u> if $f_1f_2 = f$ and $f_2f_1 = f$ where f is the identity abstraction. If an abstraction has an inverse, then it really hasn't thrown away any information about the set of clauses. For example, a 1-1 renaming of predicate symbols is an invertible abstraction. In general, a weak abstraction can throw away more information than can an ordinary abstraction.

<u>Theorem 2.2</u>. The composition of two abstractions is an abstraction. The composition of two weak abstractions is a weak abstraction. The composition of an ordinary abstraction and a weak abstraction is always a weak abstraction, but not necessarily an ordinary abstraction.

<u>Definition</u>. If f_1 and f_2 are abstractions, then their <u>union</u> f is defined by $f(C) = f_1(C) \cup f_2(C)$ for all clauses C.

<u>Theorem 2.3</u>. If f_1 and f_2 are abstractions, their union is also an abstraction. If f_1 and f_2 are weak abstractions, their union is also a weak abstraction. If f_1 is an abstraction and f_2 is a weak abstraction, their union is a weak abstraction but not necessarily an ordinary abstraction.

<u>Proof</u>. Easy.

It is not difficult to show that if f_1 and f_2 are abstractions defined in terms of literal mappings, then the union and composition of f_1 and f_2 are also defined in terms of literal mappings.

2.3 *ABSTRACTIONS OF RESOLUTION PROOFS*

We now show how abstractions can be used to guide the search for a proof of a clause C from a set S of clauses. First we show that if there is a proof of C from S, then there is an "abstracted proof" of something subsuming an abstraction of C, from abstractions of clauses in S. We then informally describe procedures which, given an abstracted proof, attempt to reconstruct the original proof.

If f is an abstraction mapping and S is a set of clauses, then we write f(S) to indicate $\cup\{f(C): C \in S\}$.

Theorem 2.4. Suppose S is a set of clauses and f is an abstraction mapping or a weak abstraction mapping for S. Suppose C' is a clause derivable from S by resolution and D' ε f(C'). Then there is a clause B' derivable from f(S) by resolution, such that B' subsumes D'.

Proof. By induction on the depth of the proof of C'. If C' ε S, the theorem is true since we can choose B' to be D'. Suppose C' is a resolvent of C1 and C2, where C1 and C2 can be derived from S by proofs of depth less than the depth of the proof of C'. Suppose D1 is a weak abstraction of C1 such that D1 subsumes D'. Applying the theorem inductively, there must be a cluase B1 derivable from f(S) by resolution, such that B1 subsumes D1. Hence B1 subsumes D'.

Suppose D1 and D2 are abstractions or weak abstractions of C1 and C2, respectively, such that some resolvent D of D1 and D2 subsumes D'. The clauses D1 and D2 must exist, if the preceding case does not apply. Applying the theorem inductively, there must exist clauses B1 and B2 derivable from f(S) such that B1 subsumes D1 and B2 subsumes D2. It follows by the properties of subsumption that either B1 subsumes D or B2 subsumes D or some resolvent B of B1 and B2 subsumes D. Hence either B1 subsumes D' or B2 subsumes D' or some resolvent B' of B1 and B2 subsumes D'. This completes the proof. Note that the derivation of B from f(S) will have depth not more than the depth of the derivation of C from S.

Since the depth of the proof of B' from f(S) is not more than the depth of the proof of C' from S, this result can be used to give a heuristic for estimating the difficulty of deriving a clause from a set of clauses. To do this, choose an abstraction mapping f and find the smallest depth at which some such clause B' can be derived from f(S). This is a lower bound on the depth of a derivation of C' from S. This use of abstractions as heuristics may be helpful for problem solving methods based on a decomposition of a problem into subproblems.

Corollary: If S is inconsistent so is f(S).

Proof. Take C' to be NIL. Then D' is also NIL, by properties of abstraction and weak abstraction mappings. Since B' subsumes D', B' must be NIL also. Since B' is derivable from f(S), f(S) is inconsistent.

This theorem can be used to show that S is consistent, but its main value for us is in the information that a proof in f(S) can give us about the structure of a possible proof in S. Here are some examples, using the propositional abstraction.

Example 1.

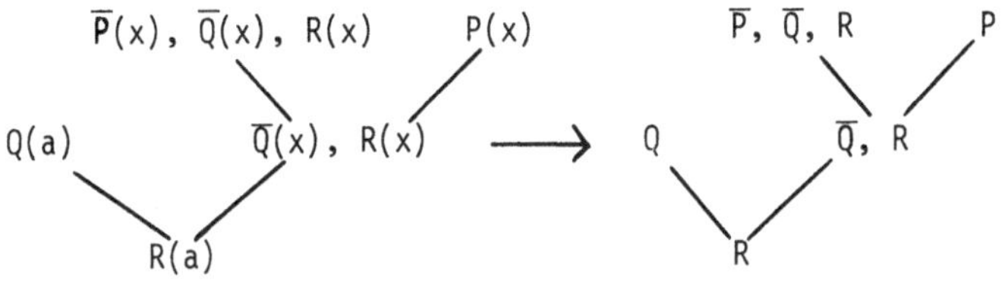

Example 2.

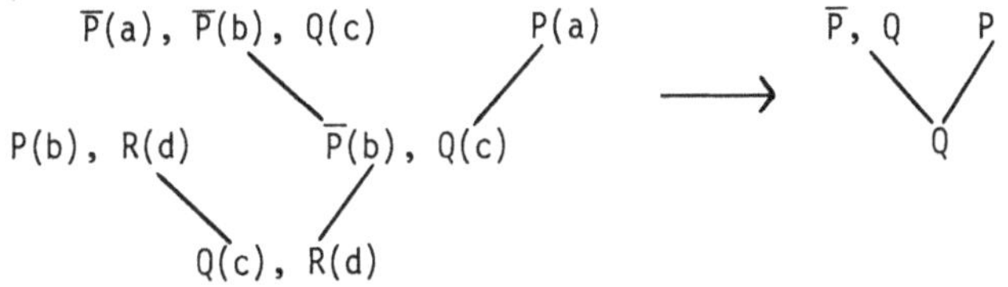

Note that we lost the literal $\overline{P}$ from {$\overline{P}$, Q} when resolving with P, even though the literal $\overline{P}$(b) remains in {$\overline{P}$(b), Q(c)}.

3. STRATEGIES BASED ON ABSTRACTION

We now show how abstractions can be used to guide the search for a proof. For simplicity, we only consider proof *trees*, i.e., resolution proofs in which no lemma is used more than once, so a lemma must be derived separately each time it is used.

3.1 *A CORRESPONDENCE BETWEEN PROOFS AND THEIR ABSTRACTIONS*

Definition. A *resolution proof tree* T is a finite binary tree in which each node is either a leaf or has a left son and a right son. Also, each node N of T is labeled with a clause CL(N) such that if N1 and N2 are the sons of N, then CL(N) is a resolvent of CL(N1) and CL(N2). We call T a proof *of C* if C is the label of the root of T. Also, we call T a proof *from S* if all leaves of T have labels in S.

Definition. Suppose T and U are proof trees and f is a weak abstraction. We write $T \underset{f}{\rightarrow} U$ if there is a mapping g from the nodes of U into the nodes of T having the following properties:

1) If N1 and N2 are brothers in U then g(N1) and g(N2) are brothers in T.
2) If N is the father of N1 and N2 in U and M is the father of g(N1) and g(N2) in T then M is a descendant of g(N). (Note that CL(M) is an ancestor of CL(g(N)) in T.) Possibly M = g(N).
3) If M is the root of T and N is the root of U then M = g(N).
4) For all nodes N of U, CL(N) subsumes an element of f(CL(g(N))). That

is, CL(N) subsumes a weak abstraction of the clause labeling g(N).

Theorem 3.1. Suppose T is a proof tree of C from S. Suppose f is a weak abstraction mapping, and D ε f(C). Then there is a proof tree U of something subsuming D from f(S), such that $T \underset{f}{\to} U$.

We may think of U as an abstracted proof which guides the search for T. If f is an m-abstraction or a bounded m-abstraction, then it can be shown that T exists with g mapping onto the nodes of T. Thus T and U have the same shape, and every resolution in T corresponds to a resolution in U. We consider instead weak abstractions and ordinary abstractions, in which a resolution in U may correspond to many resolutions in T. In some ways this is actually more useful because the search for T is broken up into smaller pieces, each piece being larger than a single resolution. However, m-abstractions and bounded m-abstractions have other advantages which we do not discuss here. We introduce m-abstractions in Section 4.

Suppose we are looking for a proof of a clause C from set S of clauses using weak abstraction f. We do this by choosing a weak abstraction D of C and generating proofs U of something subsuming D from f(S). Such proofs exist by Theorem 3.1. For each such U, we attempt to generate T such that $T \underset{f}{\to} U$. If C can be derived from S, then a proof tree T of C from S exists, so U as above exists and the method is complete. It is necessary to specify how to generate candidates for T, given U. We will give examples of this process below.

This approach can be generalized to provide more flexibility, in a way reminiscent of the abstractions of Sacerdoti [8].

Theorem 3.2. Suppose T is a proof tree of C from S1 and f is a weak abstraction mapping. Suppose D ε f(C). Suppose S2 is a set of clauses such that for all C1 ε S1, for all D1 ε f(C1), there exists D2 ε S2 such that D2 subsumes D1. Then there is a proof tree U from S2 such that $T \underset{f}{\to} U$ and such that U is a proof of something subsuming D.

The significance of this result is that there is a great deal of flexibility in choosing S2. For example, if f is the identity mapping, we can delete any subset of the literals of any of the clauses in S1. Also, we can choose clauses more general than those in S1.

The following result gives us more information about the relationship between clauses of T and clauses of U, if T is a P_1-deduction and f preserves signs of literals. In particular, positive clauses of U correspond to positive clauses of T.

Definition. A positive clause is a clause in which every literal is positive (non-negated).

Definition. A P_1-deduction is a resolution proof in which one of the parents of each clause is positive. The P_1-deduction strategy is known to be complete.

Theorem 3.3. Suppose T is a P_1-deduction of C from S and f is a weak abstraction mapping. Suppose D ε f(C). Also, suppose f preserves positive clauses. That is, if C1 is positive then so are all elements of f(C1). Then there is a proof U from f(S) such that $T \vec{f} U$ and such that U is a proof of something subsuming D. Also, if CL(N) is positive then so is CL(g(N)) for g as in the definition of $T \vec{f} U$.

Proof. Let N1 and N2 be brothers of U. Then g(N1) and g(N2) are brothers of T. Therefore exactly one of CL(g(N1)) and CL(g(N2)) will be positive. Suppose it is CL(g(N1)). Since f preserves positive clauses, CL(N1) is positive also and CL(N2) is not. Note that if S2 is some set of clauses such that for all C1 ε S1, for all D1 ε f(C1), there exists D2 ε S2 such that D2 subsumes D1, then a proof U from S2 exists as in the theorem.

3.2 *DELETION OF TAUTOLOGIES*

A disadvantage of m-abstractions and bounded m-abstractions is the necessity to include tautologies derived in the abstract search space. However, tautologies can be deleted if an appropriate combination of weak abstraction and P_1-deduction is used.

Theorem 3.4. Suppose T is a P_1-deduction of a clause C from S. Suppose f is a weak abstraction. Suppose D ε f(C). Then if proofs from f(S) are generated by P_1-deduction with tautology deletion, a proof U will be generated such that $T \vec{f} U$ and such that U is a proof of something subsuming D.

Proof. The idea is that if U1 is a proof from f(S) containing a tautology, then there is a proof U2 not containing a tautology such that $U1 \vec{i} U2$ where i is the identity abstraction. By a similar argument, we never need to generate proofs U from f(S) such that a clause in U subsumes one of its descendants in U.

A reasonable strategy, then, is to use P_1-deduction in both the original space and in the abstracted space, assuming the weak abstraction preserves positive clauses. Tautologies can be deleted in both spaces, but subsumed clauses can only be deleted in the original space (except as indicated above). Another possibility is to use locking resolution in the original space, with indices chosen so that negative literals resolve away first. This is a specialization of P_1-deduction. Therefore, P_1-deduction can be used in the abstracted space.

3.3 *DELETION OF NIL FROM ABSTRACTED CLAUSES*

If S is a set of clauses and f is a weak abstraction, frequently NIL will be an element of f(S) even if NIL $\not\in$ S. We now give conditions guaranteeing that NIL may be deleted from f(S) without affecting the completeness of the abstraction method.

Theorem 3.5. Suppose f is a weak abstraction which may be expressed as $f_1 f_2$, where f_2 deletes all literals containing certain predicate symbols and f_1 is an ordinary abstraction obtained from literal mappings. Suppose T is a proof tree of C from S, and for at least one of the input clauses B used in T, NIL $\not\in$ f(B). Then for every D ε f(C), there exists a proof tree U from f(S) such that $T \vec{f} U$ and U is a proof of something subsuming D and U does not consist entirely of the clause NIL.

The significance of this result is that if f is a weak abstraction satisfying the conditions of the theorem and some support set for S consists entirely of clauses B such that NIL $\notin$ f(B), then we need only consider abstracted proofs U which contain at least one resolution. That is, NIL can be deleted from f(S).

This property of weak abstractions can be used to reduce the search space when there are many input clauses in S. To do this, we choose f to be a weak abstraction which deletes all predicate symbols except a small set. This small set is chosen to include at least one predicate symbol occurring in each clause in some set-of-support for S. Typically the set-of-support will be some of the clauses derived from the particular theorem to be proved, and will be small. Most clauses in S will be mapped to NIL by f, and so need not be considered by above remarks. Thus the search space will be small. By using a sequence of abstractions which delete fewer and fewer predicate symbols, a proof from S can be found. The method of using a sequence of abstractions will be illustrated in the Tower of Hanoi example.

3.4 *A SEARCH STRATEGY*

We now indicate in more detail how abstracted proofs can be used to guide the search for proofs in the original space. Suppose S is a set of clauses, f is a weak abstraction, and U is a proof from f(S). We want to find proofs T from S such that $T \underset{f}{\rightarrow} U$. To do this, with each node N of U we keep two sets of clauses, the A-clauses and B-clauses, denoted by A-CL(N) and B-CL(N), respectively. Initially, A-CL(N) = {C ε S: CL(N) ε f(C)} for leaves N. (Recall that CL(N) is the label of node N.) Also, B-CL(N) = $\emptyset$ for leaves and non-leaves N initially, and A-CL(N) = $\emptyset$ for non-leaves N. If N is the father of N1 and N2, it is permissible to add to A-CL(N) resolvents C of elements of B-CL(N1) and B-CL(N2) such that CL(N) subsumes some element of f(C). If C ε A-CL(N) for some node N of U, then it is permissible to add to B-CL(N) any clause C1 satisfying the following conditions:

1) CL(N) subsumes some element of f(C1).
2) C1 can be derived from S $\cup$ {C} by resolution.
3) The clause C actually is used in some such proof of C1. (This is not really necessary for completeness.)

In particular, any clause in A-CL(N) can be added to B-CL(N). If f preserves positive clauses then we can also require that if CL(N) is positive and N is not the root of U, all elements of B-CL(N) are positive. This is particularly useful if P_1-deduction is done in both the original and the abstracted space.

The general strategy is the following: To find proofs of C from S using weak abstraction f, choose D ε f(C) and generate proofs U of something subsuming D from f(S). For each such proof U, attempt to generate T such that $T \underset{f}{\rightarrow} U$ using the above rules. If B-CL(N) ≠ ∅ for the root N of U, then such a proof T has been generated. Generate such proofs T until a proof of C (or something subsuming C) has been found, that is, C or something subsuming C is in B-CL(N).

In some cases, as the following examples show, it helps to generate the proofs T depth-first. That is, at each step we attempt to do resolutions to generate new elements of A-CL(N) or B-CL(N) for nodes N nearest to the root of U. Each step of U represents a subproblem which may require many resolutions in T. Note that many proofs U will typically exist, with many shared nodes between them. The strategy should make use of the fact that making progress towards obtaining a proof T1 such that $T1 \underset{f}{\rightarrow} U1$ may also contribute to obtaining a proof T2 such that $T2 \underset{f}{\rightarrow} U2$.

3.5 *EXAMPLES*

Cycle Example

We now consider an example which illustrates the use of weak abstractions. Suppose there are four locations P, Q, R, S arranged in a cycle. There are three ways (f, g, and h) to get from P to Q, from Q to R, from R to S, and from S to P. Also, there is one way (k) to go backwards. There is a barrier between P and Q which can only be crossed in the forward direction if condition A is true. The problem is to go around the cycle 3 times, starting at P. There is a choice of two states (a and b) to start in. The predicate P(x, y) means you are at location P, having gone around the cycle a number of times indicated by x, using a sequence of moves indicated by y. The predicates Q, R, and S are defined similarly. Translating this into a set of clauses, making minor modifications, we obtain the following set S1 of clauses:

A(x) ∧ P(x, y) ⊃ Q(x, fy)	Q(x, y) ⊃ P(x, ky)
A(x) ∧ P(x, y) ⊃ Q(x, gy)	
A(x) ∧ P(x, y) ⊃ Q(x, hy)	
Q(x, y) ⊃ R(x, fy)	R(x, y) ⊃ Q(x, ky)
Q(x, y) ⊃ R(x, gy)	
Q(x, y) ⊃ R(x, hy)	
R(x, y) ⊃ S(x, fy)	S(x, y) ⊃ R(x, ky)
R(x, y) ⊃ S(x, gy)	
R(x, y) ⊃ S(x, hy)	
S(x, y) ⊃ P(sx, fy)	P(sx, y) ⊃ S(x, ky)
S(x, y) ⊃ P(sx, gy)	
S(x, y) ⊃ P(sx, hy)	

P(a, a)	A(a)	A(sb)	A(ssa)
P(b, b)	A(b)		A(ssb)
$\overline{P}$(sssb, x)	$\overline{P}$(sssa, x)		

Let a_n be the number of positive unit clauses obtained from P(b, b) at depth n using P_1 deduction, counting the input clauses as depth 0. Then $a_0 = 1$, $a_1 = 0$, $a_2 = 3$, $a_3 = 12$, and $a_n = 3a_{n-1} + 3a_{n-2}$ for $4 \leq n \leq 13$. It follows that a_{13} is 7296561 and so there are this many positive unit clauses derived at depth 13. A proof of NIL exists at depth 14; in fact, there are 531441 such proofs at depth 14.

Let f be the weak abstraction which deletes the second argument of P, Q, R, and S and deletes all literals containing the predicate symbol A. Deleting NIL from f(S1), which is permissible, we obtain the following abstracted set of clauses:

$P(x) \supset Q(x)$ $\quad$ $Q(x) \supset P(x)$

$Q(x) \supset R(x)$ $\quad$ $R(x) \supset Q(x)$

$R(x) \supset S(x)$ $\quad$ $S(x) \supset R(x)$

$S(x) \supset P(sx)$ $\quad$ $P(sx) \supset S(x)$

$P(a)$ $\quad$ $\overline{P}(sssa)$

$P(b)$ $\quad$ $\overline{P}(sssb)$

Using P_1-deduction from this set of clauses, we easily obtain two proofs at depth 13, one starting from P(a) and one from P(b). Suppose we choose the proof from P(a) and attempt to obtain a proof from S1. There will be three subproblems to solve: derive A(a), A(sa), and A(ssa). The second subgoal cannot be achieved. Therefore we choose the proof from P(b). All subgoals can now be achieved, and a proof at depth 14 can be obtained using depth-first search. The subgoals are present in S1, but in general this may not be true. Deriving a subgoal may be a difficult problem, to be solved using other abstraction mappings. Also, if the subgoals share variables, then when one of them is solved it will restrict the possible solutions of other subgoals. This should be taken into consideration. Note the similarity of weak abstractions to abstractions considered by Sacerdoti [8], in which preconditions of actions are deleted.

Tower of Hanoi Example

This example illustrates the use of more than one level of abstraction. Let S be the following set of clauses, where D(x, y, z, w, s) means disk 1 is on peg x, disk 2 is on peg y, disk 3 is on peg z, and disk 4 (the largest disk) is on peg w, in situation s. Also, M(i, j, s) is the situation that results from moving disk i to peg j in situation s.

D(1,1,1,2,s)⊃D(1,1,1,3,M(4,3,s)) $\quad$ D(1,2,x,y,s)⊃D(1,3,x,y,M(2,3,s))

D(1,1,1,3,s)⊃D(1,1,1,2,M(4,2,s)) $\quad$ D(1,3,x,y,s)⊃D(1,2,x,y,M(2,2,s))

D(2,2,2,1,s)⊃D(2,2,2,3,M(4,3,s)) $\quad$ D(2,1,x,y,s)⊃D(2,3,x,y,M(2,3,s))

D(2,2,2,3,s)⊃D(2,2,2,1,M(4,1,s)) $\quad$ D(2,3,x,y,s)⊃D(2,1,x,y,M(2,1,s))

D(3,3,3,1,s)⊃D(3,3,3,2,M(4,2,s)) $\quad$ D(3,1,x,y,s)⊃D(3,2,x,y,M(2,2,s))

D(3,3,3,2,s)⊃D(3,3,3,1,M(4,1,s)) $\quad$ D(3,2,x,y,s)⊃D(3,1,x,y,M(2,1,s))

D(1,1,2,x,s)⊃D(1,1,3,x,M(3,3,s)) $\quad$ D(1,x,y,z,s)⊃D(2,x,y,z,M(1,2,s))

D(1,1,3,x,s)⊂D(1,1,2,x,M(3,2,s)) $\quad$ D(2,x,y,z,s)⊃D(1,x,y,z,M(1,1,s))

$D(2,2,1,x,s) \supset D(2,2,3,x,M(3,3,s))$

$D(2,2,3,x,s) \supset D(2,2,1,x,M(3,1,s))$

$D(3,3,1,x,s) \supset D(3,3,2,x,M(3,2,s))$

$D(3,3,2,x,s) \supset D(3,3,1,x,M(3,1,s))$

$D(1,1,1,1,so)$

$\overline{D}(2,2,2,2,s)$

$D(1,x,y,z,s) \supset D(3,x,y,z,M(1,3,s))$

$D(3,x,y,z,s) \supset D(1,x,y,z,M(1,1,s))$

$D(2,x,y,z,s) \supset D(3,x,y,z,M(1,3,s))$

$D(3,x,y,z,s) \supset D(2,x,y,z,M(1,2,s))$

Let abstraction f_4 delete all but the fourth argument of "D", let f_3 delete all but the third and fourth arguments, let f_2 delete all but the second, third, and fourth arguments, and let f_1 delete only the last (fifth) argument. Then $f_4(S)$ is the following set of clauses, after elimination of tautologies:

$D(2) \supset D(3)$	$D(1) \supset D(2)$
$D(3) \supset D(2)$	$D(2) \supset D(1)$
$D(1) \supset D(3)$	$D(1)$
$D(3) \supset D(1)$	$\overline{D}(2)$

From these clauses a refutation is easily obtained at depth 2 using P_1-deduction. Now, $f_3(S)$ is the following set of clauses, after deletion of tautologies:

$D(1,2) \supset D(1,3)$	$D(2,x) \supset D(3,x)$
$D(1,3) \supset D(1,2)$	$D(3,x) \supset D(2,x)$
$D(2,1) \supset D(2,3)$	$D(1,x) \supset D(3,x)$
$D(2,3) \supset D(2,1)$	$D(3,x) \supset D(1,x)$
$D(3,1) \supset D(3,2)$	$D(1,x) \supset D(2,x)$
$D(3,2) \supset D(3,1)$	$D(2,x) \supset D(1,x)$
$D(1,1)$	
$\overline{D}(2,2)$	

Notice that f_4 can be expressed as the composition $f \circ f_3$ where f deletes the first argument of D(x, y). Hence if T is a P_1-deduction of NIL from $f_3(S)$ then there is a P_1-deduction U of NIL from $f_4(S)$ such that $T \vec{f} U$. Therefore, the P_1-deduction found from $f_4(S)$ can be used to guide the search for a P_1-deduction T of NIL from $f_3(S)$. The clause $D(1) \supset D(2)$ used in the proof from $f_4(S)$ can only correspond to the clause $D(3, 1) \supset D(3, 2)$ in $f_3(S)$, hence we look for proofs involving this clause. Therefore, we have as a subgoal to derive the positive unit clause D(3, 1). Continuing in this way, the following proof from $f_3(S)$ is easily found:

1.	$D(1, 1)$	given
2.	$D(1, x) \supset D(3, x)$	given
3.	$D(3, 1)$	1, 2
4.	$D(3, 1) \supset D(3, 2)$	given
5.	$D(3, 2)$	3, 4
6.	$D(3, x) \supset D(2, x)$	given
7.	$D(2, 2)$	5, 6
8.	$\overline{D}(2, 2)$	given
9.	NIL	7, 8

In a similar way, a P1-deduction of NIL from $f_2(S)$ is found, then a P1-deduction of NIL from $f_1(S)$ is found, and finally a P1-deduction of NIL from S is found at depth 16. This again illustrates the use of abstraction in splitting a proof up into subgoals. Also, this illustrates the use of more than one level of abstraction. This proof can be found by a fairly mechanical application of the general abstraction strategy. Note how abstractions are closely related to the "differences" of GPS of Newell, Shaw, and Simon [5].

For comparison, suppose P_1-deduction and breadth first search were done directly from S. It is easy to see that in any situation three moves are possible, so there would be 3^{15} sequences of moves of length 15. Hence there would be 3^{15} or about 14,000,000 positive unit clauses at depth 15 and a refutation at depth 16. A better way to solve this problem would be to leave off the situation variable and insert it when a proof has been found. Using this approach, which is really an application of abstraction, the search space would be much smaller since there are only 81 allowable configurations of the disks. Therefore, there would be at most 81 positive unit clauses at each level. However, the use of several levels of abstraction would reduce the search space to some extent even over this method.

3.6 *EQUALITY*

Abstraction is compatible in a natural way with clauses including sets of equations. The idea is to choose a semantic abstraction (or a weak abstraction based on it) in which all the equations are true. In such an abstraction, all equations will have abstractions of the form a = a for domain elements a. These abstractions will not contribute to proofs in the abstracted space. When looking for proofs in the original space, the equations can be added using any paramodulation or simplification strategy for equality.

<u>Definition</u>. Suppose E is a set of equations. An <u>E-paramodulation</u> is a paramodulation performed on a clause C using an equation $t_1 = t_2$ which is a logical consequence of E.

<u>Definition</u>. Suppose S is a set of clauses and E is a set of equations. A <u>resolution-paramodulation proof tree</u> T from (S, E) is like a resolution proof except that E-paramodulations are also permitted. That is, if N is the father of nodes N1 and N2 in T, then CL(N) is either a resolvent or an E-paramodulant of CL(N1) and CL(N2). Note that if E is true in an interpretation I, then so is any logical consequence $t_1 = t_2$ of E, hence all abstractions of $t_1 = t_2$ will be of the form a = a for the semantic abstraction obtained from I.

<u>Definition</u>. If T and U are resolution-paramodulation proof trees and f is a weak abstraction, then $T \underset{f}{\rightarrow} U$ is defined as before.

<u>Theorem 3.6</u>. Suppose f is a semantic abstraction obtained from an interpretation in which all the equations in E are true. Suppose T is a resolution-paramodulation proof tree from (S, E). Then there is a resolution proof tree U from f(S)

such that $T \underset{f}{\rightarrow} U$. (Note that the equations are not used in U.) Also, if T is a proof of C then for every $D \in f(C)$, there is such a proof U of something subsuming D. A similar result can be shown for certain weak abstractions.

This result can be used to obtain an abstraction strategy for equality. The strategy is almost the same as before: given proof U from f(S), we attempt to construct proof T from $S \cup E$ such that $T \underset{f}{\rightarrow} U$ and such that T is a proof of the desired clause (or something subsuming it). We keep A-clauses and B-clauses with each node of U. Initially, A-CL(N) = $\{C \in S: CL(N) \in f(C)\}$ for leaves N of U and A-CL(N) = $\emptyset$ for other nodes of U. Also, B-CL(N) = $\emptyset$ for all nodes of U initially. Elements of A-CL(N) are generated by resolution as before. Elements of B-CL(N) are generated as follows: Suppose N is the father of nodes N1 and N2 in U. Then it is permissible to add to B-CL(N) all clauses C1 satisfying the following properties:

1. CL(N) subsumes an element of f(C).
2. For some clause $C \in$ A-CL(N), C1 can be derived from $S \cup E \cup \{C\}$ using resolution and E-paramodulation. Thus the only equations that may be used for paramodulation are those that are derived solely from E.
3. The clause C must actually be used in some such proof of C1.

Also, if P_1-deduction is used in the original and abstracted spaces, and f preserves positive clauses, then we can require all elements of B-CL(N) to be positive if CL(N) is positive unless N is the root. The advantage of this approach is that powerful strategies for equality are known [3]. The use of an abstraction in which all equations are true has an intuitive appeal, because it corresponds to the use of nontrivial semantic information as a guide to the theorem proving process.

4. M-RESOLUTIONS AND M-ABSTRACTIONS

Definition. A multiset M is a set S together with a function g mapping S into the set of positive integers. We refer to S as Set(M) and for $x \in S$, g(x) is denoted by mult(x, M). By convention, mult(x, M) = 0 if $x \notin S$.

Intuitively, a multiset is a set in which elements can occur more than once. For $x \in S$, mult(x, M) tells "how many times" x occurs in M. We write M as $\{n_1{*}x_1, \ldots, n_k{*}x_k\}$ where mult(x, M) = $\sum_{j:\, x_j = x} n_j$. Also, 1*x is written as x.

We often regard an ordinary set A as a multiset M in which each element of A occurs exactly once, and in which no other elements occur.

The size |M| of a multiset M = $\{n_1{*}x_1, \ldots, n_k{*}x_k\}$ is defined to be $\sum_{i=1}^{k} n_i$.

Definition. If M1 and M2 are multisets, then their union M1 ⊎ M2 is defined by mult(x, M1 ⊎ M2) = mult(x, M1) + mult(x, M2). Their intersection M1 ⩀ M2 is defined by mult(x, M1 ⩀ M2) = min(mult(x, M1), mult(x, M2)). Their difference M1 - M2 is defined by mult(x, M1 - M2) = max(0, mult(x, M1) - mult(x, M2)). Sometimes we write ⊎ as ∪. Note that Set(M1 ⊎ M2) = Set(M1) ∪ Set(M2).

Definition. If M1 and M2 are multisets, then we write M1 ⊂ M2 (M1 is a sub-multiset of M2) if for all x, $mult(x, M1) \leq mult(x, M2)$.

Definition. If M is a multiset and g is a mapping from Set(M) into a set N, then $g(M) = \biguplus_{x \in M}\{g(x)\}$. Thus $mult(y, g(M)) = \sum_{x, f(x)=y} mult(x, M)$, and $|g(M)| = |M|$.

Note that for multisets M1 and M2, $g(M1 - M2) = g(M1) - g(M2)$ if $M2 \subset M1$. This is not true of ordinary sets, however.

Definition. A multiclause (or m-clause) is a multiset of literals. That is, with each literal in the clause, a multiplicity is kept, which is a positive integer telling how many times the literal occurs in the multiclause. We can write a multiclause by writing each element the number of times it occurs in the multiclause. Thus {P, P, Q} is a multiclause in which the multiplicity of P is 2 and the multiplicity of Q is 1.

Definition. If C is a multiclause and α is a substitution, then $C\alpha$ is $\{L\alpha : L \in C\}$ where $L\alpha$ is counted the right number of times. That is, $mult(L1, C\alpha) = \sum_{L \in C, L\alpha = L1} mult(L, C)$. Thus $|C\alpha| = |C|$, and if $C = \{L_1, L_2, \ldots, L_n\}$ then $C\alpha = \{L_1\alpha, L_2\alpha, \ldots, L_n\alpha\}$.

Example. Suppose C is $\{\bar{P}(x), \bar{P}(c), Q(x)\}$. Suppose α is $\{x \leftarrow c\}$. That is, α replaces x by c. Then $C\alpha$ is $\{\bar{P}(c), \bar{P}(c), Q(c)\}$. Note that the literal $\bar{P}(c)$ occurs twice in $C\alpha$.

Definition. Suppose C1 and C2 are multiclauses. Suppose $A1 \subset C1$ and $A2 \subset C2$. (This means that every literal occurs no more times in A1 than in C1, and similarly for A2.) Suppose there exist substitutions $\alpha 1$ and $\alpha 2$ such that for some literal L, $Set(A1\alpha 1) = \{L\}$ and $Set(A2\alpha 2) = \{\bar{L}\}$. Let $\alpha 1$ and $\alpha 2$ be most general such substitutions. Then $(C1 - A1)\alpha 1 \uplus (C2 - A2)\alpha 2$ is an m-resolvent of C1 and C2. (Recall the definition of C1 - A1 for multisets C1 and A1, and similarly for C2 - A2.)

Examples. Suppose C1 is $\{\bar{P}(a), \bar{P}(x)\}$ and C2 is $\{P(a)\}$. Then $\{\bar{P}(x)\}$, $\{\bar{P}(a)\}$ and NIL (the empty multiset) are m-resolvents of C1 and C2.

Suppose C1 is $\{\bar{P}, \bar{P}\}$ and C2 is {P, Q, Q}. Then the following are the m-resolvents of C1 and C2:

$\{\bar{P}, Q, Q\}$

{Q, Q}.

Ordinary clauses can be viewed as multiclauses in which the multiplicity of each literal in the clause is 1. We have the following results concerning the relation of m-resolution to ordinary resolution.

Theorem 4.1. Suppose C3 is an ordinary resolvent of clauses C1 and C2. Suppose D1 and D2 are m-clauses such that Set(D1) = C1 and Set(D2) = C2. Then there is an m-resolvent D3 of D1 and D2 such that Set(D3) = C3.

Theorem 4.2. Suppose clause C is derivable from set S of clauses by ordinary resolution. Then there is an m-clause D derivable from S by m-resolution such that Set(D) = C. (In the derivation of D, we consider the clauses of S to be m-clauses). Note that if C = NIL then D = NIL also.

Theorem 4.3. Suppose m-clause D3 is an m-resolvent of m-clauses D1 and D2.

Then some ordinary resolvent of Set(D1) and Set(D2) subsumes Set(D3).

Theorem 4.4. Suppose S is a set of clauses, and m-clause D is derivable from S by m-resolution. (For this derivation, we consider the clauses of S to be m-clauses.) Then there is an ordinary clause C derivable from S by ordinary resolution, such that C subsumes Set(D).

Definition. An m-abstraction is a mapping f from multiclauses to (ordinary) sets of multiclauses, satisfying the following properties:

1. If C3 is an m-resolvent of C1 and C2, and D3 ε f(C3), then there exist D1 ε f(C1) and D2 ε f(C2) such that D3 is an instance of an m-resolvent of D1 and D2.
2. f(NIL) = {NIL}.

Notice how much simpler the properties of m-abstractions are than those of ordinary abstractions. The following result, analogous to theorem 2.1 for ordinary abstractions, shows that m-abstractions are also easy to construct.

Theorem 4.5. Suppose F is a set of mappings from literals to literals. For each $\emptyset \ \varepsilon$ F, extend $\emptyset$ to a mapping from multiclauses to multiclauses. Suppose that for all $\emptyset \ \varepsilon$ F, $\emptyset(\bar{L}) = \overline{\emptyset(L)}$ for all literals L. Suppose also that if multiclause D is an instance of multiclause C, then for all $\emptyset_2 \ \varepsilon$ F there exists $\emptyset_1 \ \varepsilon$ F such that $\emptyset_2$(D) is an instance of $\emptyset_1$(C). Define mapping f on multiclauses by f(C) = {$\emptyset$(C):$\emptyset \ \varepsilon$ F}. Then f is an m-abstraction mapping. (Note that f(C) is an ordinary set of multiclauses.)

Proof: Similar to the proof of Theorem 2.1.

If f is an m-abstraction as in this theorem, then we say f is defined in terms of literal mappings.

4.1 *EXAMPLES OF M-ABSTRACTIONS*

It is possible to obtain m-abstractions from the abstractions presented in Section 2.1 by counting each literal the right number of times.

We can define the composition f_1f_2 of m-abstractions f_1 and f_2, and show as before that it is an m-abstraction if f_1 and f_2 are. Also, the union of two m-abstractions is an m-abstraction. Moreover, it is easy to show that if f_1 and f_2 are m-abstractions defined in terms of literal mappings, then the union and composition of f_1 and f_2 are also defined in terms of literal mappings.

4.2 *M-ABSTRACTIONS OF M-RESOLUTION PROOFS*

The significance of m-abstractions is that they map m-resolution proofs onto proofs having the same shape. This can be an advantage in search strategies based on m-abstractions.

Theorem 4.6. Suppose V2 is an m-resolution proof of m-clause C' from set S of multiclauses. Suppose f is an m-abstraction mapping, and D' ε f(C'). Then there is an m-resolution proof V1 of an m-clause D1 from f(S) such that D' is an instance of

D1 and such that V1 and V2 have the same shape.

This is a much better result than for ordinary abstractions. Also, the depth of V1 is the same as the depth of V2.

Example

Using the propositional m-abstraction, we have the following abstraction of the following proof:

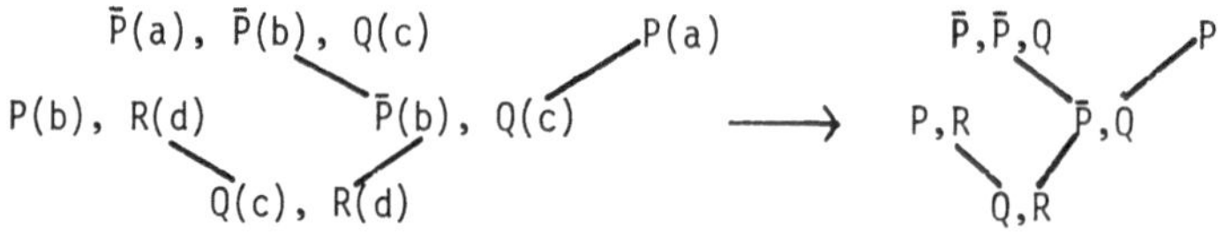

5. BOUNDED M-CLAUSES

One disadvantage of m-clauses is that there are so many of them. The set of ordinary clauses over k distinct predicate symbols is finite, but the set of m-clauses over k distinct predicate symbols is infinite. This could result in a larger search space for the various abstraction-based theorem proving strategies. It is possible to overcome this problem to some degree, while retaining the advantages of m-abstractions. The general idea is to keep less information about how many occurrences of a literal there are in an m-clause. For example, we may specify that a certain literal occurs at least twice in an m-clause. For details of this approach, see [6].

Acknowledgements

This research was supported in part by the National Science Foundation under grants MCS 77-22830 and MCS 79-04897. The comments of the referees helped in the presentation of this material.

References

[1] Boyer, R. S. Locking, a restriction of resolution, Ph.D. Thesis, University of Texas at Austin (1971).

[2] Kling, R. E. A paradigm for reasoning by analogy, Artificial Intelligence 2(1971), pp. 147-178.

[3] Lankford, D. S. Canonical algebraic simplification in computational logic. Report No. ATP-25, Southwestern University, Georgetown, Texas (1975).

[4] Munyer, J. C. Towards the use of analogy in deductive tasks, University of California at Santa Cruz (1979).

[5] Newell, A. and Simon, H. GPS, a program that simulates human thought, Computers and Thought, E. A. Feigenbaum and J. Feldman (Eds.), McGraw-Hill, New York, 1963, pp. 279-293.

[6] Plaisted, D. Theorem proving with abstraction, part II, Report No. UIUCDCS-R-79-965, University of Illinois, Urbana, Illinois (1979).

[7] Robinson, J. A. Automatic deduction with hyper-resolution, Internat. J. Comput. Math. 1 (1965), pp. 227-234.

[8] Sacerdoti, E. Planning in a hierarchy of abstraction spaces, Artificial Intelligence 5 (1974), pp. 115-135.

TRANSFORMING MATINGS INTO NATURAL DEDUCTION PROOFS

Peter B. Andrews
Mathematics Department
Carnegie-Mellon University
Pittsburgh, Pennsylvania 15213, U.S.A.

Abstract

A procedure is given for transforming refutation matings into natural deduction proofs. Thus a theorem proving system which establishes the validity of a theorem by the general matings approach can apply this procedure to obtain a comprehensible proof of the theorem without further search. This illuminates the close relationship between matings and proofs, and serves as a step toward a synthesis between apparently quite different approaches to automated theorem proving.

From a refutation mating the system constructs a plan for a theorem, describing appropriate replications of quantifiers, substitutions, and matchings of atoms. Skolem functions play a useful role in refutation matings, but terms involving such functions are replaced by appropriate variables when plans are constructed. Once a plan has been constructed, the system constructs a proof outine, or fragmentary proof, on the basis of the structure of the theorem and general principles for constructing natural deduction proofs. In a proof outline certain lines (planned lines) are justified not by rules of inference, but by plans. The outline is filled in by applying transformation rules, which add additional lines to the proof, justify certain planned lines, and sometimes create new planned lines. The linkage between plans and the proof is maintained by keeping track of the ancestries of wffs and quantifiers in the proof. Using the plans and other information about the proof, the system controls the application of transformation rules. For example, $\forall xA(x)$ is instantiated to $A(t)$ in the proof if the plan requires that t be substituted for a variable corresponding to x. Special problems arise in constructing natural proofs of existentially complex theorems, so they are proved in alternative forms.

§1. Introduction

In [1] and [2] we argued that proofs and refutations of wffs induce matings of the literal-occurrences in these wffs, that these matings embody the essential logical structure of the proofs (and the wffs proved by them), and that it is often fruitful to carry out the search for a proof in the context of a search for a suitable mating. We

---

This work is supported by National Science Foundation Grant MCS78-01462.

also argued in [2] that once one has found the essential ingredients of a proof, one should be able to present the proof in whatever format is most congenial to the reader. In pursuit of this program, we shall here describe how refutation matings [2] can be transformed into proofs presented in natural deduction style. It is hoped that this description will not only be useful in itself, but will also illuminate the close relationship between matings and proofs, and facilitate comparisons between search procedures involving matings and those involving proofs. The close relationship between matings and proofs is also illuminated by the work of Bibel and Schreiber [4] & [5], in which it is shown how eliminating redundancies from a systematic search for a proof can lead naturally to what is essentially a search for a refutation mating. Of course, many of the ideas which we use below, particularly those involving variables and Skolemization, had their origin in the work of Herbrand [8].

Much of our description of the transformation from a refutation mating to a natural deduction proof will sound like a traditional recipe for constructing a natural deduction proof. However, we shall always have a mating to guide the process, so that once the mating has been found, no additional search is required, and heuristics are needed only to make decisions concerning the arrangement of the proof, or to apply methods for increasing its simplicity, elegance, or conceptual clarity. We contemplate the eventual development of a system where the searches for a natural deduction proof and for the associated mating are carried out more or less simultaneously, so that human insights into theorem proving can readily be incorporated into the process, and so that information about compatibility of substitutions and fundamental logical structure acquired in the search for a mating can be made available to a human using the system interactively to prove a complex theorem. For the present, however, we shall assume that the refutation mating is found first.

The system described below is being implemented by Dale Miller and Eve Cohen. Experience with the working system will doubtless suggest possible refinements. The present paper should provide a general overview of the system, but certain details and examples are omitted due to space limitations. We shall assume familiarity with [2], and use the notations therein.

§2. Natural Deduction Proofs

We shall be concerned with a formulation of first order logic with function symbols in which the primitive connectives are $\sim$, $\wedge$, $\vee$, $\supset$, and $\perp$ (falsehood), and both $\forall$ and $\exists$ are quantifiers. $\perp$ plays the role of a contradiction in indirect proofs. $\equiv$ is treated as a defined expression. Other defined expressions may also occur. For example, $x \subseteq y$ might be a defined expression, which could be replaced when appropriate by $\forall t[t \epsilon x \supset t \epsilon y]$.

We now give an abbreviated and informal description of a system of natural deduction in which our proofs will be presented. A natural deduction proof consists of a sequence of lines, each of the form

$(\alpha)\ \mathcal{H} \vdash A$ justification.

Here α is a label for the line, $\mathcal{H}$ is a finite (possibly empty) set of wffs called hypotheses for the line, and A is a wff called the assertion of the line. When a line such as that above occurs in a proof, we may use α as an abbreviation for A elsewhere in the proof. $\mathcal{H} \vdash A$ may be read as "A is derivable from $\mathcal{H}$". If B is a wff, $\mathcal{H}, B \vdash A$ means $\mathcal{H} \cup \{B\} \vdash A$. We may say that $\underline{x}$ is not free in $\mathcal{H}$ when we mean that $\underline{x}$ is not free in any member of $\mathcal{H}$.

The rules of inference are stated below. Some redundant rules are included so that proofs can be presented in a relatively natural and efficient manner. In presenting these rules, we use M and N as notations for wffs which may be "empty". Thus $M \vee A$ is A when M is empty.

Hypothesis Rule (Hyp): Infer $\mathcal{H}, A \vdash A$.

Deduction Rule (Ded): From $\mathcal{H}, A \vdash B$ infer $\mathcal{H} \vdash A \supset B$.

Rule of Propositional Calculus (Rule P): From $\mathcal{H}_1 \vdash A_1, \ldots,$ and $\mathcal{H}_n \vdash A_n$ infer $\mathcal{H}_1 \cup \ldots \cup \mathcal{H}_n \vdash B$, provided that $[[A_1 \wedge \ldots \wedge A_n] \supset B]$ is tautologous.

Negation Rule (Neg): From $\mathcal{H} \vdash A$ infer $\mathcal{H} \vdash B$, where A is $\sim\forall\underline{x}C$, $\sim\exists\underline{x}C$, $\forall\underline{x}\sim C$, or $\exists\underline{x}\sim C$, and B is $\exists\underline{x}\sim C$, $\forall\underline{x}\sim C$, $\sim\exists\underline{x}C$, or $\sim\forall\underline{x}C$, respectively.

Rule of Indirect Proof (IP): From $\mathcal{H}, \sim A \vdash \bot$ infer $\mathcal{H} \vdash A$.

Rule of Cases (Cases): From $\mathcal{H} \vdash A \vee B$ and $\mathcal{H}, A \vdash C$ and $\mathcal{H}, B \vdash C$ infer $\mathcal{H} \vdash C$.

Rule of Alphabetic Change of Bound Variables ($\alpha\beta$)

Rule of Definition (Def): Eliminate or introduce a definition.

Universal Generalization ($\forall$G): From $\mathcal{H} \vdash M \vee A \vee N$ infer $\mathcal{H} \vdash M \vee \forall\underline{x}A \vee N$, provided that $\underline{x}$ is not free in $\mathcal{H}$, M, or N.

Existential Generalization ($\exists$G): Let $A(\underline{x})$ be a wff and let t be a term which is free for $\underline{x}$ in $A(\underline{x})$. (t may occur in $A(\underline{x})$.) From $\mathcal{H} \vdash M \vee A(t) \vee N$ infer $\mathcal{H} \vdash M \vee \exists\underline{x}A(\underline{x}) \vee N$.

Universal Instantiation ($\forall$I): From $\mathcal{H} \vdash \forall\underline{x}A(\underline{x})$ infer $\mathcal{H} \vdash A(t)$, provided that t is a term free for $\underline{x}$ in $A(\underline{x})$.

Rule C: From $\mathcal{H} \vdash \exists\underline{x}A$ and $\mathcal{H}, A \vdash B$ infer $\mathcal{H} \vdash B$, when $\underline{x}$ is not free in $\mathcal{H}$ or in B.

When one has inferred a wff of the form $\exists\underline{x}A$, one often "existentially instantiates" to obtain the wff A, which asserts that x (a free variable of A) is an entity of the sort whose existence is asserted by ∃xA. We simply regard A as an additional hypothesis, and Rule C can be used to eliminate it at an appropriate time.

To illustrate a question about styles of proofs, we consider how one should prove ∀u∀v∀w[Puv ∨ Pvw] ⊃ ∃x∀yPxy. One proof is:

(1)	1 ⊢	∀u∀v∀w[Puv ∨ Pvw]	hyp
(2)	1 ⊢	Puv ∨ Pvw	∀I
(3)	1 ⊢	Puv ∨ ∀w Pvw	∀G
(4)	1 ⊢	Puv ∨ ∃v∀w Pvw	∃G
(5)	1 ⊢	∀v Puv ∨ ∃v∀wPvw	∀G
(6)	1 ⊢	∃u∀v Puv ∨ ∃v∀wPvw	∃G
(7)	1 ⊢	∃x∀yPxy ∨ ∃x∀yPxy	$\alpha\beta$
(8)	1 ⊢	∃x∀yPxy	Rule P
(9)	⊢	1 ⊃ 8	Ded

Actually, this technique of proving ∃xC by first proving ∃xC ∨ ∃xC is not really very enlightening from an intuitive point of view. We shall say that a wff is existentially complex (relative to a plan for a proof) iff it has the form ∃xC(x) and our proof plan does not provide a term t such that we can prove C(t) from the relevant hypotheses. We shall avoid giving direct proofs of existentially complex wffs. Instead, we shall prove them by indirect or contrapositive methods.

§3. Plans

An occurrence of a quantifier or a wff in a wff A is said to be positive(+) [negative(-)] in A iff it is in the scope of an even [odd] number of occurrences of ~ when all wf parts [B ⊃ C] of A are regarded as abbreviations for [~B ∨ C]. We use ∃∀x as a notation in our metalanguage for a quantifier which is either ∀x or ∃x. An occurrence of a quantifier is essentially universal [existential] in A iff it is positive in A and universal [existential], or it is negative in A and existential [universal]. A variable is essentially universal [existential] iff it occurs in a quantifier which is essentially universal [existential].

A normal wff has distinct variables in distinct quantifier-occurrences, and a wff in negation normal form (nnf) has atomic scopes for all negations.

In order to develop a systematic way of describing sequences of quantifier duplications, let D be a normal wff containing a subwff M of the form ∃∀xK. Let n ≥ 2 and for each i ≤ n let M_i be the result of replacing each bound variable y in M by a variable $\underline{y}_i$ obtained from $\underline{y}$ by attaching i as (an additional) subscript. Replace the given occurrence of M in D by $M_1 \wedge \ldots \wedge M_n$ if M is $\forall \underline{x}K$, and by $M_1 \vee \ldots \vee M_n$ if M is $\exists \underline{x}K$. We say that the resulting wff is obtained from D by an n-replication of $\exists\!\!\!/\,\underline{x}$.

A replication scheme $\mathcal{R}$ for a normal wff D is a sequence $<(\underline{z}_1,k_1), \ldots, (\underline{z}_r,k_r)>$ of pairs such that $\underline{z}_i$ is a variable and k_i is an integer ≥ 2 for each i. We write $\mathcal{R}(D)$ for the result of sequentially performing a k_i-replication of $\exists\!\!\!/\,\underline{z}_i$ for i = 1, . . . , n.

Example: D is $\forall x \exists y \forall z Pxyz$ and $\mathcal{R}$ is $<(x,2), (y_1,2), (y_2,3)>$.

$\mathcal{R}(D)$ is $\forall x_1[\exists y_{11} \forall z_{11} Px_1y_{11}z_{11} \vee \exists y_{12} \forall z_{12} Px_1y_{12}z_{12}]$
$\wedge \forall x_2 [\exists y_{21} \forall z_{21} Px_2y_{21}z_{21} \vee \exists y_{22} \forall z_{22} Px_2y_{22}z_{22} \vee \exists y_{23} \forall z_{23} Px_2y_{23}z_{23}]$

Let $\mathcal{A}(D)$ be the set of occurrences of atoms (atom-occurrences) in D.

Definition A *mating $\mathcal{M}$ for a normal wff B of first order logic is a binary relation on $\mathcal{A}(B)$ such that there is a substitution θ for the essentially existential variables of B such that if L and K are members of $\mathcal{A}(B)$ and $L\mathcal{M}K$, then L is positive in B, K is negative in B, and $\theta L = \theta K$ (i.e., L and K are occurrences of atoms which become the same atom under θ). When such a substitution θ exists, there is an essentially unique most general such substitution, which we call the substitution associated with $\mathcal{M}$.

We let $\natural F$ be the wff obtained from a wff F by deleting all of its quantifiers.

Definition A *mating $\mathcal{M}$ for a normal wff B is a proof-*mating iff $\natural B$ is true with respect to every assignment of truth values to atoms which gives the same values to *mated atoms (atoms with occurrences L and K such that $L\mathcal{M}K$).

If A is any normal wff, let A′ be a normal wff obtained by eliminating all definitions from A.

Definition A plan for (a proof of) A is a quadruple $<A', \mathcal{R}, \mathcal{M}, \theta>$, where $\mathcal{R}$ is a replication scheme for A′ which replicates only essentially existential quantifiers of A′, $\mathcal{M}$ is a proof-*mating of $\mathcal{R}(A')$, and θ is the substitution associated with $\mathcal{M}$; in addition, θ must satisfy a condition which we shall not state here, but which reflects the use of Skolem functions in the construction of plans as described below. A′ is called the simple plan formula, and $\mathcal{R}(A')$ is called the expanded plan formula. There is a natural one-many correspondence between the subformulas and quantifiers of A′ and those of $\mathcal{R}(A')$.

To construct a proof for a wff A, we first construct an initial plan for A, from which other plans are constructed as the work progresses. To construct the initial plan, we rename the variables of A so that A is normal, temporarily relabel the free variables of A as constants, and proceed as follows:

1) Let B be the wff obtained from A′ by replacing wf parts of the form [M ⊃ N] by [~M ∨ N], and let C be an nnf of ~B.

2) Let D be the universal sentence in nnf obtained by Skolemizing to eliminate the existential quantifiers from C. This should be done so that when a wf part $\exists \underline{y} M(\underline{y})$ is replaced by $M(f\underline{x}^1 \ldots \underline{x}^n)$, the arguments $\underline{x}^1, \ldots, \underline{x}^n$ of f are the variables $\underline{x}^i$ such that there is an occurrence of $\forall \underline{x}^i$ in C with $\exists \underline{y} M(\underline{y})$ in its scope. (This method of Skolemization differs from that in [2], but we cannot discuss this interesting matter here.)

3) As discussed in [2], find a refutation mating $\mathfrak{M}$ and the associated substitution θ_m for some amplification G of D. We may assume there is a replication scheme $\mathcal{R}$ for D such that G is $\not\!\mathcal{R}(D)$. Note that $\mathcal{R}$ can also be applied to A′. From the natural one-to-one correspondence between $\mathcal{A}(A')$ and $\mathcal{A}(D)$ one easily obtains such a correspondence between $\mathcal{A}(\mathcal{R}(A'))$ and $\mathcal{A}(G)$. For each $K \in \mathcal{A}(\mathcal{R}(A'))$, let τK be the corresponding element of $\mathcal{A}(G)$.
Let $\mathfrak{N} = \{(L,K) \in \mathcal{A}(\mathcal{R}(A'))^2 \mid \sim\tau K \text{ is mated to } \tau L \text{ by } \mathfrak{M}\}$.

4) Let the essentially universal variables of $\mathcal{R}(A')$ serve as names for the terms which replace them (in a sense which becomes obvious when one considers examples) in $\theta_m G$. θ_m can be represented by a set of <u>substitution</u> <u>components</u> (pairs of terms) $<\underline{y}, \theta_m \underline{y}>$, where $\underline{y}$ is a universal variable of $\mathcal{R}(D)$. In each substitution term, replace each named subterm by its name, thus eliminating all Skolem functions from the term $\theta_m \underline{y}$. By these changes θ_m is modified to a new substitution, which we shall call θ_n.

5) $<A', \mathcal{R}, \mathfrak{N}, \theta_n>$ is the initial plan for A.

§4. Construction of a Natural Deduction Proof

The construction of a proof involves starting with a trivial <u>proof</u> <u>outline</u>, which is progressively modified until a proof is obtained. Transformation rules, to be discussed in §5, are used to modify proof outlines, using the plans and other information associated with proof outlines. A proof outline is like a natural deduction proof, except that some of the lines are justified not by rules of inference, but by plans. Thus a line in a proof outline may have the form:

(j) $\mathcal{H} \vdash B$ Plan k.

Hypotheses introduced by rule P-Choose of §5 are called special, and all others are ordinary. Plan k is a plan for the wff $[H_1 \wedge \ldots \wedge H_n \supset B]$, where $H_1, \ldots, H_n$ are the ordinary hypotheses in $\mathcal{H}$.

For example, a proof of the trivial theorem $\forall x[Px \wedge Qx] \supset \exists yPy$ might be obtained in stages with these proof outlines:

Stage 1

(100)		$\vdash \forall x[Px \wedge Qx] \supset \exists yPy$	Plan 1

Stage 2

(1)	1	$\vdash \forall x[Px \wedge Qx]$	hyp
(99)	1	$\vdash \exists yPy$	Plan 1
(100)		$\vdash \forall x[Px \wedge Qx] \supset \exists yPy$	Ded:99

Stage 3

(1)	1	$\vdash \forall x[Px \wedge Qx]$	hyp
(98)	1	$\vdash Py$	Plan 2
(99)	1	$\vdash \exists yPy$	$\exists$G:98
(100)		$\vdash \forall x[Px \wedge Qx] \supset \exists yPy$	Ded:99

Plan 2 is a plan for $\forall x[Px \wedge Qx] \supset Py$, which has been obtained in a systematic way from Plan 1. Using Plan 2, the system decides to instantiate $\forall x$ in (1) with y, obtaining

Stage 4

(1)	1	$\vdash \forall x[Px \wedge Qx]$	hyp
(2)	1	$\vdash Py \wedge Qy$	$\forall$I : 1
(98)	1	$\vdash Py$	Plan 2
(99)	1	$\vdash \exists yPy$	$\exists$G : 98
(100)		$\vdash \forall x[Px \wedge Qx] \supset \exists yPy$	Ded : 100

With the help of Plan 2 it is now observed that (98) follows by Rule P from (2), so the justification for (98) is changed to Rule P: 2, and the proof is complete.

Lines justified by plans are called planned lines. Lines which must still be used to infer other lines are called active; in general, when a line is created it becomes active, and the lines from which it is obtained may become inactive. Every active line is sponsored by one or more planned lines. Active lines are intended to contribute to the proofs of their sponsors. The sponsor of a line may change as work on the proof progresses. To simplify technical descriptions, we shall speak as though each active line had a unique sponsor, and multiple copies were made of lines with more than one sponsor. Actually, only one copy of any line will appear in a proof, but information associated with the line will be maintained for each sponsor.

In general, an atom-occurrence or quantifier in an active line has a primitive ancestor in the simple plan formula of the plan for the sponsoring line, and one or more possible ancestors in the expanded plan formula. These relations are established in the following way. Suppose K is an occurrence of a quantifier or atom in a wff B which has been derived from certain hypotheses. In general, the ancestry of K can be traced to certain occurrences of quantifiers or atoms in the ordinary hypotheses. (Of course, if one infers p ∨ q from p, q has no ancestor in the wff p. If one infers p from p ∨ p, both occurrences of p in p ∨ p are ancestors of p in the following line.) These correspond to occurrences of quantifiers or subformulas in the simple plan formula for the sponsoring line, which are called primitive ancestors of K. (If K contains a definition, its primitive ancestors may be non-atomic subformulas.) The possible ancestors of K are chosen from among the occurrences of quantifiers and atoms in the expanded plan formula which correspond (under the natural many-one correspondence) to the primitive ancestors of K. As the proof progresses and quantifiers are instantiated, the set of possible ancestors of (descendants of) K shrinks to reflect the fact that one can now indicate more precisely what part of the expanded plan formula corresponds to K. If K has only one possible ancestor, it is called the actual ancestor of K. Ancestries of quantifiers can be determined from their variables, which are changed only by replications.

If K and L are atom-occurrences in active or planned lines which have actual [possible] ancestors K_1 and L_1 in the plan, and K_1 and L_1 are mated by $\mathcal{M}$, we say that K and L are mated [possibly mated].

Using the plans, the induced matings, and other information about the proof outline, the system determines which transformation rules may be applied to it. If more than one rule is applicable, a choice must be made. In some cases the choices will affect the arrangement of the final proof, and heuristics designed to produce "natural" proofs may be used to make the choices.

After each transformation rule has been applied, the system checks to see whether a planned line has now been proved, or can be inferred by Rules P and $\alpha\beta$ from active lines. By using information from the plan it can do this efficiently.

§5. Transformation Rules

Transformation rules are divided into deducing rules and planning rules. Deducing rules simply cause one of the logical rules P, Neg, ∀I, or Def of §2 to be applied so that one or more lines are added to the proof outline. Inferred lines become active, and take as sponsors those lines which are sponsors of all the lines from which they were inferred. In general it is obvious in each case how to assign one or more ancestors in the inferring lines to each occurrence of an atom or quantifier in the inferred line, and to decide whether the inferring lines become inactive. We shall state only the deducing rule D∀, which applies the logical rule ∀I:

D∀: From $\forall \underline{x} A(\underline{x})$ infer A(t), where t is a term such that the substitution in the plan for the sponsoring line contains a substitution component $<\underline{x}_i, t>$, where $\exists\!\!\!/ \underline{x}_i$ is a possible ancestor of $\forall \underline{x}$. The possible ancestors of the occurrences of quantifiers and atoms in A(t) are those possible ancestors of the corresponding entities in $A(\underline{x})$ which are in the scope of $\exists\!\!\!/ \underline{x}_i$ in the expanded plan formula. After applying this rule, delete $\exists\!\!\!/ \underline{x}_i$ from the list of possible ancestors of $\forall \underline{x}$. If the list is now empty, the line asserting $\forall \underline{x} A(\underline{x})$ becomes inactive (with respect to the sponsoring line under consideration); otherwise, it remains active.

Clearly the plan plays a crucial role in deciding how to apply Rule ∀I. It plays a similar role with regard to Rule P. The system infers B from $A_1, \ldots, A_n$ only if the induced *mating of $[[A_1 \wedge \ldots \wedge A_n] \supset B]$ is a proof-*mating.

Next we discuss planning rules, which alter the proof outline so that it has different planned lines. In the description of each planning rule below, we are given a proof outline containing a line

$\mathcal{H} \vdash C$ Plan a,

where the wff C is to be specified. For convenience we shall always write Plan a as $<\mathcal{H}_o' \supset C', \mathcal{R}, \mathcal{n}, \theta>$, where $\mathcal{H}_o$ is the set of ordinary hypotheses in $\mathcal{H}$, $'$ still denotes the elimination of definitions, and $\mathcal{H}_o'$ is understood as the conjunction of the hypotheses in the set $\mathcal{H}_o'$ when appropriate. If $\mathcal{H}_o$ is empty, $\mathcal{H}_o' \supset C'$ is simply C'. In each case we shall create a line of the form

$\mathcal{G} \vdash D$ Plan b,

where $\mathcal{G}$ and D are obtained in a simple way from $\mathcal{H}$ and C, so that there is a natural one-to-one correspondence between $\mathcal{A}(\mathcal{G}_o' \supset D')$ and a subset of $\mathcal{A}(\mathcal{H}_o' \supset C')$. It will be understood that Plan b is $< \mathcal{G}_o' \supset D', \mathcal{R}, \mathcal{n}_1, \theta_1 >$, where the replication sequence $\mathcal{R}$ is the same as for Plan a, members of $\mathcal{A}(\mathcal{R}(\mathcal{G}_o' \supset D'))$ are *mated by $\mathcal{n}_1$ iff the corresponding members of $\mathcal{A}(\mathcal{R}(\mathcal{H}_o' \supset C'))$ are *mated by $\mathcal{n}$, and θ_1 is obtained from θ by deleting any substitution components $<\underline{x}, t>$ such that no quantifier $\exists\!\!\!/ \underline{x}$ occurs in $\mathcal{R}(\mathcal{G}_o' \supset D')$.

P⊃ : Replace the line

(α) $\mathcal{H} \vdash A \supset B$ Plan a

by

$(\gamma)\ \mathcal{H}, A \vdash A$ hyp
$(\beta)\ \mathcal{H}, A \vdash B$ Plan b
$(\alpha)\ \mathcal{H} \vdash A \supset B$ Ded

The new active line (γ) is sponsored by line (β).

<u>P∧</u> : Replace the line

$(\alpha)\ \mathcal{H} \vdash A \wedge B$ Plan a
by
$(\gamma)\ \mathcal{H} \vdash A$ Plan b
$(\beta)\ \mathcal{H} \vdash B$ Plan c
$(\alpha)\ \mathcal{H} \vdash A \wedge B$ Rule P: γ,β

Plan c is obtained from Plan a in a manner analogous to that used to obtain Plan b. Active lines originally sponsored by (α) are now sponsored by both (β) and (γ).

<u>P∨1- P∨4</u>:

These four rules for dealing with disjunctions each involve replacing the line
$(\alpha)\ \mathcal{H} \vdash A \vee B$ Plan a
by certain lines, as specified below for each rule:

<u>P∨1</u>:
$(\delta)\ \mathcal{H}, \sim A \vdash \sim A$ hyp
$(\gamma)\ \mathcal{H}, \sim A \vdash B$ Plan b
$(\beta)\ \mathcal{H} \vdash \sim A \supset B$ Ded
$(\alpha)\ \mathcal{H} \vdash A \vee B$ Rule P

<u>P∨3</u>:
$(\beta)\ \mathcal{H} \vdash A$ Plan b
$(\alpha)\ \mathcal{H} \vdash A \vee B$ Rule P

P∨3 is to be used only when examination shows that $<\mathcal{H}_0' \supset A', \mathcal{R}, \mathcal{N}_1, \theta_1>$ is in fact a plan for $\mathcal{H}_0 \supset A$. Two other rules, called P∨2 and P∨4, are obtained from those above by interchanging A and B.

<u>P∀</u>: Replace the line

$(\alpha)\ \mathcal{H} \vdash \forall\underline{x}A$ Plan a
by
$(\beta)\ \mathcal{H} \vdash A$ Plan b
$(\alpha)\ \mathcal{H} \vdash \forall\underline{x}\ A$ ∀G

Because of the way variables are handled, $\underline{x}$ will not occur free in $\mathcal{H}$.

<u>P∃</u>: Suppose the current plan outline contains a line

$(\alpha)\ \mathcal{H} \vdash \exists\underline{x}A(\underline{x})$ Plan a,

where $\exists\underline{x}A(\underline{x})'$ is not existentially complex with respect to Plan a, and let $t = \theta\underline{x}$, the term substituted for $\underline{x}$ by θ. Replace this line by:

(β) $\mathcal{H} \vdash A(t)$	Plan b
(α) $\mathcal{H} \vdash \exists\underline{x}A(\underline{x})$	$\exists$G

Additional planning rules replace the problem of proving $\mathcal{H} \vdash C$ by that of proving $\mathcal{H} \vdash C*$, where C* is $[\sim C \supset \bot]$ (producing an indirect proof), or a contrapositive of C, or is obtained from C by Rules Neg or Def of §2.

The rules we have discussed so far have been directed toward improving the proof outline on the basis of the form of the wff to be proved in a planned line. The next two rules focus on wffs which are asserted in active lines.

P-Cases: Suppose the current proof outline contains the lines

(κ) $\mathcal{H} \vdash A \vee B$	Rule x
(α) $\mathcal{H} \vdash C$	Plan a

where line (κ) is active and sponsored by line (α). Replace line (α) by the lines:

(ϵ) $\mathcal{H}, A \vdash A$	hyp (Case 1)
(δ) $\mathcal{H}, A \vdash C$	Plan b
(γ) $\mathcal{H}, B \vdash B$	hyp (Case 2)
(β) $\mathcal{H}, B \vdash C$	Plan c
(α) $\mathcal{H} \vdash C$	Cases: κ,δ,β

(κ) becomes inactive. All active lines originally sponsored by (α) are now sponsored by both (β) and (δ). (ϵ) is sponsored by (δ) and (γ) by (β). For this rule Plan b is obtained from Plan a in a slightly more complicated way than that described above. Recall that Plan a is $< \mathcal{H}_0' \supset C', \mathcal{R}, \mathcal{N}, \theta >$; Plan b is to be $< \mathcal{H}_0' \wedge A' \supset C', \mathcal{R}, \mathcal{N}_1, \theta >$, where $\mathcal{N}_1$ is obtained from $\mathcal{N}$ by giving to atoms of A′ the same mates as their ancestors in $\mathcal{H}_0'$ have. This ancestry information is obtained from the corresponding information about parts of A in line κ. (Note that $\mathcal{H}_0' \wedge A'$ may not be a normal wff; however, it is convenient to extend the definition of a plan so that this fact may be ignored.) Plan c is obtained from Plan a in a similar way.

P-Choose: Suppose the current proof outline contains the lines

(κ) $\mathcal{H} \vdash \exists\underline{x}B(\underline{x})$	Rule x
(α) $\mathcal{H} \vdash C$	Plan a

where (κ) is active and sponsored by (α). Let $\exists\!\!\!/\,\underline{x}_i$ be the actual ancestor of $\exists\underline{x}$. Because of the way variables are handled, $\underline{x}_i$ will not occur free in $\mathcal{H}$ or in C. Replace line (α) by the lines

(γ) $\mathcal{H}$, $B(x_i) \vdash B(x_i)$ hyp (choose x_i)
(β) $\mathcal{H}$, $B(\underline{x}_i) \vdash C$ Plan a
(α) $\mathcal{H} \vdash C$ Rule C: κ,β

(γ), and all active lines originally sponsored by (α), are sponsored by (β). (κ) becomes inactive. Hypothesis $B(\underline{x}_i)$ in (γ) is special. We do not need to incorporate special hypotheses into a plan for a line because in a sense the information represented by the special hypotheses is already present in the plan. When we derive $\exists\underline{x}B$, and then assume B to reflect the fact that from a logical point of view we are making an assumption about $\underline{x}$, we are performing a quantifier elimination which is already implicit in our understanding of the plan for $\mathcal{H} \vdash C$, so a plan for $[\mathcal{H}_0 \wedge B] \supset C$ would be redundant.

References

[1] Peter B. Andrews, "Refutations by Matings", IEEE Transactions on Computers C-25 (1976), 801-807.

[2] Peter B. Andrews, "General Matings", Proceedings of the Fourth Workshop on Automated Deduction, Austin, Texas, February 1-3, 1979, 19-25. (Condensation of [3].)

[3] Peter B. Andrews, "Theorem Proving via General Matings", Journal of the ACM (to appear).

[4] W. Bibel, "An Approach to a Systematic Theorem Proving Procedure in First-Order Logic", Computing 12 (1974), 43-55.

[5] W. Bibel and J. Schreiber, "Proof Search in a Gentzen-like System of First Order Logic", Proceedings of the International Computing Symposium 1975, edited by E. Gelenbe and D. Potier, North-Holland Publishing Company, 1975, 205-212.

[6] W. Bibel, "A Syntactic Connection between Proof Procedures and Refutation Procedures", Theoretical Computer Science 3rd GI Conf., Lecture Notes in Computer Science 48, Springer, 1978, 215-224.

[7] Wolfgang Bibel, "A Comparative Study of Several Proof Procedures", Universitat Karlsruhe, (August 1979).

[8] Jacques Herbrand, "Recherches sur la Theorie de la Demonstration", Travaux de la Societe des Sciences et des Lettres de Varsovie, Classe III sciences mathematiques et physiques, no. 33 (1930). Translated in [9].

[9] Jacques Herbrand, Logical Writings, edited by Warren D. Goldfarb, Harvard University Press, 1971, 312pp.

ANALYSIS OF DEPENDENCIES TO IMPROVE THE BEHAVIOUR OF LOGIC PROGRAMS.

Maurice Bruynooghe
Afdeling Toegepaste Wiskunde en Programmatie
Katholieke Universiteit Leuven
B-3030 Heverlee/Belgium

Abstract

Traditionally, backtracking uses a total order over the derivation steps. On failure, it returns to the most recent state. We consider states as a set of derivation steps. For each step, we save a 'inputset' validating the derivation. This results in a partial order over the derivation steps and allows a more accurate 'intelligent' backtracking.

1. Introduction

The goal statement of a logic program comprises a set of problems (procedure calls) which have to be solved simultaneously. For each problem, there exists a set of alternative rules (procedure definitions). The successful application of a rule to a problem reduces the problem to a (possibly empty) set of subproblems. It is common practice to use a depth first search combined with backtracking to explore the search space (and- or tree). However, backtracking ignores the dependencies between different subproblems and very systematically explores the complete search space. This results in a sometimes stupid behaviour when backtracking reacts on the failure to solve a subproblem with an attempt to find another solution for a subproblem which is completely irrelevant to the detected failure, i.e. the behaviour condemned by Sussman and McDermott [8] of a robot attempting to pick up an object with the left hand after a trial with the right hand failed because the object was to hot. In this paper, we show how an anlysis of the dependencies between different subproblems can be used to obtain a substantially better search behaviour.

2. Logic programs [5], [6]

A logic program comprises a set of procedures and a goal statement. A goal statement has the form $\leftarrow A_1, \ldots, A_n$ $(n \geq 1)$ with the symbols A_i procedurecalls ('literals'). A procedure has the form $B \leftarrow A_1, \ldots, A_n$ $(n \geq 0)$. The heading B and the calls A_i of the body are again literals.

A procedure can be used to execute a call when the heading matches that call ('resolution' [7]). To execute a goal statement $\leftarrow A_1, \ldots, A_n$, a call A_i is selected, (in PROLOG [3], [9], the leftmost). A procedure $B \leftarrow B_1, \ldots, B_m$ matching that call with substitution θ is taken and a new goal statement $\leftarrow (A_1, \ldots, A_{i-1}, B_1, \ldots, B_m, A_{i+1}, \ldots, A_n)\,\theta$ is derived. A solution is obtained when an empty goal statement is derived. The answer consists of the composition of the successive substitutions.

Backtracking is necessary because some calls A_i match different procedures.

3. Backtracking

The computation consists of a sequence of states or goal statements starting with the given goal statement. The successor of a state is derived by applying a procedure on the selected call. Once all solutions derivable from a given state are obtained, the backtracking algorithm returns to the previous state and attempts to apply a yet untried procedure on the call selected in that state. This search process does not consider the dependencies between the different calls of the successive goal statements. It simply assumes that each step depends on the whole state.

To improve the search, we consider the state not as a whole but as a set of derivation steps. Each step being the result of applying a procedure on a call and resulting in a substitution and in the creation of instances of the calls in the body of the applied procedure. Having refined the concept of state, we can imagine that a new step of the computation only depends on a subset of the previous steps. Let us call this subset the inputset. The inputset of a derivation step - an attempt to apply a definition on a call - consists of all previous steps containing information (substitutions - created calls) which are

- either necessary to generate the substitutions resulting from the unification between the call and the heading.
- or necessary to conclude that the definition is inadequate for the given call.

In other words, the inputset contains all steps crucial to the validity of the computed results of the performed derivation step.

To exploit the information about the inputset, we have to save it, i.e. for each executed call, we not only need the set of available (yet untried) definitions and the substitution resulting from the applied definition but also the inputsets of the applied and eliminated definitions.

Assuming we can determine the inputsets, we can write an algorithm for 'intelligent' backtracking.

Algorithm

Repeat
- select a unexecuted call.
- if there are some definitions available
  then
  . select a definition and remove it from the set of available definitions
  . unify the call with the heading of the selected definition and determine the inputset
  . if the unification is successful
    then . store the substitution and the inputset of the applied definition
         . create instances of the calls in the body of the applied procedure and initialise their sets of available definitions
    else ('shadow' backtracking)
         . add the selected definition with its inputset to the set of eliminated defintions.

  else (exhaustion of the available definitions - 'deep' backtracking)
  . the set of 'suspects' for this failure is the union of the inputsets of the eliminated definitions.
    Undoing one of the steps in this set will at least return one of the eliminated definitions to the set of available definitions and will eventually cure the failure.
  . to obtain a systematic generation of the potential solutions, it is necessary to select as 'culprit' the 'last' one in the set of suspects. As 'last' one we mean any of the suspects which is not in the inputsets (of the applied and eliminated definitions) of the other suspects.

the derivation performed by the 'culprit' is deleted (substitution - created calls), the applied definition is added to the set of eliminated definitions with its inputset extended by all other suspects, indeed the incompatibility of the substitution generated by this step with the other suspects caused the rejection of the definition.
All computations partly based on the computation performed by the culprit are no longer valid and must be deleted :
.. The application of a definition on a call must be undone and the definition returns to the set of available definitions when the culprit belongs to the inputset (In turn, other steps based on this deleted steps need to be deleted)
.. An eliminated definition returns to the set of available definitions when the culprit (or another deleted step) belongs to the inputset.

until a solution is found.

<u>Note on the selection of calls</u>

Usually, the same call will be selected after shadow backtracking while, after deep backtracking, the call associated with the culprit will be selected.

<u>Differences with the naïve backtracking algorithm</u>

The naïve algorithm simply returns to the previous call on deep backtracking. However, all attempts to find a solution are doomed to fail as long as all information contributing to the first failure remains in the system. It means that the naïve algorithm finally will return to a possible culprit.

The intelligent algorithm not only returns immediately to the culprit, it also saves the still valid information about some eliminated and applied definitions. The naïve algorithm deletes this information and, eventually, will rediscover the unsuitability of the eliminated procedures and has to recompute the substitutions of the applied definitions.

<u>A partial order</u>

The naïve algorithm defines a total order over the steps of the computation i.e. the order of their execution. The intelligent algorithm defines only a partial order i.e. a step must be executed after the steps in the inputsets of the applied and eliminated definitions. As a result, a particular state can be obtained by different orders of execution. It is even possible and meaningful to execute different steps in parallel (see further). Selection of the culprit has to be such that the partial order <u>over the steps</u> remains.

<u>A simple example</u>

We can already give a simple example, even before discussing the determination of inputsets. It is easy to convince oneself that the given inputsets are <u>sufficient</u> to perform the given steps and thus result in a correct behaviour of the algorithm.

```
   <- FP(u,v,w)
a. FP(x,y,z) <-- Q(x), T(z), R(y,y), P(x,y)
b. Q(A) <--
c. Q(B) <--
d. R(A,A) <--
e. R(s,t) <-- SR(s,t)
f. SR(B,A) <--
g. P(B,A) <--
```

h. P(A,B) <--
i. T(C) <--
j. T(D) <--

Applying definition (a) on the only call FP(u,v,w) results in the substitution {x <- u, y<- v, z <- w} and the creation of the subgoals Q(x), T(z), R(y,y) and P(x,y). Execution of Q(x) with FP as inputset gives, with definition (b), the substitution {u <- A}. In a similar way we can use FP as inputset to apply definition (i) on T(z) - which gives w <- C - and to apply definition (d) on R(y,y) - which gives v <- A.
Applying definition (g) on P(x,y) with FP and Q as inputset results in a failure, indeed x does not unify with B because of the existing components x <- u and u <- A. Applying definition (h) on the same call, but with FP and R as inputset also results in a failure. The set of available definitions for the call P being exhausted, we need deep backtracking. The set of suspects consists of FP, Q and R. Q and R can be selected as culprit. Suppose we select R; its applied definition (d) becomes eliminated due to the other suspects i.e. FP and Q. Undoing the execution of R affects the computations based on R; i.e. the elimination of definition (h) for the call P(x,y). This definition becomes again available, however the other definition (g) remains eliminated.
Applying definition (e) on R(y,y) with inputset FP results in the substitution {s <- v, t <- v)} and the subgoal SR(s,t). An attempt to solve SR with R and FP in the inputset and using the only definition (f) fails. Deep backtracking with FP and R as suspects results in undoing the call R : the definition (e) is eliminted and the inputset consists of FP. R has exhausted its set of available definitions and we again have deep backtracking. The suspects are FP and Q. Q must be the culprit and definition (b) is eliminated by FP (inputset). Now definition (d) becomes again available for call R (but not e!) and definition (g) for call P. Also the execution of call T(z) remains intact although executed after Q. The state of the computation is summarized in Fig 1.

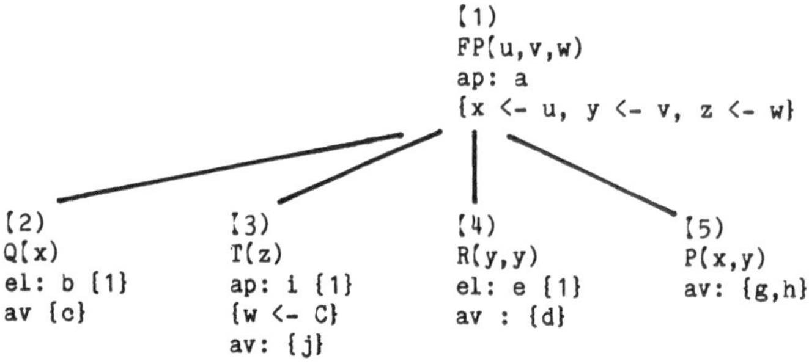

Fig 1. The computations takes a new start from the call Q(x). A substantial amount of information about the calls T and R has been saved. This picture represents the and-tree of subproblems. The calls are labeled by numbers. ap: i {1} means that definition (i) is applied with call (1) i.e. FP, as inputset. el: b {1} means that definition (b) is eliminated and the inputset consists of call (1). av: {g,h} means that definitions (g) and (h) are available.

This new attempt starting with call Q will lead to a solution. The naïve algorithm will not only loose the information about the calls T and R, it will also make a completely irrelevant attempt starting with procedure (j) for call T(z). Remark that Q(x) and T(z) both do not share variables with the failing call R(y,y). It is the <u>history</u> of previous failures (P(x,y)) which makes Q(x) a valid backtrackingpoint and T(z) an irrelevant one. We conclude that any serious attempt to apply 'intelligent' backtracking needs to consider the history of the successive failures.

4. Determination of the inputsets

Taking the inputsets to big does not result in an incorrect behaviour , it only makes the backtracking and thus the search for a solution less accurate.

As an extreme, we could take all previously executed calls as inputset. In this case, the algorithm will always select the most recent call as culprit (the other suspects are all in the inputset of the most recent one) and the algorithm reduces to the naïve backtracking algorithm.

At least, the inputset contains the father of the considered call and the inputset of the procedure applied to it (a.o. the other ancestors). Indeed, these elements are responsible for the merely existence of the call. It also means that the unification algorithm at least has access to the substitutions associated with these elements.

Which other information is needed by the unification algorithm is a question with different possible answers. We consider two seemingly interesting answers.

a. When the unification algorithm has to unify a variable x with a term t, it will create the component x <- t unless there exists a component x <- t' in which case the algorithm has to unify t' with t. Thus, whenever the unification algorithm accesses the binding of a variable, it adds the call generating that binding and the inputset of the procedure applied on that call to the inputset of the call being executed. Some non-determinism is possible when unification fails. For example, the failure of the unification between P(x,y) and P(A,B) can be detected, either by accessing a binding x <- B or by accessing a binding y <- A.

b. When unifying x with t we can generate x <- t even when a component x <- t' exists but the generator does not belong to the inputset (to the ancestors). However in that case, we can have different substitutions for the same variable and we have to verify their compatibility. (Maybe by adding a special subgoal Equal (t,t') having both generators as its initial inputset). This is a rather complex task - t and t' can contain variables which in turn can get different bindings - but this approach has some advantages :
   - As a result of unification, the inputset solely consists of the ancestors. This gives more independance between subgoals.
   - The behaviour is less dependent on the order of selection of the calls. Indeed most conflicts arise from inconsistencies between different substitutions. The set of suspects consists of the generators (and their inputsets) of these conflicting substitutions and - as far as the solution of earlier conflicts allows - any of these generators can be choosen as culprit. With method a, the last generator, the one completing the conflicting set, will read the already existing substitutions and the unification will fail. As a result, the last generator is the culprit (in fact shadow backtracking). Method b is more neutral to the order in which the generators are executed and can choose as culprit the one which gives the most promising (smallest) search space (especially interesting where the user is unaware of the desired control i.e. theorem proving).
   - This approach is attractive in the context of parallellism : some processes performing unification, only using the information of the ancestors; other processes checking (read only!) the consistency of different substitutions.

Optimising the manipulation of the inputsets

The construction of the inputsets is such that, whenever a call (b) is in the inputset of the procedure applied on a call (c), also the elements in the inputset of the procedure applied on (b) belong to the inputset of the procedure applied on (c). This is in fact a transitive relationship between the different steps of the execution. Without loss of information, we can delete any call (a) from the inputset of a call (c) when (a) also belongs to the inputset of a definition applied on another call belonging to the inputset of (c). Indeed, computing the closure of the relation restores the original information. This not only reduces the amount of information to be stored but also simplifies the computation of the culprit. Indeed, only the explicitly listed elements can be last in the set of suspects and are possible culprits. It is not necessary to compute the whole set of suspects.

With both of the above methods, the inputset becomes initialised by the father of the call. With method a, it is extended with the generator of a substitution x <- t whenever the unification algorithm needs that substitution and the generator is not yet in the *implicit* inputset.

When all definitions of a particular call are exhausted (deep failure), it is sufficient to take as suspects the explicitly known elements of the inputsets of the eliminated procedures (the set of the original algorithm can be derived by transitive closure). The inputset of the culprit 's rejected definition need only be extended with the other possible culprits (again the transitive closure gives all elements). The invalid information can be removed by initialising a set of deleted steps with the culprit. Any *eliminated* procedure with an element of the set of deleted steps on its inputset returns to the set of available procedures. Any step with an element of the set of deleted steps in the inputset of the *applied* procedure is undone (substitution removed, definition becomes available) and added to the set of deleted steps.

With method b, when an inconsistency is detected, the set of suspects consists of all generators of substitutions contributing to the inconsistency. The further treatment is similar to the case of deep failure.

Reduction of the number of calls

If the set of available definitions is empty and the explicit inputset of the eliminated and applied definitions consists of only one and the same element, then we can discard that call and its substitution can be added to the substitution of the only call in its inputsets. This will not substantially change the backtracking behaviour because selection of that call as culprit immediately results in another failure followed by deep backtracking and, in the new set of suspects, the call is only replaced by the single element of its inputsets. Of course, the discarded call is again initialised when the call absorbing the discarded call is undone by the backtracking.

A simplification

A simpler but less accurate version is obtained by giving the eliminated and applied procedures the same inputset, namely the union of their respective inputsets. For each call, only one inputset needs to be stored. This approach results in the method described in [1].

5. Example

As an example, we take the eight queensproblem, however, to simplify the exposition, we restrict ourself to three queens and a square board of side three. The observation that each row and each column need to have exactly one queen allows us to write a configuration as a list of numbers where the i th number gives the columnnumber of the queen on the i th row. In our case a possible solution is a permutation of the numbers 1 to 3. To be a solution, two queens have to be on different diagonals. Solutions can be computed by the goal statement

<- Perm (1.2.3. Nil, q), Safe (q)

(1.2.3. Nil stands for the term .(1, .(2, . (3, Nil)))) where Perm expresses the binary permutation relation and safe the unary relation describing all configurations with the queens on different diagonals.

Using a ternary relation Del(e, l_1, l_2) which is true when the list l_2 is derived from the list l_1 by removal of the element e, we can define the permutation as follows.

```
a.  Perm (Nil, Nil) <-
b.  Perm (x.y, u.v) <- Del(u, x.y, w), Perm(w, v)
c.  Del (x, x.y, y) <-
d.  Del (u, x.y, x.v) <- Del(u, y, v)
```

Using a ternary relation Check (p, l, d) to compare a queen p with the queens of list l with d the distance between the row of p and the row of the first queen on l and another ternary relation Nodiag (p, q, d) to compare p with queen q and d the distance between the rows of p and q, we can write Safe as :

```
e.  Safe (Nil) <-
f.  Safe (p.q) <- Check (p, q, 1), Safe (q)
g.  Check (p, Nil, n) <-
h.  Check (p, q.r, n) <- Nodiag (p, q, n), m = n+1, Check (p, r, m)
```

For simplicity we assume Nodiag is given by an explicit enumeration of the tupples satisfying the relation. This program is remarkable simple because the generation of possible solutions (Perm) is written independant of the diagonaltest (Safe), however with a strictly left to right execution order and naïve backtracking, the behaviour will be poor. The program will generate all possible permutations. When Safe for example detects that the first and second queen are on the same diagonal, the program will only move the second queen after all configurations with another 3th, 4th, ... nth queen have been exhausted.

Rewriting the program such that parts of Safe are moved inside the permutation and the compatibility of the queens 1, ..., i is checked before the i+1 th queen is computed will make the program less transparent. (A more complex logic component cfr [6]).

As we will illustrate, our intelligent backtracking will immediately react on a failure with moving the right queen. Using method a for determining the inputsets, the computation up to the point where Safe detects the first conflict is given in Fig. 2. The call Nodiag (p_9, q_9, n_9) fails due to the values of p_9, q_9 and n_9 (the positions of queens 1 and 2). The substitutions causing the failures are generated by the steps (1) and (2). However (1) is in the inputset of (2) and (2) is the only possible culprit. The situation after backtracking is given in Fig. 3. A new position for the second and following queens has to be computed but part of the execution of Safe has been saved. The algorithm reacts on the detected failure with the replacement of one of the involved queens.

Remark that different calls are associated with the same computation step. This will be a general situation for logic programs because large parts are

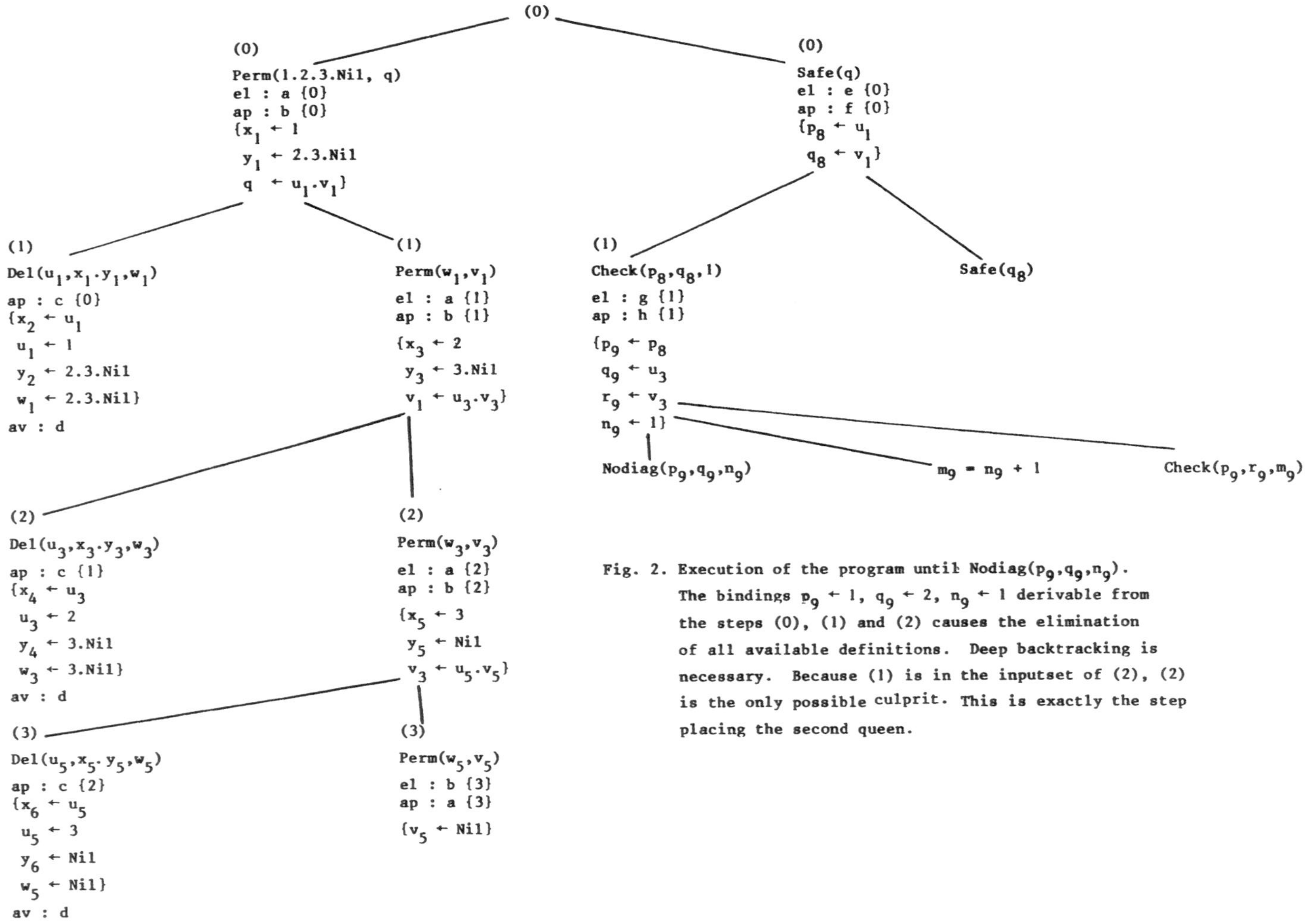

Fig. 2. Execution of the program until $Nodiag(p_9,q_9,n_9)$. The bindings $p_9 \leftarrow 1$, $q_9 \leftarrow 2$, $n_9 \leftarrow 1$ derivable from the steps (0), (1) and (2) causes the elimination of all available definitions. Deep backtracking is necessary. Because (1) is in the inputset of (2), (2) is the only possible culprit. This is exactly the step placing the second queen.

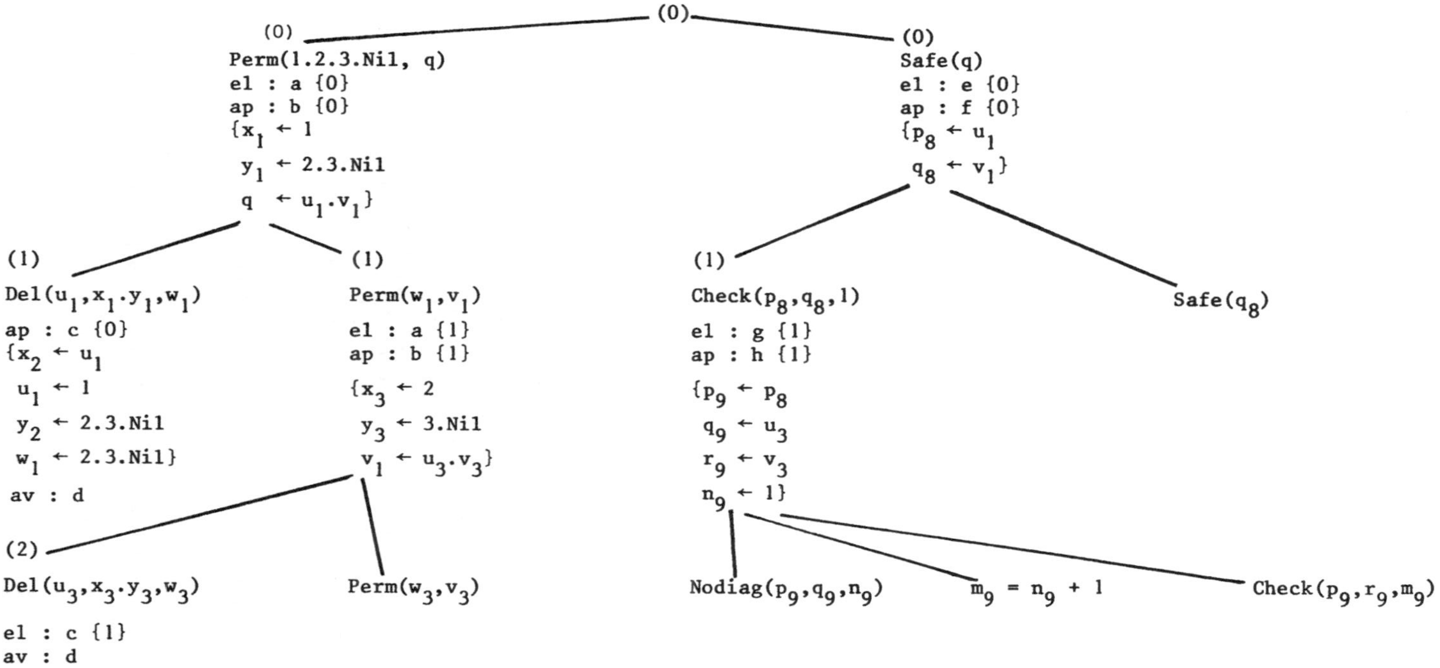

Fig. 3 : The situation after deep backtracking. The elimination of procedure (c) of (2) is due to (1). The execution can take a new start. Remark that part of the execution of Safe has been saved.

determinate i.e. only one definition matching a call. This largely simplifies the task of manipulating the inputsets.

Notes

1. The behaviour still depends on the order in which the calls on Nodiag are generated. This can be avoided by pursuing the Safe computation until all failing calls are detected. Backtracking needs to remove all the found conflicts.
2. This example is typical for a large class of generate and test programs where the separation of the generate part and the test part gives a simple transparant logic, but, with naïve backtracking, an extremely inefficient behaviour.

6. Finding all solutions.

Up to now, we were only handling failures, however, at some point, a solution will be found and the user can be eager to know more solutions. A straightforward way to obtain another solution follows from the statement that the found solution must be rejected, in other words that there is a conflict with all calls composing the solution as set of suspects. The disadvantage of this approach is that nonessential dependencies are created between subproblems which are independant of each other. This slows down the further search process.

Using an example we will indicate how the information about the inputsets allows to find independant subproblems and to speed up the process of finding all solutions.

```
   <- P(A.B.C.D.E.F.Nil, z)
a. P(l, p(x,y)) <- Split (l, l1, l2), Member (x, l1), Member (y, l2)
b. Split (Nil, Nil, Nil) <-
c. Split (x.Nil, x.Nil, Nil) <-
d. Split (x.y.l, x.l1, y.l2) <- Split (l, l1, l2)
e. Member (x, y.z) <- Member (x, z)
f. Member (x, x.y) <-
```

The derivation of a first solution is given in Fig. 4. The conflict induced by the rejection of that solution has steps (2) and (4) as possible culprits. Choosing (4) as culprit makes (4) dependent on (2). The systematic rejection of all solutions imposes step by step a total order over the steps of the execution. The computation degrades to the naïve backtrackingalgorithm and, although the subproblems Member (x,l_1) and Member (y,l_2) are independent of each other, we shall compute all solutions of Member (y,l_2) for each solution of Member (x,l_1). To avoid this duplication of effort on Member (y,l_2), we want to compute its solutions only once and to represent them as lemma's. We can use this lemmas to get a new set of solutions of the total problem for each new solution of Member (x,l_1)

In the remainder of this section, we will briefly outline when such savings are possible and feasible. We start with some definitions related to a solution of a problem.

A feasible set of subtrees of a solution is such that the steps occurring in these subtrees are not in the inputsets of steps outside the subtree (e.g. the subtrees with Member (x_1,z_1) and Member (y,l_2) as root but not the subtree with Split (l,l_1,l_2) as root). In other words, the set of calls in the roots

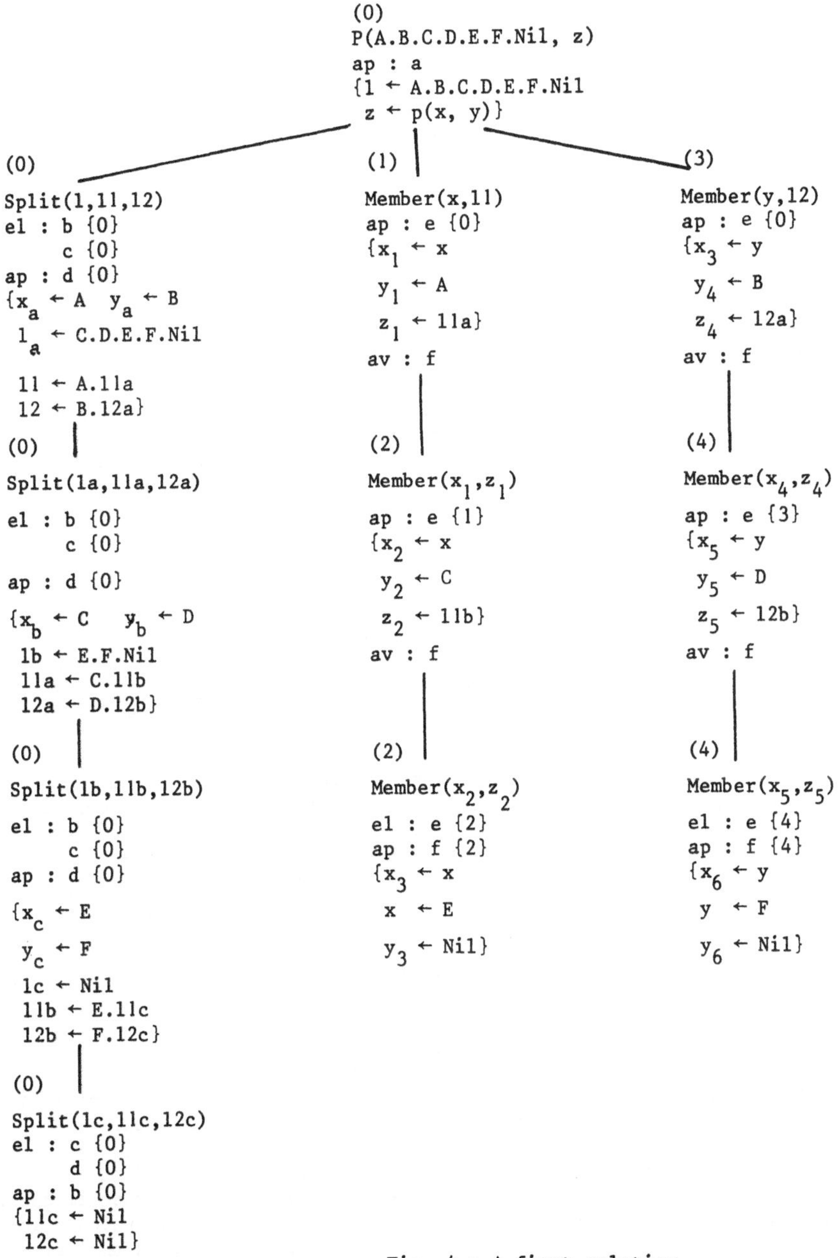

Fig. 4 : A first solution.

could have appeared in an intermediate goal statement leading to the solution.

A set of subtrees is nondeterminate if it contains a call with a nonempty set of available definitions.

The use of lemmas can be useful if we can distinguish two disjoint nondeterminate feasible set of subtrees (e.g. the subtree with Member (x,l_1) as root and the subtree with Member (y,l_2) as root). Then we isolate the subproblem defined by one set of subtrees and compute all its solutions. We represent each solution by a lemma and determine the validityset of the lemma's : all steps outside the subproblem which occur somewhere in the inputset of a step leading to the solution of the isolated subproblem (e.g. the solutions of Member (y,l_2) are y <- F, y <- D and y <- B, the validityset consists of step (0)).

The effort was worth while if the validityset still does not contain steps belonging to the solution of the other feasible nondeterminate set of subtrees. We then replace the roots of the set of subtrees by a single call whose solution is given by the first lemma and which has the other lemmas as available definitions. Then we reject the first solution and select the call using lemma's as culprit. This results in one solution for each lemma. (e.g. the solutions z <- p(E,F), z <- p(E,D) and (z <- p(E,B)). As next culprit we select a step from the other feasible nondeterminate set of subtrees. (e.g. step (2)). In other words, we start by searching a new solution for the associated subproblem. Once found (e.g. x <- C) we again apply the lemma's to obtain a new set of

solutions (e.g. z <- p(C,F), z <- p(C,D) and z <- p(C,B)).

We have to remove the lemmas and to replace the call by the original set of calls once it is necessary to choose as culprit a step belonging to the validityset of the lemmas.

This approach gives the computation a flavour of parallellism where it is feasible i.e. where it is possible to distinguish independent subproblems each (eventually) having different solutions. In those situations this approach in fact makes the crossproduct of both solutionsets. The information in the inputsets allows to detect potentially independent subproblems.

7. Final remarks

We have presented a framework for purely syntactical analysis of the dependencies between subgoals. It shows the possibilities, but also the price to be paid, for a more intelligent behaviour which moves the burden of the control (algorithm = logic + control [6]) from the user to the execution mechanism.

It is important to realise that the reaction on a failure not only depends on the failure state but also on the history of previous failures.

The method of determining the inputsets is likely to depend on the purpose of the system. In the context of theorem proving, where the user is unable to handle control at all, one wants the most intelligent system which carefully analyses the state and the impact of the choice of a culprit on the size of the search space, whereas, for the execution of logic programs, it is more important to obtain a simple and fast execution mechanism using only a limited amount of space.

Our method can be considered a kind of 'truth maintenance' [4] for resolution theorem provers.

8. Acknowledgments.

This paper is extracted from [2]. For this research, which was supported by the Belgian 'Nationaal Fonds voor Wetenschappelijk Onderzoek', we are especially indebted to R. Kowalski and Y.D. Willems.

9. References

[1] Bruynooghe, M.
Intelligent backtracking for an interpreter of Horn Clause logic programs.
Report CW 16, Afdeling Toegepaste Wiskunde en Programmatie,
K.U.Leuven, Belgium, september 1978.

[2] Bruynooghe, M.
Naar een betere beheersing van de uitvoering van programma's in de logika der Horn-uitdrukkingen. (In Dutch).
Doctoral dissertation, Afdeling Toegepaste Wiskunde en Programmatie,
K.U.Leuven, Belgium, may 1979.

[3] Colmerauer, A., Kanoui, H., Roussel, P. and Pasero, R.
Un système de communication homme-machine en francais.
Groupe d'Intelligence Artificielle, U.E.R. de Luminy,
Université d' Aix-Marseille, 1973.

[4] Doyle, J.
A truth maintenance system.
Artificial Intelligence 12 (1979), 231-272.

[5] Kowalski, R.A.
Predicate logic as a programming language.
Information Processing 74, North-Holland Pub. Co., Amsterdam,
1974, pp. 569-574.

[6] Kowalski, R.A.
Algorithm = logic + control.
Comm. ACM 22, 7 (july 1979), 424-436.

[7] Robinson, J.A.
Logic : form and function.
Edinburgh University Press 1979.

[8] Sussman, G.J. and McDermott, D.V.
Why conniving is better than planning.
AI Memo 255A, MIT, april 1972.

[9] Warren, D., Pereira, L.M. and Pereira, F.
PROLOG- The language and its implementation compared with LISP.
Proc. symp. on Art. Intell. and Programming Languages;
SIGPLAN Notices (ACM) 12, 8; SIGART Newsletters (ACM) 64
(august 1977), pp. 109-115.

SELECTIVE BACKTRACKING FOR LOGIC PROGRAMS

Luís Moniz Pereira
António Porto

Departamento de Informática
Universidade Nova de Lisboa
1899 Lisboa, Portugal

Abstract

We present a selective backtracking method for Horn clause programs, as applied to Prolog (2)(6)(11)(12), a programming language based on first-order predicate calculus (3)(4), developed at the university of Marseille (10).
This method is based on the general method expounded in (7) for backtracking intelligently in AND/OR trees. It consists, essentially, in avoiding backtracking to any goal whose alternative solutions cannot possibly prevent the repetition of the failures which caused backtracking.
This is a renewed version of an earlier report (8), which was spurred by the work of Bruynooghe (1).
In (9) we present an implementation of a selective backtracking interpreter using the methods discussed in this paper.

1. Introduction

The desire for control over resolution theorem provers (5) has always flourished. General search strategies, however, have remained too general. On the other hand, the standard backtracking methods remain plagued by the inefficiency of looking for remedy where it cannot be found.

This paper provides a strategy for backtracking selectively in top-down executions of Horn clause programs, which avoids searching irrelevant branches of the corresponding AND/OR tree. It consists, essentially, in avoiding backtracking to any goal whose alternative solutions cannot possibly prevent the repetition of the failures which caused backtracking.

This more intelligent backtracking strategy is based on the general method expounded first in (7) for backtracking intelligently in AND/OR trees, and is presented here as applied to Prolog (2)(6)(11)(12), a programming language based on first-order predicate calculus (3)(4), developed at the university of Marseille (10). The adaptation of the techniques developed here to resolution-logic theorem provers is, we believe, straightforward, if its executions can be modelled by an AND/OR tree.

This paper is a renewed version of an earlier report (8), that was spurred by the work of Bruynooghe (1), although improving it in several respects.

Our method has been tested in its implemented form. The implementation, which further elucidates our strategy, is presented in (9).

2. Prolog AND/OR trees

To solve a goal a clause must be found whose head matches the goal and whose body goals (if any) can all be solved. Execution of a goal generates an AND/OR tree with that goal as the root.

The clauses that match a goal give rise to OR branches at that goal. Each such OR branch is split into AND branches. One such branch leads to the node where unification between the goal and the clause head is achieved. Each remaining AND branch leads to a goal in the body of the clause. The unification node is itself an AND of the unifications of corresponding arguments of the goal and head, where each argument may be a compound term giving rise to further AND branching.

3. Backtracking selectively in Prolog AND/OR trees

In the execution of a Prolog program, a goal G is said to fail when G is selected for activation and there are no alternative clauses that G can match.

Standard backtracking consists in going back to reactivate the goal activated just before G.

Selective backtracking consists in updating the current set of goals selected as backtrack goals, by analysing the reasons for the failure of G, going back to reactivate the most recently activated goal present in that set, and deleting it from the set.

Whenever an activated goal G fails, the following are the sole goals selected for inclusion in the set of backtrack goals - these are the only goals that may prevent repetition of the same failure:

- The parent goal of G, since an alternative solution for it will avoid reactivation of G. It is an avoiding goal for G.

- All goals after whose reactivation the arguments of G may be modified so as to allow G to match some clause head which it did not match in its current form. These are the modifying goals for G.

Why are these the only goals selected for backtracking when activation of G has failed?

First, the goals that necessarily avoid G are its ancestors. Selecting only the parent of G is, however, sufficient for selecting all its ancestors, because if and when the parent fails its own parent will be selected as a backtrack goal, and so on, up to the top goal if necessary.

Second, let us discuss which are the only goals that may prevent G from failing again. Since G failed, for every clause for G either its head did not match G or some goal S subsequent to G failed. Backtrack goals that may prevent S from failing again have assumedly been selected already, when S failed. So the only possibly new backtrack goals that may prevent G from failing are those after whose reactivation the arguments of G may be modified so as to allow G to match some clause head that did not match the current failed G. If G fails on first activation it means no matching clause was found; in this case the parent goal and the modifying goals for G are clearly the only backtrack goals that may allow solution of the top goal. This is the basis for the recursive argument given above.

How are the modifying goals for G obtained?

The modifying goals for G are all those selected at each failed attempted match of G with a clause head, because each clause provides an independent possibility for solving G.

This is the OR rule.

For each failed match, since unification is an AND of several matching conditions, every individual condition that failed must be modified to allow unification to succeed. For every individual failed condition there is a (possibly empty) set of modifying goals. To allow modification of all failed conditions backtracking must occur up to the least recent among the most recent goals in each set (if one set is empty no modifying goals exist for a failed condition, and thus for the whole match). Because there may be no alternatives left at that least recent goal, it may not be able to modify the failed conditions depending on it. Thus, the least recent of the most recent remaining modifying goals for those conditions must also be selected for backtracking (again, if no modifying goals remain for one of those conditions, no more goals should be selected). This argument may be repeated until some condition has no more modifying goals. No modifying goals for any other condition should be considered: if some condition cannot be modified the match will always fail, regardless of modifications of other conditions; on the other hand, if one or more conditions are modified but some others are not, a new failure of the goal G will then select a set of modifying goals for any remaining failed conditions.

So, the set of modifying goals selected for a failed attempted match is the least of the sets of modifying goals for individual failed conditions, according to the following (lexicographic) order among the sets:

- The empty set is the least of all sets.

- If two sets have different most recent goals, the least set is the one with the least recent of them.

- The order between two sets with equal most recent goals is the order between those same sets with the most recent goal deleted from them.

This is the AND rule.

What are the modifying goals for a failed matching condition?

Since every failed condition is a unification conflict between two different constant names (ie. constants or functors), the modifying goals for the conflict are the modifying goals for each constant name, because changing either of them may solve the conflict.

The modifying goals for a conflicting constant name are readily obtained if, during execution, each individual binding of a constant name in a goal is tagged with its set of modifying goals (a detailed analysis of all the various binding possibilities is carried out in a subsequent section). In such taggings the reference to a goal may be its number, goals being numbered in the order they are activated. On backtracking, both tagging and goal numbering are undone.

Because any goal is an avoiding goal for its subgoals (since it is their ancestor), no goal is a modifying goal for its subgoals. Thus tagging should only be performed when a goal is solved, not during unification of a goal with a clause head.

Finally, a more expediently implemented but less precise selective backtracking can be obtained if, for each failed goal, no analysis is made of which are the failed conditions in each failed match. In that case, all modifying goals for every goal term are selected as backtrack goals, ie. failure is not associated with any particular goal term but with all of them.

4. On the possible types of unification conflicts

In this section we concentrate on the different types of conflict that may cause failure of the matching of a goal with a clause head.

We distinguish two cases.

First case. One of the constant names textually occurs at the conflicting position in the clause head.

e.g. (1) ←p(a)
p(b)← (2) ←p(X) [X=...=a]
p(b)← (X bound to a) (3) ←p(X,X)
p(a,b)←

For the match to succeed the other constant name must change. If this other constant name is also textually present in the goal, as in (1), no modifying goals exist for the conflict, meaning that it cannot be solved. If the second constant name is not textual, then it is obtained through some variable, as in (2). In this case, the modifying goals are those for the binding of the variable. When, in particular, the second constant name, binding a variable, is textually present in the clause head and is transmitted to the conflicting position through a multiple occurrence of that goal variable, as in (3), the conflict is irrevocable.

Second case. No constant name textually occurs at the conflicting position in the clause head.

There is a variable at that position, which must occur somewhere else in the clause head to receive the mismatching constant name from the goal itself, either textually or by a bound variable.

e.g. (1) ←p(a,b)
p(X,X)← (2) ←p(a,Y) [Y=...=b]
p(X,X)←

(3) ←p(Z,Y) [Z=...=a]
p(X,X)← [Y=...=b] (4) ←p(Y,Y,b)
p(a,X,X)←

In this case, the conflict is really between two constant names referred to by the goal. Accordingly, the conflict may be solved if either of them changes, and the modifying goals are the ones for each constant name. If, in particular, both conflicting constant names are textually present in the goal or head, as in (1) and (4), the conflict is irrevocable.

In all cases, a conflict is irrevocable if both conflicting constant names are textually present in the unification, and it is revocable if at least one conflicting constant name is not textually present in the unification.

Irrevocable conflicts can only be avoided, whereas revocable conflicts can either be avoided or modified by the undoing or redoing of a binding.

5. On the dependency of bindings on goal matches

We next examine how the presence of a constant name in the binding of a variable depends on goal matches.

The presence of a constant name as part of the binding of a textual variable in a goal G depends solely on the goals whose matches have transmitted that part of the binding. All of these, except the ancestors of G, are modifying goals for that constant name present in G.

1) If in a match a variable is bound to some non-variable term directly, ie. without any intervening variables, then all the constant names part of that binding will be tagged with the matching goal, when it is solved.

e.g.

```
←p(X)
p(f(Y,a))←q(Y)
```

'f' and 'a' will be tagged with goal 'p'

2) The remaining case is where a variable becomes dependent, by unification, on other variables, even though the latter may have not yet acquired a value.

2.1) The simplest subcase is when a variable occurring in the clause head becomes dependent on the actual or future binding (possibly in the same match) of a matching variable in the goal.

Because the goal is only an avoiding goal for any failed goal in the body of the matching clause, reference to the goal should not appear in the tagging of any constant name binding to those variables.

e.g.

```
(1)  ←p(X)           [ X=...=a ]      (2)  ←p(X,X)
      p(Y)←q(Y)                              p(a,Y)←q(Y)

(3)  ←p(X,X)
      p(Z,Y)←r(Z),q(Y)
              r(a)←
```

In all examples, reference to goal 'p' will not appear in the tagging of 'a' bound to Y.

The particular case, as in (3), where two variables in the clause head are unified through the same multiple occurring variable in the goal is already catered for.

2.2) Another subcase is where a goal variable becomes dependent on some future binding (possibly in the same match) of a matching variable in the head.

2.2.1) If there is a single occurrence of the variable in the head, then the future binding can only be made in the execution of a subgoal. Any failure due to the transmission of that binding, via the goal variable, will select that subgoal for backtracking. But that subgoal, through its parent, will reactivate all its ancestors, including the current goal, if need be. Consequently, the dependency of the goal variable on the current goal needs no tagging.

e.g.

```
←p(X)
 p(Y)←r(Y)
       r(a)←
```

reference to goal 'p' will not appear in the tagging of 'a' bound to X.

2.2.2) If there is a multiple occurrence of the head variable, it is linking two or more terms in the goal.

2.2.2.1) If both terms are ground no tagging occurs.

e.g.

```
←p(X,a)      ( X=...=a )
 p(Y,Y)←
```

2.2.2.2) If a constant name in one of the terms is bound to some variable in the other term it will be tagged with the goal, in that binding.

e.g.

```
←p(X,a)
 p(Y,Y)←
```

2.2.2.3) If a binding is between two free variables but, upon solution of the goal, they have become bound to some non-variable term by a subgoal, then reference to the goal should not be put in the tagging of that term, which already refers to the subgoal.

e.g.

```
←p(X,Y)
 p(Z,Z)←q(Z)
         q(a)←
```

reference to goal 'p' will not be put in the tagging of 'a' bound to X or Y.

On the other hand, if both variables are still free after the goal is solved, any subsequent binding of a constant name to them will have to be tagged with a reference to the goal. This is easily achieved if both variables are tagged with a reference to the goal.

e.g.

```
←p(X,Y)
 p(Z,Z)←
```

reference to goal 'p' will be tagged to X and Y.

To sum up:

Because parent goals of failed goals are always selected as backtrack goals, the goal dependencies created by simple transmission of bindings up and/or down the tree (through chains of ancestors possibly linked by common variables at brother goals) should not be noted explicitly.

Any solved goal in whose match a goal variable was bound to a textual non-variable term must be retained as a modifying goal for any constant name which then became part of the binding of that variable.

The only other case in which a solved goal must also be retained as a modifying goal is when two (or more) still free variables in the goal have been unified to one another in the matching of the goal.

6. <u>Selective backtracking examples</u>

6.1 Map colouring example

This example is a program for colouring any planar map with at most four colours, such that no two adjacent regions have the same colour (proved to be always possible).

The program consists of a complete list of pairs of different colours, taken from a collection of four. These constitute the admissible pairs of colours for regions next to each other.

```
next(blue,yellow)←
next(blue,red)←
next(blue,green)←
next(yellow,blue)←
next(yellow,red)←
next(yellow,green)←
next(red,blue)←
next(red,yellow)←
next(red,green)←
next(green,blue)←
next(green,yellow)←
next(green,red)←
```

To obtain a colouring of the map below, we give as a goal to the program all the pairs of regions that are next to each other. This can be done systematically by first pairing region 1 with higher numbered regions next to it, then region 2, etc.

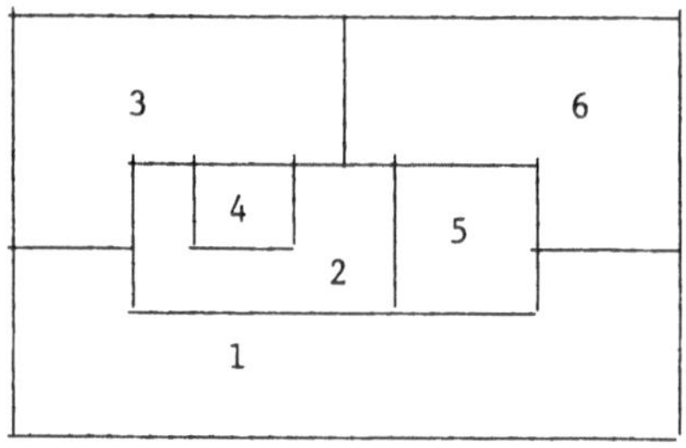


```
goal(R1,R2,R3,R4,R5,R6)←next(R1,R2),next(R1,R3),next(R1,R5),next(R1,R6),
                        next(R2,R3),next(R2,R4),next(R2,R5),next(R2,R6),
                        next(R3,R4),next(R3,R6),
                        next(R5,R6)
```

The effect of selective backtracking is best seen in the partial trace below of an execution using standard backtracking.

To help to follow the trace we have replaced the calls in the goal, which are of the form next(R_i,R_j), by calls of the form next(i:R_i,j:R_j).

In the trace, each procedure call is displayed with the current value of its arguments, and is preceded by a '-' sign; each time execution of a goal is successfully completed the call is displayed again, with the new current values of its arguments, this time preceded by a '+' sign.

On the right-hand side of each failed goal an arrowed line points to the goal to which standard backtracking returns. On the left-hand side of the trace another arrowed line points from the first failed goal to the point further down in the trace that would correspond to an execution with selective backtracking. The segment of trace jumped over corresponds exactly to the useless computation of standard backtracking, as compared with the selective version.

```
- goal(R1,R2,R3,R4,R5,R6)
- next(1:R1,2:R2)
+ next(1:blue,2:yellow)
- next(1:blue,3:R3)
+ next(1:blue,3:yellow)
- next(1:blue,5:R5)
+ next(1:blue,5:yellow)
- next(1:blue,6:R6)
+ next(1:blue,6:yellow)
- next(2:yellow,3:yellow)
+ next(1:blue,6:red)
- next(2:yellow,3:yellow)
+ next(1:blue,6:green)
- next(2:yellow,3:yellow)
+ next(1:blue,5:red)
- next(1:blue,6:R6)
+ next(1:blue,6:yellow)
- next(2:yellow,3:yellow)
+ next(1:blue,6:red)
- next(2:yellow,3:yellow)
+ next(1:blue,6:green)
- next(2:yellow,3:yellow)
+ next(1:blue,5:green)
- next(1:blue,6:R6)
+ next(1:blue,6:yellow)
- next(2:yellow,3:yellow)
+ next(1:blue,6:red)
- next(2:yellow,3:yellow)
+ next(1:blue,6:green)
- next(2:yellow,3:yellow)
+ next(1:blue,3:red)
- next(1:blue,5:R5)
+ next(1:blue,5:yellow)
- next(1:blue,6:R6)
+ next(1:blue,6:yellow)
- next(2:yellow,3:red)
+ next(2:yellow,3:red)
```

6.2 Relational database example

This example deals with a database of students who take courses, professors who teach courses, and courses held on certain weekdays and rooms. Our example query is

"Is there a student such that a professor teaches him two different courses in the same room?"

```
query(S,P)←student(S,C1),      1
           course(C1,D1,R),    2
           professor(P,C1),    3
           student(S,C2),      4
           course(C2,D2,R),    5
           professor(P,C2),    6
           C1≠C2               7
```

The example database follows.

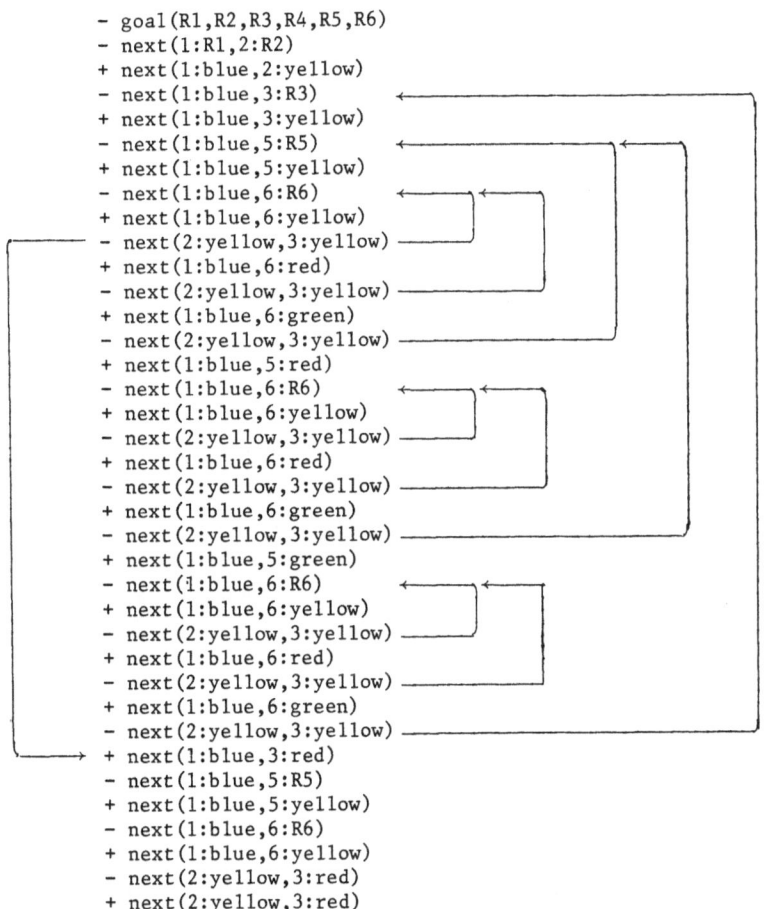

```
student(robert,prolog)←

student(john,music)←
student(john,prolog)←
student(john,surf)←

student(mary,science)←
student(mary,art)←
student(mary,physics)←

professor(luis,prolog)←
professor(luis,surf)←

professor(antonio,prolog)←

professor(eureka,music)←
professor(eureka,art)←
professor(eureka,science)←
professor(eureka,physics)←

course(prolog,monday,room1)←
course(prolog,friday,room1)←

course(surf,sunday,beach)←

course(maths,tuesday,room1)←
course(maths,friday,room2)←

course(science,thursday,room1)←
course(science,friday,room2)←

course(art,tuesday,room1)←

course(physics,thursday,room3)←
course(physics,saturday,room2)←
```

The first student and course picked up in goal 1 are robert and prolog, which are again picked up in goal 4. Execution fails at 7, where C1 and C2 are tested to be different. Both C1 and C2 contribute to failure. The binding of C1 (prolog) was obtained at 1, and that of C2 (prolog) at 4. So the set of backtrack goals is (0,1,4), where 0 is the parent goal. Consequently, backtracking jumps to 4 over 6 and 5.

Goal 4 fails since robert takes no more courses. An analysis would now ensue of the failed matches of goal 4. However, in any sized database, that analysis is too expensive. A simple solution to this problem consists in doing no analysis at all, and in taking as modifying goals all the modifying goals associated with the bindings of all the arguments of the goal. Certainly no solutions are lost in doing so, and probably any one constant name in the bindings would contribute with the most ancient modifying goal in some clause, given the diversity of the database.

To continue with the example, at goal 4 backtrack goal 1, corresponding to the binding of robert to S, is re-selected. Since C2 is still free it renders no modifying goal. The set of backtrack goals becomes (0,1).

Backtracking now jumps to 1 over 3 and 2. This time john and music are picked up. Failure occurs again at 7 but, after backtracking is made to 4, another solution is found for C2, namely prolog. However, the professor of music is not a professor of prolog, so 6 fails. Because the bindings of P and C2 have originated at 3 and 4, respectively, the set is updated to (0,3,4).

And so on and so back and forth.

It should be noted that no other ordering of the goals within the query could prevent selective backtracking from being advantageous over standard backtracking.

6.3 Non-attacking chessboard queens example

The problem consists in placing n chess queens on a nxn chessboard such that no two queens attack one another. The program is general, only the input changes according to the number of queens. For four queens it is

```
←perm(4.3.2.1.nil,L),pair(4.3.2.1.nil,L,Q),safe(Q)
```

The position of a queen is a pair of coordinates. 'perm' generates a permutation of row positions for the four queens, thereby ensuring that no two queens occupy the same row. 'pair' pairs each element of the list L of row positions with a different column, thereby ensuring that no two queens occupy the same column. 'safe' inspects whether any two queens are on the same diagonal.

The program follows.

```
perm(nil,nil)←
perm(X.Y,U.V)←delete(U,X.Y,W),perm(W,V)

delete(X,X.Y,Y)←
delete(U,X.Y,X.V)←delete(U,Y,V)

pair(nil,nil,nil)←
pair(X.Y,U.V,p(X,U).W)←pair(Y,V,W)

safe(nil)←
safe(Q.R)←check(Q,R),safe(R)

check(Q,nil)←
check(Q,R.S)←not_on_diagonal(Q,R),check(Q,S)

not_on_diagonal(p(C1,R1),p(C2,R2))←minus(C1,C2,C),
                                    minus(R1,R2,R),
                                    C≠R,
                                    minus(R2,R1,NR),
                                    C≠NR
```

'safe' uses 'check' to check each pair against all the following pairs in the list. This is accomplished by subtracting corresponding coordinates with the predicate 'minus'.

Backtracking occurs when two queens are on the same diagonal. Standard backtracking would generate the next permutation, even though it may not alter the coordinates of the conflicting queens. Selective backtracking will return to the point in the generation of the permutation where one of the conflicting queens is assigned a different row if possible.

The first permutation being 4.3.2.1.nil, the first two pairs, p(4,4) and p(3,3), are the first to conflict. Standard backtracking would then permute the rows of the queens on columns 1 and 2 to no avail. Selective backtracking begins by detecting that the conflicting constants binding C and R (1 and 1) at goal C≠R were obtained at minus(C1,C2,C) and minus(R1,R2,R), and selects them as backtrack goals. Since there are no alternatives for these calls, they fail. Which are the modifying goals for the current values of C1,C2,R1 and R2? These values were transmitted through a chain of ancestors up to the first call to 'safe', but such dependencies are implicit. The two first calls to 'pair' originated bindings responsible for the transmission of the values, so they are modifying goals. Also, the presence of the values for R1 and R2 in those two first calls to 'pair' depends on the two first calls to 'delete', where they were transmitted from the second to the first argument of the first clause head for 'delete'; these must therefore be also modifying goals. The transmission of the values from perm(4.3.2.1.nil,L) down to the two first calls to 'delete' and then up again is again implicitly accounted for.

To sum up, the successive backtrack goals generated upon failure of C≠R are: first, minus(4,3,C), minus(4,3,R) and the parent not_on_diagonal(p(4,4),p(3,3)); next, on failure of minus(4,3,R), the second call to 'pair', pair(3.2.1.nil,3.2.1.nil,W), and the second call to 'delete', delete(U,3.2.1.nil,W); next, on failure of the call minus(4,3,C), the first call to 'pair', pair(4.3.2.1.nil,4.3.2.1.nil,Q), and the first call to 'delete', delete(U,4.3.2.1.nil,W).

Finally, everytime one of these backtrack goals fails because there are no more clauses for it, its parent goal is included as a backtrack goal, and an analysis takes place to find reasons for any failure of the goal in matching the clauses. Thus, when 'not_on_diagonal' fails, there being no failure in entering its single clause, only its parent, 'check', is selected. But 'check' also fails, so an analysis takes place to examine why its first clause failed to match the goal. The reason is the conflict between the principal functor '.' in the binding of R and 'nil'. A modifying goal for that conflict is the goal in whose match that functor was obtained. It turns out to be the second call to 'pair', already retained as a backtrack goal. The fact is that all constant components of p(4,4).W were obtained in the match of the second call of 'pair'. Similarly, analysis of the failure of 'safe', ie. the conflict between the binding of Q and 'nil', singles out the first call of 'pair' as the culprit, already retained as a backtrack goal.

References

[1] Bruynooghe, M.
Intelligent backtracking for Horn clause logic programs
Colloquium on Mathematical Logic in Programming
Salgotarjan, Hungary 1978

[2] Coelho, H. ; Cotta, J.C. ; Pereira, L.M.
How to solve it with Prolog
Laboratório Nacional de Engenharia Civil, Lisboa 1979

[3] Kowalski, R.A.
Predicate logic as a programming language
IFIP 74, pp 569-574, North-Holland Publ. Co. 1974

[4] Kowalski, R.A.
Logic for problem solving
North-Holland Publ. Co. 1980

[5] Loveland, D.
Automated theorem proving: a logical basis
North-Holland Publ. Co. 1978

[6] Pereira, L.M. ; Pereira, F.C.N. ; Warren, D.H.D.
User's guide to DECsystem-10 Prolog
Laboratório Nacional de Engenharia Civil, Lisboa 1978

[7] Pereira, L.M.
Backtracking intelligently in AND/OR trees
Departamento de Informática
Universidade Nova de Lisboa, Lisboa 1979

(8) Pereira, L.M. ; Porto, A.
Intelligent backtracking and sidetracking in Horn clause programs - the theory
Departamento de Informática
Universidade Nova de Lisboa, Lisboa 1979

(9) Pereira, L.M. ; Porto, A.
An interpreter of logic programs using selective backtracking
Paper submitted to the Workshop on Logic Programming, organized by the von Neumann Computer Society, Budapest July 1980

(10) Roussel, P.
Prolog: manuel de référence et d' utilisation
Groupe d' Intelligence Artificielle,
Université d' Aix-Marseille II 1975

(11) Warren, D.H.D.
Implementing Prolog, Parts I and II
Department of Artificial Intelligence
Edinburgh University 1977

(12) Warren, D.H.D. ; Pereira, L.M. ; Pereira, F.C.N.
Prolog, the language and its implementation compared with Lisp
ACM Symposium on Artificial Intelligence and Programming Languages,
Sigart Newsletter no. 64, and Sigplan Notices vol. 12, no. 8, August 1977

CANONICAL FORMS AND UNIFICATION

Jean-Marie Hullot
INRIA and SRI International

Abstract

Fay has described in [2,3] a complete T-unification for equational theories T which possess a complete set of reductions as defined by Knuth & Bendix [12]. This algorithm relies essentially on using the narrowing process defined by Lankford [13]. In this paper, we first study the relations between narrowing and unification and we give a new version of Fay's algorithm. We then show how to eliminate many redundancies in this algorithm and give a sufficient condition for the termination of the algorithm. In a last part, we show how to extend the previous results to various kinds of canonical term rewriting systems.

1. Introduction.

In this paper, we are interested in unification problems arising in equational theories. More precisely, we study the case of an equational theory T which may be embedded in complete set of reductions (or canonical term rewriting system) $\mathcal{R}$ as defined by Knuth & Bendix [12]. Let us write $\rightarrow$ the reduction relation associated to $\mathcal{R}$ and $\mathcal{R}(M)$ the (unique) $\rightarrow$-normal form of any term M. A decision procedure for T-equality is known via $\rightarrow$-normal form:

$$M =_T N \Leftrightarrow \mathcal{R}(M) = \mathcal{R}(N).$$

We would like also to be able to solve equations in T, that is to T-unify two given terms; this is a quite difficult problem. Take for instance the canonical term rewriting sytem reduced to the single equation:

$$f(f(x,y),z) \rightarrow f(x,f(y,z)).$$

Solving equations in the corresponding equational theory is the problem of associative unification which has only recently shown to be decidable by Makanin [21].

A general result has been obtained by Fay who describes in [2,3] such a process to resolve equations in T. This algorithm is not in general a decision procedure for T-unification since the termination of the process is not insured. However one can organize this algorithm in such a manner that it can be used as a semi-decision procedure for T-unification in the following sense: if the two terms are unifiable, then a solution will be found in a finite time. Moreover Fay has shown that this algorithm produces a complete set of T-unifiers. Improvements over Fay's algorithm have been given by Lankford. An analogous result has been found by Huet who has given in [4] a unification algorithm for λ-calculus which relies essentially on the existence of a canonical form for λ-conversion.

In this paper we shall present a new version of Fay's algorithm. In a second step we shall eliminate many redundancies in our algorithm and sufficient conditions will be given insuring the termination of the algorithm.

New kinds of canonical term rewriting systems have been defined by Lankford & Ballantyne [16,17,18], Huet [6,7] and Peterson & Stickel [24]. Fay conjectures in [3] that his algorithm extends to the case of permutative reductions [17] and Lankford & Ballantyne use such an extension for associative and commutative theories in the appendix of [19]. We give in this paper extensions of Fay's algorithm for Huet's and Peterson & Stickel's canonical term rewriting systems.

2. An overview of first order terms.

We give a brief survey on first order terms. Our definitions and notations are consistent with those of Huet [6] and Huet & Oppen [9].

2.1. *First order terms.*

Let $\mathcal{V}$ be a denumerable set of elements called *variables*, and C a finite or denumerable set of elements called *constants* with $\mathcal{V} \cap C = \emptyset$. Elements in C are graded by an *arity function* α: $C \mapsto \mathcal{N}$ where $\mathcal{N}$ is the set of integers. The *set of terms* $\mathcal{T}$ is the smallest set containing $\mathcal{V}$ and closed by the operations:

$$M_1, \ldots, M_{\alpha(f)} \mapsto f(M_1, \ldots, M_{\alpha(f)})$$

for every f in C. If $\alpha(f) = 0$ we abbreviate $f()$ in f. For every M in $\mathcal{T}$, $\mathcal{V}(M)$ will denote the *set of variables of* M.

2.2. *Substitutions.*

Definition. A *substitution* is a mapping σ from $\mathcal{V}$ to $\mathcal{V}$ with $\sigma(x) = x$ almost everywhere. Substitutions are classically extended as morphisms of $\mathcal{T}$. For every substitution σ, we define the *set of variables affected by* σ or *domain of* σ by $\mathcal{D}(\sigma) = \{x|\sigma(x) \neq x\}$ and the *set of variables introduced by* σ by $\mathcal{I}(\sigma) = \bigcup_{x \in \mathcal{D}(\sigma)} \mathcal{V}(\sigma(x))$. For $V \subset \mathcal{V}$, we define the *restriction of* σ *to* V by $(\sigma \uparrow V)(x) = \sigma(x)$ if $x \in V$ and $(\sigma \uparrow V)(x) = x$ otherwise.

We define the preorder $\leq$ of *subsumption* in $\mathcal{T}$ by $M \leq N \Leftrightarrow \exists\sigma \quad \sigma(M) = N$. If such a σ exists, its restriction to $\mathcal{V}(M)$ is unique and we call it *the match of N by M*. Finally, we define: $M \equiv N \Leftrightarrow M \leq N \;\; \& \;\; N \leq M$, $\equiv$ is *variable renaming*. We extend $\leq$ to substitutions by: $\sigma \leq \sigma' \Leftrightarrow \forall x \quad \sigma(x) \leq \sigma'(x)$.

We say that two terms M and M' are *unifiable* iff $\exists\sigma \quad \sigma(M) = \sigma(M')$. Let us denote by $\mathcal{U}_\emptyset(M, M')$ the set of all unifiers of M and M'. If two terms are unifiable, then there exists a minimum unifier, that is $\exists\sigma \in \mathcal{U}_\emptyset(M, M')$, $\forall\theta \in \mathcal{U}_\emptyset(M, M')$, $\sigma \leq \theta$. This element is unique up to variable renaming and may be found by the unification algorithm [5,22,23,27]. Furthermore, note that it is always possible to impose to the minimum unifier σ the condition $\mathcal{D}(\sigma) \cap \mathcal{I}(\sigma) = \emptyset$. We will always choose such a minimum unifier.

2.3. *Occurrences.*

In order to have a formal way to deal with subterm of a term we define the *occurrences* that is sequences of integers denoting an access path in a term. Let N_+^* be the set of finite sequences of positive integers, Λ the empty sequence and $\cdot$ the operation of concatenation on sequences. The elements of N_+^* are called *occurrences* and we denote them by u, v, w. The set of occurrences is partially ordered by the prefix ordering: $u \preceq v \Leftrightarrow \exists w \quad v = u \cdot w$ in this case, we define $v/u = w$. If $u \not\preceq v$ and $v \not\preceq u$ we say that u and v are *disjoint*, and we write $u \mid v$. Finally, we define $u \prec v$ iff $u \preceq v$ and $u \neq v$. For any term M, we define its *set of occurrences* $\mathcal{O}(M)$ as a finite subset of N_+^* as follows:

(i) $\Lambda \in \mathcal{O}(M)$,
(ii) $u \in \mathcal{O}(M_i) \Rightarrow i \cdot u \in \mathcal{O}(F(M_1, \ldots, M_n)) \quad \forall i \quad 1 \leq i \leq n$.

If $u \in \mathcal{O}(M)$, we define the *subterm of* M at u as the term M/u, and for every M' the *replacement in* M *of* M' *at* u *as the term* $M[u \leftarrow M']$, by:

(i) $M/\Lambda = M$,
(ii) $F(M_1, \ldots, M_n)/i \cdot u = M_i/u$,
(iii) $M[\Lambda \leftarrow M'] = M'$,
(iv) $F(M_1, \ldots, M_n)[i \cdot u \leftarrow M'] = F(M_1, \ldots, M_i[u \leftarrow M'], \ldots, M_n)$.

In order to distinguish between variable and non-variable occurrences we define $\overline{\mathcal{O}}(M)$ as $\{u \in \mathcal{O}(M)|M/u \notin \mathcal{V}\}$, and $\mathcal{F}r(M) = \{u \in \mathcal{O}(M)|M/u \in \mathcal{V}\}$.

3. Reduction, narrowing and unification.

We call an *equation* any pair of terms. An equation will be denoted by $M = N$. Let T be any finite set of equations. We define the relation $\doteq_T$ on T as the compatible, stable, symmetric closure of T. We define the equality in the *equational theory* T or T-*equality* to be the congruence generated by T, that is $\overset{*}{\doteq}_T$. It will be denoted by $=_T$.

Example. The equational theory of groups is generated by the set of equations:

$$x + 0 = x;$$
$$x + (-x) = 0;$$
$$(x + y) + z = x + (y + z).$$

We are interested in *unifying terms in the equational theory* T; given two terms M and M', we shall say that a substitution σ is a T-unifier of M and M' iff $\sigma(M) =_T \sigma(M')$. $\mathcal{U}_T(M, M')$ will denote the set of all T-unifiers of M and M'. When dealing with ordinary unification, we have seen that two unifiable terms have a minimum unifier. This is no longer the case with T-unification. However, this notion generalizes to the one of *complete set of* T-*unifiers* [4,5,25].

Definition. Let M and M' be two terms and W be a finite set of variables containing $V = \mathcal{V}(M) \cup \mathcal{V}(M')$. We say that a set of substitutions Σ is a *complete set of unifiers of* M *and* M' *away from* W iff:

(i) $\forall \sigma \in \Sigma \quad \mathcal{D}(\sigma) \subset V \quad \& \quad \mathcal{I}(\sigma) \cap W = \emptyset$,
(ii) $\Sigma \subset \mathcal{U}_T(M, M')$,
(iii) $\forall \sigma \in \mathcal{U}_T(M, M') \quad \exists \theta \in \Sigma \quad \theta \leq_T \sigma \, [\![V]\!]$.

where $\leq_T$ is the preorder defined on T by $M \leq_T M'$ iff $M' =_T \sigma(M)$ and $\leq_T \ [\![V]\!]$ is the following extension of $\leq_T$ to substitutions: $\sigma \leq \sigma' \, [\![V]\!]$ iff there exists a substitution ρ such that $\sigma'(x) =_T \rho(\sigma(x))$ for all x in V. $CSU_T(M, M', W)$ will denote the set of all such Σ's. In addition Σ is said to be *minimal* iff it satisfies the further condition:

(iv) $\forall \sigma, \sigma' \in \Sigma \quad \sigma \neq \sigma' \Rightarrow \sigma \not\leq_T \sigma' \, [\![V]\!]$.

A discussion of general properties of complete set of unifiers may be found in [5]. Remark that the introduction of the set W is motivated by technical reasons: it is a way to avoid conflicts when choosing new variables.

A T-unification algorithm is *complete* if it generates a complete set of T-unifiers for all T-unifiable input terms. Complete unification algorithm are known for various theories including commutativity, associativity [25], idempotence [26], associativity and commutativity [29], associativity commutativity and idempotence [20] and abelian group theory [14]. Note that we do not require any termination property.

A very general result on T-unification problems has been obtained by Fay who describes in [2,3] such a T-unification algorithm in the case where the equational theory T may be described by a *complete set of reductions* as defined by Knuth & Bendix [12]. We recall that a *term rewriting system* $\mathcal{R}$ is a set of pairs of terms $\gamma_k \to \delta_k$, such that $\mathcal{V}(\gamma_k) \subseteq \mathcal{V}(\delta_k)$. We say that term $M \to_{\mathcal{R}}$-reduces to term N at occurrence u and we write $M \to_{\mathcal{R}} N$ iff:

$$\exists \gamma_k \to \delta_k \in \mathcal{R} \quad \exists \sigma \quad \exists u \in \mathcal{O}(M) \quad M/u = \sigma(\gamma_k) \quad \& \quad N = M[u \leftarrow \sigma(\delta_k)].$$

Sometimes we will write: $M \to_{[u,k]} N$. We say that a term M is in $\to_{\mathcal{R}}$-*normal form* iff $\not\exists N$, $M \to_{\mathcal{R}} N$. If $M \overset{*}{\to}_{\mathcal{R}} N$ and N is in $\to_{\mathcal{R}}$-normal form, we say that N is a $\to_{\mathcal{R}}$-*normal form of* M. We say that a substitution σ is in $\to_{\mathcal{R}}$-*normal form* iff $\forall x \in \mathcal{D}(\sigma)$, $\sigma(x)$ is in $\to_{\mathcal{R}}$-normal form.

A term rewriting system $\mathcal{R}$ is said to be a *complete set of reductions* or a *canonical term rewriting system* iff:

(a) $\rightarrow_{\mathcal{R}}$ is nœtherian, that is, there does not exist infinite derivation $M_1 \rightarrow M_2 \cdots$

(b) $\rightarrow_{\mathcal{R}}$ is confluent, that is $\forall M, M_1, M_2$, such that $M \overset{*}{\rightarrow}_{\mathcal{R}} M_1$ and $M \overset{*}{\rightarrow}_{\mathcal{R}} M_2$ then $\exists M'$ such that $M_1 \overset{*}{\rightarrow}_{\mathcal{R}} M'$ and $M_2 \overset{*}{\rightarrow}_{\mathcal{R}} M'$.

Note that if $\mathcal{R}$ is a canonical term rewriting system, each term M admits a unique $\rightarrow_{\mathcal{R}}$-normal form we shall denote by $\mathcal{R}(M)$. Thus a decision procedure for T-equality is known via this normal form. More precisely, we have:

$$M =_T N \Leftrightarrow \mathcal{R}(M) = \mathcal{R}(N).$$

Knuth & Bendix give in [12] a way to decide if a finite and nœtherian term rewriting system $\mathcal{R}$ is canonical. A detailed study may be found also in [6].

It is not possible to define a canonical term rewriting system for a theory T which includes for instance a commutativity axiom (condition (a) never holds with such an axiom). Differents ways have been proposed to extend the notion of canonical term rewriting systems as we shall see in section 5.

Our aim in this section is to give a new version of Fay's algorithm.

Notation. $\rightarrow$ will be $\rightarrow_{\mathcal{R}}$ where $\mathcal{R}$ is any canonical term rewriting system defining an equational theory T.

3.1. Narrowing.

The basic idea in Fay's algorithm is to combine ordinary unification and "narrowing". Informally narrowing a term is applying to it the minimum substitution such that the resulting term is not in $\rightarrow$-normal form and then reducing it one step. Slagle [28] has first noticed the difficulties arising when working with terms which are "narrowable" and thus considered in its work only sets of clauses which are fully "narrowed". Lankford in [13] was first to show the interest of the iteration of narrowing and to give applications of such a "narrowing procedure". We define below a relation on $\mathcal{T}$ we call the *narrowing relation*. A step of derivation using this relation is very close to the notion of *immediate narrowing* of Lankford.

Definition. Let M be a term and V be a finite set of variables containing $\mathcal{V}(M)$. Assume there exists a non-variable subterm of M, say M_1, which is unifiable with the left part of some rule $\gamma_k \rightarrow \delta_k$ in $\mathcal{R}$. More formally:

$$\exists u \in \overline{O}(M), \quad \exists \gamma_k \rightarrow \delta_k \in \mathcal{R}, \quad \mathcal{U}_{\emptyset}(M/u, \gamma_k) \neq \emptyset,$$

where we assume $\gamma_k \rightarrow \delta_k$ is renamed so that $\mathcal{V}(\gamma_k) \cap V = \emptyset$.
Let σ the minimum unifier of $M_1 = M/u$ and γ_k. We say that σ is a *narrowing substitution of M away from V*. $NS(M, V)$ will denote the finite set of such substitutions.
Let us now consider the term M obtained from $\sigma(M)$ in replacing $\sigma(M_1)$ by $\sigma(\delta_k)$, that is:

$$M' = \sigma(M)[u \leftarrow \sigma(\delta_k)] = \sigma(M[u \leftarrow \delta_k]).$$

We say that *M is narrowable in M' at occurrence u using rule $\gamma_k \rightarrow \delta_k$* and we write:

$$M \rightsquigarrow_{[u,k,\sigma]} M'.$$

$\rightsquigarrow$ is called *the narrowing relation on $\mathcal{T}$*.

Notations. Sometimes we will abbreviate $\rightsquigarrow_{[u,k,\sigma]}$ in $\rightsquigarrow_{[u,k]}$ or $\rightsquigarrow_{[\sigma]}$.

Remark 1. Note that our definition is not exactly the one used by Lankford, Fay or Slagle. More precisely, we do not assume that M is normalized and we do not normalize M', so that we have:

$$\rightarrow \subseteq \rightsquigarrow.$$

The motivation behind this modification is that we want to express very precisely correspondances between reduction and narrowing in order to find sufficient conditions insuring the termination of the narrowing process.

Remark 2. The condition $\mathcal{V}(\gamma_k) \cap V = \emptyset$ insures that $\sigma(M)$ is reducible by $\rightarrow$ using $\gamma_k \rightarrow \delta_k$. Note that M' may contain variables which are not in V; these variables come from $\mathcal{V}(\gamma_k)$. In the next section, when we shall study the iteration of the narrowing process we will have to make explicit the renaming of variables of the γ_k's at each step of the iteration in order to avoid conflict between the names of all these "new" variables. In practice the problem is simplified in using the GENSYM operator of LISP for instance.

Example. Let us consider the following canonical term rewriting system $\mathcal{R}$:

$$\mathcal{R} = \{f(x,x) \rightarrow x\},$$

and the term $M = f(x_1, f(y_1, z_1))$. M is narrowable at occurrence Λ since it is unifiable with $f(x,x)$:

$$M \rightsquigarrow_{[\Lambda]} f(y_1, z_1), \quad \sigma = \{(x \leftarrow f(y_1, z_1)), (x_1 \leftarrow f(y_1, z_1))\},$$

and at occurrence 1 for the subterm $f(y_1, z_1)$ is unifiable with $f(x,x)$:

$$M \rightsquigarrow_{[1]} f(x_1, x), \quad \sigma = \{(y_1 \leftarrow x), (z_1 \leftarrow x)\}.$$

Remark 3. Let us now have a look back to the initial problem we were interested in: given two terms P and Q, find a substitution σ such that:

$$\sigma(P) =_T \sigma(Q), \tag{1}$$

or equivalently:

$$\mathcal{R}(\sigma(P)) = \mathcal{R}(\sigma(Q)). \tag{2}$$

Assume the problem has a solution, say σ, then there are two cases: either σ is a unifier of P and Q in the usual sense in which case one can find a minimum unifier of P and Q in using the unification algorithm. Or σ does not unify P and Q, in which case at least one of $\sigma(P)$ and $\sigma(Q)$ is $\rightarrow$-reducible, since otherwise it would be impossible to have (2). In this case at least one of P and Q is narrowable.

3.2. Narrowing and reduction.

Using the notations of remark 3 of the previous section, the equivalence between (1) and (2) gives a particular interest in the study of $\rightarrow$-derivation issuing from $\sigma(M)$ where σ is any substitution and M is any term. The aim of the following theorem is to show how one can make correspond any such derivation to a $\rightsquigarrow$-derivation and conversely. In other words, we shall show that any $\rightarrow$-derivation issuing from $\sigma(M)$ may be "projected" on a $\rightsquigarrow$-derivation issuing from M. And conversely any $\rightsquigarrow$-derivation issuing from M may be considered as the "projection" of a certain class of $\rightarrow$-derivations.

Theorem 1. *Let M be any term, V be a finite set of variables containing $\mathcal{V}(M)$, and η be a normalized substitution with $\mathcal{D}(\eta) \subseteq \mathcal{V}(M)$. Consider any $\rightarrow$-derivation issuing from $\eta(M)$:*

$$\eta(M) = N_0 \rightarrow_{[u_0,k_0]} N_1 \rightarrow_{[u_1,k_1]} N_2 \rightarrow \cdots \rightarrow_{[u_{n-1},k_{n-1}]} N_n. \tag{1}$$

There exists an associated $\rightsquigarrow$-derivation issuing from M:

$$M = M_0 \rightsquigarrow_{[u_0,k_0,\sigma_0]} M_1 \rightsquigarrow_{[u_1,k_1,\sigma_1]} M_2 \rightsquigarrow \cdots \rightsquigarrow_{[u_{n-1},k_{n-1},\sigma_{n-1}]} M_n, \tag{2}$$

and for each i, $0 \le i \le n$, a substitution η_i and a finite set of variables V_i such that:

(i) $\mathcal{D}(\eta_i) \subseteq V_i$,
(ii) η_i *is normalized,*
(iii) $(\eta \uparrow V) = (\eta_i\theta_i \uparrow V)$,
(iv) $\eta_i(M_i) = N_i$,

where $\theta_0 = \epsilon$ and $\theta_{i+1} = \sigma_{i+1}\theta_i$.

Conversely, to each $\rightsquigarrow$-derivation (2) and every η such that $\theta_n \le \eta[\![V]\!]$, we can associate a $\rightarrow$-derivation (1).

As usual the motivation behind the introduction of the V_i is technical: it is a way to avoid conflicts between "new" and initial variables. As one can see in the following proof, it is not totally trivial to deal in a mathematical way with this problem of renaming. In a certain way we have to give all the details of a "garbage collecting" process.

Proof. $\Rightarrow$ By induction on i. For $i = 0$ it is obvious, taking $\eta_0 = \eta$ and $V_0 = V \cup \mathcal{D}(\eta)$. Let us assume (i) to (iv) hold for i.
Since $M_i \rightsquigarrow_{[u_i,k_i]} M_{i+1}$, we have:

$$\exists\sigma \quad \sigma(\gamma_{k_i}) = M_i/u_i,$$

where γ_{k_i} is renamed so that $\mathcal{D}(\sigma) \cap V_i = \emptyset$.
From assumptions (ii) and (iv) for i, we get $u_i \in \overline{O}(M_i)$ and therefore:

$$\eta_i(M_i/u_i) = \sigma(\gamma_{k_i}).$$

Let us consider $\rho = \eta_i \cup \sigma$. We have:

$$\rho(M_i/u_i) = \rho(\gamma_{k_i}),$$

and thus:

$$M_i \rightsquigarrow_{[u_i,k_i,\sigma_i]} M_{i+1},$$

where σ_i is the minimum unifier of M_i/u_{i+1} and γ_{k_i}. We have $\sigma_i \le \rho$, and thus there exists a substitution η' such that $\rho = \eta'\sigma_i$. Therefore:

$$\eta_i = ((\eta'\sigma_i) \uparrow V_i).$$

Now, let:

$$V_{i+1} = (V_i \cup \mathcal{I}(\sigma_i)) - \mathcal{D}(\sigma_i),$$

and:

$$\eta_{i+1} = \eta' \uparrow V_{i+1}.$$

We get (i) and:

$$\eta_i = (\eta_{i+1}\sigma_i) \uparrow V_i. \tag{*}$$

(remember that we impose $\mathcal{D}(\sigma_i) \cap \mathcal{I}(\sigma_i) = \emptyset$).
Now, let us consider x in V_{i+1}. There are two cases:

a) $x \in \mathcal{I}(\sigma_i)$; then $\exists y \in \mathcal{D}(\eta_i)$ such that $x \in \mathcal{V}(\sigma_i(y))$, and $\eta_i(y) = \eta_{i+1}(\sigma_i(y))$ normalized implies $\eta_{i+1}(x)$ normalized.
b) otherwise $\sigma_i(x) = x$ since $x \notin \mathcal{D}(\sigma_{i+1})$ and therefore $\eta_{i+1}(x) = \eta_i(x)$ is normalized, which proves (ii).

We now assume (iii) for i:

$$\eta \uparrow V = \eta_i\theta_i \uparrow V,$$

and show it fo $i+1$. From (*) above, we get:

$$(\eta_i\theta_i) \uparrow V = (((\eta_{i+1}\sigma_i) \uparrow V_i)\theta_i) \uparrow V.$$

From the definition of θ_i, we get $\mathcal{I}(\theta_i) \subseteq V_i$ and $V \subseteq V_i \cup \mathcal{D}(\theta_i)$. The above expression simplifies therefore to:

$$(\eta_{i+1}\sigma_i\theta_i) \uparrow V = (\eta_{i+1}\sigma_i) \uparrow V,$$

proving (iii).

Finally we get easily $\mathcal{V}(M_i) \subseteq V_i$, from which we get:

$$\eta_{i+1}(M_{i+1}) = \eta_{i+1}\sigma_i(M_i[u_i \leftarrow \delta_{k_i}]) = \eta_i(M_i[u_{i+1} \leftarrow \delta_{k_{i+1}}) = N_{i+1},$$

proving (iv).

Note that because of (iii) every $\theta_i \uparrow V$ is normalized.

$\Leftarrow$ Conversely, let us consider any $\rightsquigarrow$-derivation (2) and any substitution η such that $\theta_n \leq \eta[\![V]\!]$. Let ρ be such that $\eta \uparrow V = (\rho\theta_n) \uparrow V$. We define substitutions η_i for $0 \leq i \leq n-1$ by:

$$\eta_i = \rho\sigma_n\sigma_{n-1}\cdots\sigma_i,$$

and substitution η_n as being ρ. With $N_i = \eta_i(M_i)$, it is easy to show by induction on i, that:

$$\eta(M) = N_0 \rightarrow_{[u_0,k_0]} N_1 \rightarrow_{[u_1,k_1]} N_2 \rightarrow \cdots \rightarrow_{[u_{n-1},k_{n-1}]} N_n.$$

Now:

$$N_0 = \eta_0(M_0) = \eta_0(M) = \eta_n\theta_n(M) = \eta(M),$$

since $\mathcal{V}(M) \subseteq V$, which establishes the $\Leftarrow$-part. ∎

3.3. Narrowing and unification.

In this section we will describe a non-deterministic T-unification algorithm. Before giving this algorithm we prove two key lemmas about the connection between narrowing and T-unification.

Let us consider two terms P and Q. In order to find T-unifiability properties of these two terms we will have to iterate the narrowing process on both P and Q in parallel. It will simplify matters to iterate the narrowing process on the single term $M = H(P,Q)$ where H is a "new" function symbol, that is $H \notin C$, playing the rôle of cartesian product.

In lemma 1, we show how to combine narrowing and ordinary unification to build T-unifiers. This lemma will show the correctness of the unification algorithm.

Lemma 1. *Let us consider any $\rightsquigarrow$-derivation:*

$$M = H(P,Q) = M_0 \rightsquigarrow M_1 = H(P_1,Q_1) \rightsquigarrow \cdots \rightsquigarrow M_n = H(P_n,Q_n),$$

such that P_n and Q_n are unifiable say by substitution σ. Then $\sigma\theta_n$ is a T-unifier of P and Q, where θ_n is the composition of substitutions along the derivation, as defined in theorem 1.

Proof. Using the $\Leftarrow$ part of the previous theorem with $\eta = \rho_n$, we can associate to this $\rightsquigarrow$-derivation the following $\rightarrow$-derivation:

$$\theta_n(M) = N_0 \rightarrow N_1 \rightarrow N_2 \rightarrow \cdots \rightarrow N_n = H(N_n^P, N_n^Q),$$

and thus, we have:

$$\theta_n(P) \rightarrow^* N_n^P \quad \& \quad \theta_n(Q) \rightarrow^* N_n^Q.$$

Moreover, since $\eta_n = \epsilon$ in this case, we have:

$$N_n^P = P_n \quad \& \quad N_n^Q = Q_n,$$

thus:

$$\sigma\theta_n(P) =_T \sigma\theta_n(Q),$$

since these two terms are $\rightarrow$-reducible to the same term. ∎

In lemma 2 we show that any T-unifier may be reached in such a way. This lemma will be used when showing the completeness of the T-unification algorithm.

Lemma 2. *Let P and Q be two terms which are T-unifiable, ρ be any T-unifier and V be a finite set of variables containing $\mathcal{V}(P) \cup \mathcal{V}(Q)$. Then there exists a $\leadsto$-derivation:*

$$M = H(P,Q) = M_0 \leadsto M_1 = H(P_1, Q_1) \leadsto \cdots \leadsto M_n = H(P_n, Q_n),$$

such that P_n and Q_n are unifiable. Let μ be the minimum unifier of P_n and Q_n we have:

$$\mu\theta_n \leq_T \rho \,[\![V]\!].$$

Moreover we are allowed to restrict our attention to $\leadsto$-derivations such that: $\forall i,\ 0 \leq i \leq n$, $\theta_i \uparrow V$ is normalized.

Proof. We have $\rho(P) =_T \rho(Q)$ thus with $\eta = \mathcal{R}(\rho)$, where $(\mathcal{R}(\rho))(x) = \mathcal{R}(\rho(x))$, $\eta(P) =_T \eta(Q)$ that is these two terms have a same normal form which we call R. Then we have:

$$\eta(M) = H(\eta(P), \eta(Q)) = N_0 \to \cdots \to N_n = H(R, R).$$

The corresponding $\leadsto$-derivation is such that:

$$\eta_n(M_n) = H(\eta_n(P_n), \eta_n(Q_n)) = N_n = H(R, R).$$

Thus η_n is a unifier of P_n and Q_n. Let μ be the minimum unifier, we have: $\exists\xi \quad \xi\mu = \eta_n$, therefore:

$$(\xi\mu\theta_n \uparrow V) = (\eta_n\theta_n \uparrow V) = (\eta \uparrow V) =_T (\rho \uparrow V),$$

that is:

$$\mu\theta_n \leq_T \rho \,[\![V]\!],$$

which proves the lemma. ∎

We are now ready to describe how to build a complete set of T-unifiers for two terms.

Theorem 2. *Let T be the equational theory defined by a canonical term rewriting system $\mathcal{R}$. Let P and Q be two terms, M be $H(P,Q)$ where H is a new function symbol ($H \notin C$), and V be a finite set of variables containing $\mathcal{V}(M)$. Let Σ be the set of all substitutions σ such that σ is in Σ iff there exists a $\leadsto$-derivation:*

$$M = H(P,Q) = M_0 \leadsto M_1 = H(P_1, Q_1) \leadsto \cdots \leadsto M_n = H(P_n, Q_n),$$

such that P_n and Q_n are unifiable, θ_n is normalized and $\sigma = \mu\theta_n$ where μ is the minimum unifier of P_n and Q_n. Then Σ is a complete set of T-unifiers of P and Q away from V.

Proof. Lemma 1 proves consistency and lemma 2 proves completeness. ∎

A T-unification algorithm follows from the construction of theorem 2: enumerate all elements of Σ. Essentially this algorithm is the same as Fay's; however we do not normalize terms at each step. Note that, althougt this set may be infinite, one can organize the enumeration in such a way that if two terms P and Q are T-unifiable, then a T-unifier will be produced in a finite number of steps. Thus this algorithm gives a semi-decision procedure for T-unifiability. In the following section we shall study how to refine this algorithm in order to eliminate some redundancies. Moreover a sufficient condition for the termination of the construction will be given.

Note also that this algorithm does not enumerate a minimal set of unifiers (even when such a set exists as one can see in the example of associativity). We will give an example in section 5 where a complete and finite T-unification algorithm is known and the algorithm described here does not even terminate.

4. Elimination of redundancies.

In this section we are interested in eliminating some redundancies in the construction of theorem 2. To achieve this aim we shall restrict our attention to special $\leadsto$-derivations. Since we have seen in theorem 1 that any $\leadsto$-derivation issuing from M is the "projection" of a $\to$-derivation issuing from $\eta(M)$ such that η is normalised, we shall first give a particular property verified by all such $\to$-derivations.

Definition. Let us consider a term N and a set of occurrences U of a proper prefix of N (e.g. $U = \overline{O}(M)$, for some $M \leq N$). We define by induction what it means for a derivation:

$$N = N_0 \rightarrow_{[u_0,k_0]} N_1 \rightarrow_{[u_1,k_1]} \cdots \rightarrow_{[u_{i-1},k_{i-1}]} N_i,$$

to be *based on* U, and we construct sets of occurrences $U_i \subset O(N_i)$, $0 \leq i \leq n$, as follows:

- the empty derivation is based on U, and $U_0 = U$,
- if the derivation above is based on U, then the derivation obtained from it by adding one step $N_i \rightarrow_{[u_i,k_i]} N_{i+1}$ is based on U iff $u_i \in U_i$, and in this case we take:

$$U_{i+1} = (U_i - \{v \in U_i | u_i \preceq v\}) \cup \{u_i \cdot v | v \in \overline{O}(\delta_{k_i})\}.$$

This definition is quite technical, but the practical meaning is easy to understand. Consider for instance the following term rewriting system:

$$f(h(x)) \rightarrow h(x); \qquad (r1)$$

$$h(h(x)) \rightarrow x; \qquad (r2)$$

$$h(a) \rightarrow a. \qquad (r3)$$

We consider terms $M = h(f(x))$ and $N = h(f(h(a)))$ that is $N = \sigma(M)$ with $\sigma(x) = h(a)$. Note that σ is not normalized (see lemma 3 below). In order to be based on $\overline{O}(M)$ a derivation issuing from N must not affect $\sigma(x)$. For instance the following derivation using rule $(r3)$ is not based on $\overline{O}(M)$:

$$N = h(f(h(a))) \rightarrow h(f(a)).$$

Thus it must affect a subterm which has a prefix in M, for instance the following step of reduction using rule $(r1)$:

$$N = h(f(h(a))) \rightarrow h(h(a)).$$

Since the affected subterm was $f(h(a))$ the definition says that we can iterate these considerations with $M_1 = h(x)$, $N_1 = h(h(a))$ and $\sigma_1(x) = h(a)$. Thus the only way to go on is:

$$h(h(a)) \rightarrow a.$$

Let us now give a lemma which shows our interest in derivations based on a set of occurrences.

Lemma 3. *Let $N = \eta(M)$, with η normalised. Every $\rightarrow$-derivation from N is based on $\overline{O}(M)$.*

Proof. Obvious. ∎

Definition. A $\rightsquigarrow$-derivation:

$$M = M_0 \rightsquigarrow_{[u_0,k_0]} M_1 \rightsquigarrow_{[u_1,k_1]} M_2 \rightsquigarrow \cdots \rightsquigarrow_{[u_{i-1},k_{i-1}]} M_i,$$

is said to be *basic* iff it is based on $\overline{O}(M)$ (in the same sense as in the previous definition for $\rightarrow$-derivation).

Let us now consider the $\rightsquigarrow$-derivation:

$$M = M_0 \rightsquigarrow_{[u_0,k_0,\sigma_0]} M_1 \rightsquigarrow_{[u_1,k_1,\sigma_1]} M_2 \cdots \rightsquigarrow_{[u_{n-1},k_{n-1},\sigma_{n-1}]} M_n,$$

associated by theorem 1 to any $\rightarrow$-derivation:

$$\eta(M) = N_0 \rightarrow_{[u_0,k_0]} N_1 \rightarrow_{[u_1,k_1]} N_2 \rightarrow \cdots \rightarrow_{[u_{n-1},k_{n-1}]} N_n,$$

such that η is normalised. Because of lemma 3 this $\rightarrow$-derivation is based on $\overline{O}(M)$, and since the sets U_i are the same for the $\rightarrow$-derivation and the $\rightsquigarrow$-derivation it follows easily that the considered $\rightsquigarrow$-derivation is basic. Thus, we have:

Theorem 3. *The $\rightsquigarrow$-derivations constructed in the $\Rightarrow$-part of theorem 1 are all basic.*
As a corollary of theorem 3 we can now give a refined version of theorem 2:

Theorem 4. *Theorem 2 holds if we consider only basic $\rightsquigarrow$-derivations.*

The main interest of this theorem is that we can give a sufficient condition for the termination of the narrowing process when we consider only basic $\rightsquigarrow$-derivations and therefore for the termination of the corresponding T-unification algorithm.

Proposition 1. *Let $\mathcal{R} = \{\gamma_k \rightarrow \delta_k\}$ be a canonical term rewriting system such that any basic $\rightsquigarrow$-derivation issuing from any of the δ_k's terminates. Then any $\rightsquigarrow$-derivation issuing from any term terminates.*

Proof. Let us consider any basic $\rightsquigarrow$-derivation:

$$M = M_0 \rightsquigarrow_{[u_0]} M_1 \rightsquigarrow \cdots \rightsquigarrow_{[u_{i-1}]} M_i \rightsquigarrow_{[u_i]} M_{i+1} \rightsquigarrow \cdots;$$

The basic idea underlying the proof is the following: at each step of the derivation, either u_i comes from $\overline{O}(M)$ and such an occurrence may be used only one time, or this step of derivation "is part" of some $\rightsquigarrow$-derivation issuing from a δ_k. More formally, we define sets of occurrences $\mathcal{G}_i$ by:

- $\mathcal{G}_0 = \overline{O}(M_0)$,
- $\mathcal{G}_{i+1} = (\mathcal{G}_i - \{u \mid u \in \mathcal{G}_i \quad \& \quad u_i \preceq u\})$ if $u_i \in \mathcal{G}_i$ and $\mathcal{G}_{i+1} = \mathcal{G}_i$ otherwise.

We define also sets $\mathcal{H}_i$. Each element of a $\mathcal{H}_i$ will be a pair which left part is an occurrence u and which right part is an integer $n(u)$. For each $\gamma_k \rightarrow \delta_k \in \mathcal{R}$ we define integer n_k to be the maximal length of a derivation issuing from δ_k.

- $\mathcal{H}_0 = \emptyset$.
- if $u_i \in \mathcal{G}_i$, $\mathcal{H}_{i+1} = (\mathcal{H}_i - \{\langle u, n(u)\rangle \mid u \in \mathcal{H}_i \quad \& \quad u_i \prec u\} \cup \{\langle u_i, n_{k_i}\rangle\}$,
- otherwise let us consider the following sequence of occurrences in N_i:

$$v_0 = \Lambda, v_1, \ldots, v_p = u_i.$$

Since $u_i \notin \mathcal{G}_i$, there exists an integer q such that $v_q \in \mathcal{G}_i$ and $v_{q+1} \notin \mathcal{G}_i$. In this case, we define: $\mathcal{H}_{i+1} = (\mathcal{H}_i - \{\langle v_{q+1}, n(v_{q+1})\rangle\}) \cup \{\langle v_{q+1}, n(v_{q+1}) - 1\rangle\}$ (note that v_{q+1} is an u_j with $j \leq i$).

Along the lines of the definition of basic $\rightsquigarrow$-derivation, it is easy to prove that, if a right part of a couple in one $\mathcal{H}_i$ reaches value 0, then no more narrowing will be possible under the corresponding left part occurrence. Moreover, one of the two following situations occurs:

(a) either $|\mathcal{G}_{i+1}| < |\mathcal{G}_i|$, $|\mathcal{H}_{i+1}| \leq |\mathcal{H}_i| + 1$,

(b) or $|\mathcal{G}_{i+1}| = |\mathcal{G}_i|$, $|\mathcal{H}_{i+1}| = |\mathcal{H}_i|$, and the right part of one of the elements of $\mathcal{H}_i$ has decreased from 1.

Thus situation (a) may occur only $|\mathcal{G}_0|$ times in the derivation. Then we have $\forall i$, $|\mathcal{H}_i| \leq |\mathcal{G}_0|$. It is then easy to prove the termination, using the decreasing of the integers in situation (b). ∎

Proposition 2. *If the hypothesis of proposition 1 holds, the construction of theorem 4 leads to a complete and finite T-unification algorithm.*

Example 1. In the case where all right parts of the rules of a canonical term rewriting system are variables, the previous proposition obviously applies. This is the case for idempotency law alone. However, in this case, a more powerful (for minimal) complete and finite T-unification algorithm is known [26].

Example 2. Another example is quasi-group theory, which can be defined by the following set of equations:

$$x * (x \setminus y) = y; \qquad (a1)$$

$$(x / y) * y = x; \qquad (a2)$$

$$x \setminus (x * y) = y; \qquad (a3)$$

$$(x * y) / y = x. \qquad (a4)$$

This set of equations can be embedded in a canonical term rewriting system $\mathcal{R}$, as shown in [11]:

$$x * (x \setminus y) \to y; \qquad (r1)$$

$$(x / y) * y \to x; \qquad (r2)$$

$$x \setminus (x * y) \to y; \qquad (r3)$$

$$(x * y) / y \to x; \qquad (r4)$$

$$(x / y) \setminus x \to y; \qquad (r5)$$

$$x / (y \setminus x) \to y. \qquad (r6)$$

Thus we obtain the first known complete and finite T-unification algorithm for quasi-group theory. Note that our result applies in the same way to all particular quasi-group with identities studied by Hullot in [11].

Example 3. This example is from Lankford [15]. Let us consider a theory T defined by a finite set of ground equations. In this case, using a lexicographic ordering to show the finite termination property, it is always possible to build a canonical term rewriting system from the equations. Moreover, since the right parts of the resulting rewrite rules are ground, no $\rightsquigarrow$-derivation is possible from these terms. Thus the narrowing process is finite and the construction of theorems 2 & 3 gives a quite elegant way to solve equations in such theories. We have implemented this equation solver as a LISP program.

5. Extensions.

Under certain conditions it is possible to define canonical term rewriting systems on equivalence classes of terms modulo permutations. This has been done for commutativity by Lankford & Ballantyne [16], for associativity and commutativity by Lankford & Ballantyne [18] and Peterson & Stickel [24]. In the case where the term rewriting sytem is left linear Huet [6,7] has given general results. We are interested in this section to extend the results of sections 3 & 4 to all these cases.

Let T be the equational theory defined by $T = \mathcal{E} \cup \mathcal{R}$ where $\mathcal{R}$ is a term rewriting system and $\mathcal{E}$ is a set of equations verifying:

$$\forall \langle \gamma, \delta \rangle \in \mathcal{E} \qquad \mathcal{V}(\gamma) = \mathcal{V}(\delta).$$

In all this section, we assume the existence of a complete $\mathcal{E}$-unification algorithm. We shall study three cases according to the three methods known to extend Knuth & Bendix's results.

First we shall study the case where $\mathcal{R}$ is a canonical term rewriting system modulo $=_\mathcal{E}$ (Huet [6]), that is $\to_\mathcal{R}$ is nœtherian in the quotient structure by $=_\mathcal{E}$, and $\to_\mathcal{R}$ is *confluent modulo* $=_\mathcal{E}$, e.g. $\forall M_1, M_2, M'_1, M'_2$, such that $M_1 =_\mathcal{E} M_2$ and $M_1 \overset{*}{\to}_\mathcal{R} M'_1$ and $M_2 \overset{*}{\to}_\mathcal{R} M'_2$, then $\exists M''_1, M''_2$ such that $M'_2 \overset{*}{\to}_\mathcal{R} M''_2$ and $M''_1 =_\mathcal{E} M''_2$. Note that if $\mathcal{E}$-equality is decidable, a decision procedure for T-equality is known: $M =_T M'$ iff $\mathcal{R}(M) =_\mathcal{E} \mathcal{R}(M')$.

For the two other cases, we have to define a new relation on $\mathcal{T}$:

$$\to_\sim = =_\mathcal{E} \cdot \to_\mathcal{R},$$

that is: to achieve one step of $\to_{\approx}$ to a term M, one has to find any term $\mathcal{E}$-equivalent to M which is $\to_{\mathcal{R}}$-simplifiable and then to achieve one step of $\to_{\mathcal{R}}$-reduction. We define also a new notion of canonical term rewriting system: we say that $\mathcal{R}$ is a *canonical term rewriting system over* $\mathcal{E}$ iff $\to_{\approx}$ is nœtherian and $\to_{\mathcal{R}}$ is *confluent over* $=_{\mathcal{E}}$, that is $\forall M,\ M_1,\ M_2$, such that $M \overset{*}{\to}_{\approx} M_1$ and $M \overset{*}{\to}_{\approx} M_2$, then $\exists M'_1,\ M'_2$ such that $M_1 \overset{*}{\to}_{\approx} M'_1$ and $M_2 \overset{*}{\to}_{\approx} M'_2$ and $M'_1 =_{\mathcal{E}} M'_2$. Note that in the case where $\mathcal{E}$-equality is decidable and $\to_{\approx}$-simplifiability is decidable and $\mathcal{R}$ is a canonical term rewriting system over $=_{\mathcal{E}}$, we have a decision procedure for T-equality: $M =_T M'$ iff $\mathcal{R}_{\approx}(M) =_{\mathcal{E}} \mathcal{R}_{\approx}(M')$ where $\mathcal{R}_{\approx}(M)$ is a $\to_{\approx}$-normal form of M which is unique modulo $=_{\mathcal{E}}$.

One difficulty with this approch is the need of a decision procedure for $\to_{\approx}$-simplifiability. In the case where all equivalences classes under $=_{\mathcal{E}}$ are finite, such a decision procedure is easily obtained by generating the equivalence class of a term and by checking $\to_{\mathcal{R}}$-simplifiability on each element of this class. This is the way used by Lankford & Ballantyne when dealing with permutative reductions [16,17,18].

Another way is given by Huet in [7]. We first give a definition.

Definition. $\to_{\mathcal{R}}$ is said to be $=_{\mathcal{E}}$-uniform iff:

$$M \to_{\mathcal{R}} N \quad \& \quad M \doteq_{\mathcal{E}} M' \Rightarrow \exists N' \quad M' \to_{\mathcal{R}} N'.$$

Proposition. *Assume $\to_{\mathcal{R}}$ is $\doteq_{\mathcal{E}}$-uniform, then for any term M, M is $\to_{\approx}$-reductible iff M is $\to$-reductible.*

Proof. Obvious ∎

Huet gives in [7] a way to decide, for any finite left linear term rewriting system $\mathcal{R}$ and any finite set of equations $\mathcal{E}$ having decidable $\mathcal{E}$-equality if $\mathcal{R}$ is a canonical term rewriting system over $=_{\mathcal{E}}$.

Another way has been introduced by Peterson & Stickel [24]. The idea is to extend $\to_{\mathcal{R}}$ in a new relation $\to_{\mathcal{R},\mathcal{E}}$.

Definition. We say that $M \to_{\mathcal{R},\mathcal{E}} N$ iff:

$$\exists \gamma \to \delta \in \mathcal{R} \quad \exists \sigma \quad \exists u \in O(M) \quad M/u =_{\mathcal{E}} \sigma(\gamma) \quad N = M[u \leftarrow \sigma(\delta)]$$

Remark that $\to_{\mathcal{R},\mathcal{E}}$-simplifiability is decidable, when $\mathcal{R}$ is finite, if T-matching is decidable. We define now a new notion of uniformity.

Definition. $\to_{\mathcal{R},\mathcal{E}}$ is said to be $\mathcal{E}$-uniform iff:

$$\forall M, N \quad M \to_{\approx} N \Rightarrow \exists P \quad M \to_{\mathcal{R},T} P.$$

Proposition. *Assume $\to_{\mathcal{R},\mathcal{E}}$ is $\mathcal{E}$-uniform, then for any term M, M $\to_{\approx}$-simplifiable iff M $\to_{\mathcal{R},\mathcal{E}}$-simplifiable.*

Proof. Obvious ∎

Peterson & Stickel give in [24] a way to decide if $\mathcal{R}$ is a canonical term rewriting system over $\mathcal{E}$ in the case where there exists a complete $\mathcal{E}$-unification algorithm and $\mathcal{R}$ is "$\mathcal{E}$-compatible" which condition is stronger that $\mathcal{E}$-uniform.

We shall extend our results to the three following cases:

(1) $\mathcal{R}$ is a canonical term rewriting system modulo $=_{\mathcal{E}}$.

(2) $\mathcal{R}$ is a canonical term rewriting system over $=_{\mathcal{E}}$ and $\to_{\mathcal{R}}$ is $=_{\mathcal{E}}$-uniform.

(3) $\mathcal{R}$ is a canonical term rewriting system over $=_{\mathcal{E}}$ and $\to_{\mathcal{R},\mathcal{E}}$ is $\mathcal{E}$-uniform.

For cases (1) and (2) we will use the same notion of narrowing as in previous section. For case (3) we shall define a new definition of narrowing using $\rightarrow_{\mathcal{R},\mathcal{E}}$ instead of $\rightarrow_{\mathcal{R}}$. Note that this notion of extended narrowing has been introduced by Lankford & Ballantyne [19] in the case of associative commutative derivations. We generalize the result to all cases covered by Peterson & Stickel's paper [23] and prove the correctness of this new T-unification process.

Note that we extend theorems 1 & 2; however it is possible to extend results of section 4 as well.

5.1. Extension to cases (1) & (2).

Lemma 4 (resp. 5) is an analogue of lemma 1 (resp.2). We use the same notations: P and Q are two terms, H is a "new" function symbol and M is $H(P,Q)$.

Lemma 4. *Let us consider any $\leadsto$-derivation:*

$$M = H(P,Q) = M_0 \leadsto M_1 = H(P_1,Q_1) \leadsto \cdots M_n = H(P_n,Q_n),$$

such that P_n and Q_n are $\mathcal{E}$-unifiable, say by substitution σ. Then $\sigma\theta_n$ is a T-unifier of P and Q.

Proof. The proof closely follows that of lemma 1. ∎

Lemma 5. *Let P and Q be two terms which are T-unifiable, ρ be any T-unifier and V be a finite set of variables containing $\mathcal{V}(P) \cup \mathcal{V}(Q)$. Then there exists a $\leadsto$-derivation:*

$$M = H(P,Q) = M_0 \leadsto M_1 = H(P_1,Q_1) \leadsto \cdots \leadsto M_n = H(P_n,Q_n),$$

such that P_n and Q_n are $\mathcal{E}$-unifiable. Let Σ be any complete set of $\mathcal{E}$-unifiers of P_n and Q_n away from $V \cup V_n$. We have:

$$\exists \mu \in \Sigma \quad \mu\theta_n \leq_T \rho[\![V]\!].$$

Moreover we are allowed to restrict our attention to $\leadsto$-derivations such that: $\forall i$, $0 \leq i \leq n$, $\theta_i \uparrow V$ is normalized.

Proof. We have $\rho(P) =_T \rho(Q)$ thus with $\eta = \mathcal{R}(\rho)$, $\eta(P) =_T \eta(Q)$. Let us consider a derivation from $\eta(M)$ to one of its normal form:

$$\eta(M) = H(\eta(P), \eta(Q)) = N_0 \rightarrow \cdots \rightarrow N_n = H(N_P, N_Q),$$

where N_P is a $\rightarrow$-normal form of $\eta(P)$ and N_Q is a $\rightarrow$-normal form of $\eta(Q)$. In the two cases we are studying, we have then $N_P =_{\mathcal{E}} N_Q$. For the corresponding $\leadsto$-derivation we have:

$$\eta_n(M_n) = H(\eta_n(P_n), \eta_n(Q_n)) = N_n = H(N_P, N_Q).$$

Thus η_n is a $\mathcal{E}$-unifier of P_n and Q_n. Let Σ be any complete set of $\mathcal{E}$-unifiers away from $V \cup V_n$. We have:

$$\exists \mu \in \Sigma \quad \mu \uparrow (V \cup V_n) \leq_{\mathcal{E}} \eta_n \uparrow (V \cup V_n),$$

then:

$$\exists \xi \quad (\xi\mu) \uparrow (V \cup V_n) =_{\mathcal{E}} \eta_n \uparrow (V \cup V_n),$$

thus:

$$(\xi\mu\theta_n) \uparrow V = ((\xi\mu) \uparrow (V \cup V_n))(\theta_n \uparrow V) =_{\mathcal{E}} ((\eta_n \uparrow (V \cup V_n))(\theta_n \uparrow V) = (\eta_n\theta_n \uparrow V)$$

and:

$$(\eta_n\theta_n \uparrow V) = (\eta \uparrow V) =_T (\rho \uparrow V),$$

that is:

$$\mu\theta_n \leq_T \rho[\![V]\!],$$

which proves the lemma. ∎

We can now give an analogue of theorem 2:

Theorem 5. *Let T be the equational theory defined by $T = \mathcal{R} \cup \mathcal{E}$ where $\mathcal{R}$ is:*

- *either a canonical term rewriting system modulo $=_{\mathcal{E}}$,*
- *or a canonical term rewriting system over $=_{\mathcal{E}}$ such that $\to_{\mathcal{R}}$ is $\doteq_{\mathcal{E}}$-uniform,*

and $\mathcal{E}$ is a set of equations defining an equational theory in which a complete $\mathcal{E}$-unification algorithm is known.

Let P and Q be two terms, M be $H(P,Q)$ where H is a new function symbol ($H \notin C$), V be a finite set of variables containing $\mathcal{V}(M)$. Let Σ be the set of all substitutions σ such that σ is in Σ iff there exists a $\rightsquigarrow$-derivation:

$$M = H(P,Q) = M_0 \rightsquigarrow M_1 = H(P_1,Q_1) \rightsquigarrow M_n = H(P_n,Q_n),$$

such that P_n and Q_n are $\mathcal{E}$-unifiable, θ_n is normalized and $\sigma = \mu\theta_n$ where μ is any element in a complete set of $\mathcal{E}$-unifiers of P_n and Q_n. Then Σ is a complete set of T-unifiers of P and Q away from V.

Proof. Lemma 3 proves consistency and lemma 4 proves completeness. ∎

5.2. Extension to case (3).

$\mathcal{R}$ is a canonical term rewriting system over $=_{\mathcal{E}}$ and $\to_{\mathcal{R},\mathcal{E}}$ is $\mathcal{E}$-uniform. In this case, we need to define a new notion of narrowing.

Definition. Let M be a term and V be a finite set of variables containing $\mathcal{V}(M)$. Assume the following situation holds:

$$\exists u \in \overline{O}(M) \quad \exists \gamma_k \to \delta_k \in \mathcal{R} \quad \mathcal{U}_{\mathcal{E}}(M/u, \gamma_k, W_k) \neq \emptyset,$$

where we assume $\gamma_k \to \delta_k$ is renamed so that $\mathcal{V}(\gamma_k) \cap V = \emptyset$ and W_k is a finite set of variables containing $\mathcal{V}(\gamma_k) \cup V$.

Let Σ be any complete set of $\mathcal{E}$-unifiers of M/u and γ_k away from W_k, then each element of Σ will be called *$\mathcal{E}$-narrowing substitution of M away from V*. $NS_{\mathcal{E}}(M,V)$ will denote the set of all such substitutions.

Let N be the term $\sigma(M[u \leftarrow \delta_k])$ where σ is any substitution in Σ. We say that M is *$\mathcal{E}$-narrowable in M* at *occurrence u using rule* $\gamma_k \to \delta_k$ and we write:

$$M \rightsquigarrow_{\mathcal{E},[u,k,\sigma]} N.$$

$\rightsquigarrow_{\mathcal{E}}$ is called *$\mathcal{E}$-narrowing relation on T*.

Notations. We will use all notations of section 3. In particular $\rightsquigarrow_{\mathcal{E}}$-derivations are defined in an obvious way.

Theorem 6. *Theorem 1 holds if we replace $\to$-derivation by $\to_{\mathcal{R},\mathcal{E}}$-derivation and $\rightsquigarrow$-derivation by $\rightsquigarrow_{\mathcal{E}}$-derivation.*

Proof. The proof follows closely the one of theorem 1, we do not give it. ∎

It is now easy to prove lemma 4 & 5 where narrowing is replaced by $\mathcal{E}$-narrowing. Finally we give an analogue of theorems 1 & 2.

Theorem 7. *Let T be the equational theory defined by $T = \mathcal{R} \cup \mathcal{E}$. $\mathcal{E}$ is a set of equations defining an equational theory in which a complete $\mathcal{E}$-unification algorithm is known. $\mathcal{R}$ is a canonical term rewriting system over $=_{\mathcal{E}}$ such that $\to_{\mathcal{R},\mathcal{E}}$ is $\mathcal{E}$-uniform. Then the result of theorem 5 holds where $\mathcal{E}$-narrowing is used instead of narrowing.*

Example. We give an example in abelian group theory. In this case $\mathcal{E}$ will be the set of two equations defining the associativity and commutativity of $+$. We list below a canonical term rewriting system for abelian group theory :

$$x + 0 \rightarrow x; \tag{r1}$$
$$x + (-x) \rightarrow 0; \tag{r2}$$
$$-0 \rightarrow 0; \tag{r3}$$
$$-(-x) \rightarrow x; \tag{r4}$$
$$-(x + y) \rightarrow (-x) + (-y). \tag{r5}$$

which appears in [24,18]. Note that we need to consider extended rules only for rule $(r2)$:

$$x + (-x) + y \rightarrow y. \tag{er2}$$

With this rule $\mathcal{E}$-compatibility is insured (see [23]), and $\mathcal{E}$-uniformity follows.

Lankford has proposed to orient rule $(r5)$ from right to left. In this case we obtain another complete set of reductions for abelian groups. Rules $(r1)$ to $(r4)$ are the same, the others are:

$$(-x) + (-y) \rightarrow -(x + y); \tag{r5'}$$
$$-((-x) + y) \rightarrow x + (-y); \tag{r6'}$$
$$x + -(y + x) \rightarrow (-y). \tag{r7'}$$

in this case we need to consider extended rules for rules $(r5')$ & $(r7')$, that is:

$$(-x) + (-y) + z \rightarrow -(x + y) + z; \tag{er5'}$$
$$x + -(y + x) + z \rightarrow (-y) + z. \tag{er7'}$$

As in the previous example $\mathcal{E}$-compatibility and $\mathcal{E}$-uniformity are then insured.

Let us now consider term $M_1 = -x_1$ (this example is from Lankford). We show that there exists an infinite $\rightsquigarrow_{\mathcal{E}}$-derivation issuing from M_1, even if we restrict ourself to basic $\rightsquigarrow_{\mathcal{E}}$-reduction, as one can define in the same way as basic $\rightsquigarrow$-reductions. We begin with the first term rewriting system . M_1 is $\mathcal{E}$-unifiable with the left part of rule $(r5)$, $\sigma = \{\langle x_1 \leftarrow x + y \rangle\}$ being a unifier. Thus, we have (after renaming):

$$M_1 = -x_1 \rightsquigarrow M_2 = (-x_2) + (-y_1).$$

In the same way, using subterm $-x_2$ we have:

$$M_2 = (-x_2) + (-y_1) \rightsquigarrow M_3 = (-x_3) + (-y_1) + (-y_2),$$

and more generally:

$$M_n = (-x_n) + (-y_1) + \cdots + (-y_{n-1}) \rightsquigarrow M_{n+1},$$

showing the existence of an infinite $\rightsquigarrow_{\mathcal{E}}$-derivation. Note that we have used only basic $\rightsquigarrow_{\mathcal{E}}$-derivations.

When dealing with the second term rewriting system, consider rule $(r6')$. We build the infinite derivation:

$$M_1 = -x_1 \rightsquigarrow M'_2 = (-x_2) + y_1 \rightsquigarrow \cdots \rightsquigarrow M'_n = (-x_n) + y_1 + \cdots + y_{n-1}.$$

Thus, none of these two canonical term rewriting systems leads to a finite T-unification algorithm with the methods described in this paper. However, there exists a complete and finite T-unification algorithm for abelian group theory as shown by Lankford [14].

Remark. Ballantyne & Lankford have shown in [1] how to solve the word problem for finitely presented commutative semigroups in using associative and commutative term rewriting systems. Thus one could be interested in solving equations in commutative semigroups in using the way described in this section. However note that we cannot expect to show the termination of the algorithm since some of these equations have infinite set of independent unifiers. Let us for instance consider the equational theory defined by $ab = a$ and let us try to unify ax and a where x is a variable. It is easy to show that $x \leftarrow b$, $x \leftarrow bb$, $x \leftarrow bbb$, ... are independent unifiers. (This example was communicated to the author by A.M. Ballantyne).

6. Conclusion.

We have shown in this paper how to improve over Fay's T-unification algorithm. In particular we have given a sufficient condition for the termination of this algorithm, proving a refined version of a conjecture by Lankford. Furthermore we have shown how to extend Fay's algorithm to equational theories defined by various kinds of canonical term rewriting systems.

7. Acknowledgments.

We thank G. Huet for his help in writing this paper and R. Shostak for his many helpful comments.

8. References.

1. Ballantyne A.M. and Lankford D.S., *New Decision Algorithms for Finitely Presented Commutative Semigroups.* Report MTP-4, Department of Mathematics, Louisiana Tech. U., May 1979.
2. Fay M., *First-order Unification in an Equational Theory.* Master Thesis, U. of California at Santa Cruz. Tech. Report 78-5-002, May 1978.
3. Fay M., *First-order Unification in an Equational Theory.* 4th Workshop on Automated Deduction, Austin, Texas, Feb. 1979, 161–167.
4. Huet G., *A Unification Algorithm for Typed Lambda Calculus.* Theoretical Computer Science, 1,1 (1975), 27–57.
5. Huet G., *Résolution d'équations dans des langages d'ordre* $1, 2, \ldots, \omega$. Thèse d'Etat, Université de Paris VII, 1976.
6. Huet G., *Confluent Reductions: Abstract Properties and Applications to Term Rewriting Systems.* 18th IEEE Symposium on Foundations of Computer Science (1977), 30–45.
7. Huet G., *Embedding Equational Theories in Complete Sets of Reductions.* Unpublished manuscript, 1979.
8. Huet G. and Lévy J.J., *Call by Need Computations in Non-Ambiguous Linear Term Rewriting Systems.* Rapport Laboria 359, IRIA, Août 1979.
9. Huet G. and Oppen D.C., *Equations and Rewrite Rules: a Survey.* In "Formal Languages: Perspectives and Open Problems". Ed. Book R., Academic press 1980.
10. Hullot J.M., *Associative-Commutative Pattern Matching.* Fifth International Joint Conference on Artificial Intelligence, Tokyo, 1979.
11. Hullot J.M., *A Catalogue of Canonical Term Rewriting Systems.* Unpublished manuscript, March 1980.
12. Knuth D. and Bendix P., *Simple Word Problems in Universal Algebras.* "Computational Problems in Abstract Algebra". Ed. Leech J., Pergamon Press, 1970, 263–297.
13. Lankford D.S., *Canonical Inference.* Report ATP-32, Departments of Mathematics and Computer Sciences, University of Texas at Austin, Dec. 1975.

14. Lankford D.S., *A Unification Algorithm for Abelian Group Theory*. Report MTP-1, Math. Dept., Louisiana Tech. U., Jan. 1979.

15. Lankford D.S., *Private Communication*. 1980.

16. Lankford D.S. and Ballantyne A.M., *Decision Procedures for Simple Equational Theories With Commutative Axioms: Complete Sets of Commutative Reductions*. Report ATP-35, Departments of Mathematics and Computer Sciences, U. of Texas at Austin, March 1977.

17. Lankford D.S. and Ballantyne A.M., *Decision Procedures for Simple Equational Theories With Permutative Axioms: Complete Sets of Permutative Reductions*. Report ATP-37, Departments of Mathematics and Computer Sciences, U. of Texas at Austin, April 1977.

18. Lankford D.S. and Ballantyne A.M., *Decision Procedures for Simple Equational Theories With Commutative-Associative Axioms: Complete Sets of Commutative-Associative Reductions*. Report ATP-39, Departments of Mathematics and Computer Sciences, U. of Texas at Austin, Aug. 1977.

19. Lankford D.S. and Ballantyne A.M., *The Refutation Completeness of Blocked Permutative Narrowing and Resolution*. Fourth Conference on Automated Deduction, Austin, Feb. 1979, 53–59.

20. Livesey M. and Siekmann J., *Unification of Sets*. Internal Report 3/76, Institut fur Informatik I, U. Karlsruhe, 1977.

21. Makanin G.S., *The Problem of Solvability of Equations in a Free Semigroup*. Akad. Nauk. SSSR, TOM 233,2 (1977).

22. Martelli A. and Montanari U., *An Efficient Unification Algorithm*. Unpublished manuscript, 1979.

23. Paterson M.S. and Wegman M.N., *Linear Unification*. J. of Computer and Systems Sciences 16 (1978), 158–167.

24. Peterson G.E. and Stickel M.E., *Complete Sets of Reductions for Equational Theories With Complete Unification Algorithms*. Tech. Report, Dept. of Computer Science, U. of Arizona, Tucson, Sept. 1977.

25. Plotkin G., *Building-in Equational Theories*. Machine Intelligence 7 (1972), 73–90.

26. Raulefs P. and Siekmann J., *Unification of Idempotent Functions*. Unpublished manuscript, 1978.

27. Robinson J.A., *A Machine-Oriented Logic Based on the Resolution Principle*. JACM 12 (1965), 32–41.

28. Slagle J.R., *Automated Theorem-Proving for Theories with Simplifiers, Commutativity and Associativity*. JACM 21 (1974), 622–642.

29. Stickel M.E., *A Complete Unification Algorithm for Associative-Commutative Functions*. 4th International Joint Conference on Artificial Intelligence, Tbilisi, 1975.

Deciding Unique Termination of Permutative Rewriting Systems: Choose Your Term Algebra Carefully

Hans-Josef Jeanrond
Computer Science Department
University of Edinburgh

Abstract: Some problems are considered related to unique termination of rewriting systems for classes of terms equal under some equational theory. It is shown that the approach of Peterson and Stickel [2] to such problems fails to cope with a rather simple equational theory which is very natural in the context of axiomatic specifications of abstract data types.

One can circumvent the problem by choosing a different axiomatic specification (with a different underlying term algebra) using only associative and commutative equations for which the techniques in [2] work nicely.

It is argued that we ought to try finding systematic ways of choosing the "right" term algebra for axiomatisations in order to be able to cope with the equational theory needed.

Some tools are presented to deal with the particular equational theory mentionned above, and some of the difficulties encountered in this approach are highlighted.

0. Introduction:

In dealing with axiomatic (equational) specifications of abstract data types one can exploit results known for rewriting systems to check the specifications for consistency: Look at axioms as rewriting rules and decide their finite termination and unique termination properties.

Knuth and Bendix [1] called a rewriting system with both these properties complete. If $R = \{l_i \to r_i \mid 1 \leq i \leq k\}$ is complete then it is "equivalent" to the equational system $E = \{l_i = r_i \mid l_i \to r_i \in R\}$ in the following sense: Two terms t_1 and t_2 are equal under E iff there exists a t such that t_1 and t_2 both derive to t by R (or t_1 and t_2 are identical).

It is this property of complete rewriting systems that makes them interesting when dealing with axiomatic definitions of abstract data types.

There are two different problems involved in going from equational systems to rewriting systems: One is to find a finitely terminating rewriting system given an equational system. (If l=r is in E, does one take l->r or r->l into R?) The other is to decide whether a finitely terminating rewriting system is uniquely terminating.

The first of these problems is aggravated by "permutative" axioms, like the commutative axiom for example. Such axioms, regarded as rewriting rules, lead trivially to non-terminating derivations. E.g. f(x,y) => f(y,x) => f(x,y) =>....

There have been various attempts to deal with axioms of this kind: See Huet [3], Lankford and Ballantyne [4], Peterson and Stickel [2]. The results are decision and semi-decision procedures for for unique termination of various classes of rewriting systems.

Peterson and Stickel split an equational theory E into two parts, a rewriting system R and an equational theory P (in which one would certainly put all the permutative axioms, but possibly others as well). They generalise the notions of unification and derivation for classes of terms equal under E. This eliminates "trivial" non-terminations like the one mentionned above.

We shall not concern ourselves here with the problems of deciding whether R is finitely terminating but always assume it is. We shall concentrate on the problems of deciding unique termiation with respect to P.

Knuth and Bendix [1] have develop ed a test for unique termination of term rewriting systems which is based on superpositions of left-hand sides of rewriting rules. Successful superpositions lead to "critical pairs" of terms that have to be checked for confluence.

The main problem with generalising the Knuth-Bendix superposition test to deal with classes of terms equal under P is to find a "sufficiently large" set of critical pairs. This problem is tackled in [2] by introducing "P-compatible enlargements" of R. This results in more rewrite rules and more superpositions to be considered. In fact, if $R^e{}_P$ is a "P-compatible enlargement" of R then it is sufficient to check the superpositions of left hand sides of rules in $R^e{}_P$ for confluence to ensure confluence of R (with respect to P in each case).

An example where this technique is successful is the case where P consists of commutative and associative axioms only.

If $P_0 = \{ F(F(t,e_1),e_2) = F(F(t,e_2),e_1) \}$ where t,e_1,e_2 are variables, F: SxEL -> S, then there is, unfortunately, no finite P_0-compatible enlargement for some R.

So the methods of [2] can not be applied successfully to P_0; neither can any other methods known to the author.

If one thinks of F as the function inserting an element into a set, then the axiom in P_0 says that the order of insertions of elements is irrelevant. So it looks as if we had not even the theoretical foundation to deal with the axiomatic specification of the data type of finite sets based on functions

EMPTY: -> SET and INSERT: SETxEL -> SET, and P_0.

However, if we choose two different "basic" functions in place of INSERT, namely

INJECT: EL -> SET , and UNION: SETxSET -> SET

such that INSERT(t,e) -> UNION(t,INJECT(e)), and take P to consist of the commutative and associative axioms for UNION, then we can apply Peterson and Stickel's machinery to decide the consistency of the axiomatisation.

This illustrates a rather unfortunate state of affairs:

When specifying an abstract data type axiomatically we have to be careful which algebra to choose. We might have to "enrich" the algebra we actually want in order to be able to use results of the underlying theory so far known. This is reminiscent of the much talked about issue of "auxiliary operations" needed for certain axiomatisations of abstract data types. But here we have as yet hardly any idea about what "auxiliary operations" one needs to introduce in any particular case to be able to use an available algorithm for deciding unique termination.

This suggests a twopronged thrust of research:

1. Find ways to deal with more equational theories P directly.

2. Find systematic ways of "enriching" algebras, as in the example above, to arrive at tractable problems.

In this paper we illustrate briefly why the concept of P-compatible enlargements fails for P_0, and, in line with suggestion (1.) develop some tools to deal with P_0 directly. No complete theory is offered, more problems are mentionned than solutions offered, but it is hoped that the concepts introduced are illuminating and will stimulate work in this field.

1. Basic Definitions:

Let T = [$\mathcal{C}$,$\mathfrak{F}$] be a many sorted algebra with a finite set of carriers $\mathcal{C}$ containing EL and S and a set of function symbols, $\mathfrak{F}$, containing F;

F:S*EL -> S

Let $t, t_1, t_2, \ldots$ be variables ranging over S,
$e, e_1, e_2, \ldots$ be variables ranging over EL.

Let V be a set of (sorted) variables containing the above variables.

Let L(T) be the <u>word algebra</u> of T,
L(T,V) = V ∪ L(T) ∪ {the set of words of L(T) where variables appear on argument positions}

The elements of L(T,V) will be referred to as <u>terms</u>.

For w∈L(T,V) let <u>V(w)</u> be the set of all variables occurring in w.

A <u>substitution</u> is a mapping $\Theta: V \to L(T,V)$ such that for almost all $x \in V$: $\Theta(x) = x$.
The <u>domain</u> of a substitution Θ is the set $dom_\Theta = \{ x \in V \mid \Theta(x) \neq x \}$.
As usual substitutions are extended homomorphically to L(T,V):
For $G(u_1, \ldots u_n) \in L(T,V)$ $\Theta(G(u_1, \ldots u_n) = G(\Theta(u_1), \ldots, \Theta(u_n))$.
<u>Composition</u>: $\mu \circ \Theta(t) = \mu(\Theta(t))$.

Let u,v∈L(T,V). u is an <u>instance</u> of v if there is a substitution Θ such that $\Theta(v)=u$.
Let $v_1, v_2 \in L(T,V)$. A substitution Θ is a <u>unifier</u> of v_1 and v_2 if $\Theta(v_1) = \Theta(v_2)$.
<u>Notation</u>: $v_1 \nabla v_2$
(Notice that we don't distinguish between terms of L(T,V) that are identical up to consistent renaming of variables, and thus are free to rename variables in v_1, v_2 such that $V(v_1) \cap V(v_2) = \emptyset$.)
μ is the <u>most general unifier</u> of v_1, v_2 if for all unifiers Θ there exists a substitution λ such that $\Theta = \lambda \circ \Theta$.
(μ is unique up to consistent renaming of variables.)
See [6] pp.32 for an algorithm to determine the most general unifier for two given terms.

A <u>rewriting rule</u> is an ordered pair of terms written $l \to r$.
Let R be a set of rewriting rules; u,v terms. Then v is an <u>immediate reduction</u> of u if there exists a rule $l \to r$ in R and a substitution Θ such that $u = u_1\Theta(l)u_2$ and $v = u_1\Theta(r)u_2$.
We refer to a sequence of immediate reductions as a <u>derivation</u>.
We use $\underset{R}{=>}$ for the immediate reduction relation and $\underset{R}{=>}^*$ for its reflexive and transitive closure, and $\underset{R}{=>}c$ for "complete" derivations, i.e. if $u \underset{R}{=>}c\ v$ then there is no rewrite rule applicable to v. In that case v is called <u>terminal</u>. We write => instead of $\underset{R}{=>}$ if the relevant R is understood from the context.

Let P be an equational theory (set of equations):
$P = \{ l_i = r_i \mid 1 \leqslant i \leqslant k,\ l_i,\ r_i \in L(T,V) \}$

Then for $u,v \in L(T,V)$ $u \mapsto v$ if there exists $l=r \in P$ such that either $u \Rightarrow v$ or $v \Rightarrow u$ by $\{ l \rightarrow r, r \rightarrow l \}$.
Let $=_P$ be the reflexive transitive closure of $\mapsto$.
$(w)_P = \{ w' \mid w' =_P w\}$ for $w \in L(T,V)$.
If P is understood we write (w) instead of $(w)_P$.
L(T,V)/P is the set of all $=_P$-classes of terms of L(T,V).

Let R be a set of rewriting rules , P an equational theory, $u,v \in L(T,V)$.
v is an <u>immediate P-reduction</u> of u if there exist $u' \in (u)_P$, $v' \in (v)_P$, such that $u' \underset{R}{\Rightarrow} v'$.

We use $\underset{P,R}{\Rightarrow}$ for the immediate P-reduction relation and $\underset{P,R}{\Rightarrow}^*$ and $\underset{P,R}{\Rightarrow}c$ as above.

For term classes we define reduction as follows: $(u)_P \underset{R}{\Rightarrow} (v)_P$ iff $u \underset{P,R}{\Rightarrow} v$.
If R is regarded as a rewriting system for classes of terms we shall call it a <u>permutative rewriting system</u>, and the derivations of term classes <u>permutative derivations</u>.

A set of rewrite rules R has the <u>Finite Termination Property</u>, (FTP), if there is no infinite derivation with respect to R from any $w \in L(T,V)$:
$w \underset{R}{\Rightarrow} w_1 \underset{R}{\Rightarrow} w_2 \underset{R}{\Rightarrow} \ldots\ldots$
It has the <u>Unique Termination Property</u> , (UTP), if for any $w \in L(T,V)$ such that $w \underset{R}{\Rightarrow}c\ u$ and $w \underset{R}{\Rightarrow}c\ v$, $u = v$.
If R has the UTP we say it is <u>confluent</u>.

The definitions for permutative rewriting systems are analogous, replacing terms by equivalence classes of terms.

If for u and v in L(T,V) there exists a w in L(T,V) such that $u \underset{R}{\Rightarrow}^* w$ and $v \underset{R}{\Rightarrow}^* w$ we write $u \downarrow v$. Analogously we shall write $(u) \downarrow (v)$ for term classes.

<u>2. Unique Termination</u>

The first step towards a decision procedure for unique termination is to "localise" the points in derivations where "forks" can occur.

<u>2.1. The Diamond Lemma</u>:

Let R be a rewriting system, $\Rightarrow\ =\ \underset{R}{\Rightarrow}$.
For $u \in L(T,V)$ let C(u) be the following predicate:
$\forall v_1,v_2 \in L(T,V)$: $u \Rightarrow v_1$ and $u \Rightarrow v_2$ implies $v_1 \downarrow v_2$
and let $C^*(u)$ be

$\forall v_1, v_2 \in L(T,V)$: u =>* v_1 and u =>* v_2 implies $v_1 \downarrow v_2$.
If C(u) holds for all u∈L(T,V) then R is called <u>locally confluent</u>.
If $C^*(u)$ holds for all u∈L(T,V) then R is confluent.

<u>Lemma</u>: Let R have the FTP (i.e. $\underset{R}{=>}$ is "noetherian" in terms of [3]). Then R is confluent iff R is locally confluent.

There are several proofs of this lemma in the literature, the most elegant in [3].

The lemma also holds if R is a permutative rewriting system.

2.2. Critical Pairs

We now concentrate on ways of deciding local confluence of rewriting systems.

The basic technique is to find a "representative" class of terms CT, called <u>critical terms</u>, with the following property: If C(u) holds for all u∈CT then C(u) holds for all u∈L(T,V).

If we can find a finite CT for a given R then we have an effective way of checking R for local confluence: Simply check whether for all u∈CT the set { u' | u $\underset{R}{=>}$c u' } is a singleton (which we can do because R is finitely terminating).

Knuth and Bendix [1] have shown that for ordinary rewriting systems (no rewriting of term classes) such a finite set CT exists and presented an algorithm of how to construct it.

The method is to consider all pairs of reductions in R, d_1: l_1 -> r_1, d_2: l_2 -> r_2, and to check whether l_1 unifies with a subterm of l_2, or vice versa. If it does, e.g. l_1= uvw, v∉V, and $\Theta(v) = \Theta(l_2)$, then $\Theta(l_1)$ is a critical term, giving rise to a <u>critical pair</u>: < $\Theta(ur_2w)$, $\Theta(r_1)$ >. If u↓v for all such critical pairs <u, v> then R is locally confluent.

A simple-minded generalisation of this method for reductions of term classes is to consider all pairs (l_1',l_2'), where $l_1' \in (l_1)$, $l_2' \in (l_2)$, when looking for critical terms. But this is not enough, as illustrated by the following example:

Example 1: Let d_1: $G(F(t_1,A)) \rightarrow r_1$, d_2: $G(F(t_2,B)) \rightarrow r_2$.
Then

$$
\begin{array}{ccc}
 & (G(F(F(T,A),B))) & \\
\theta(t_1)=F(T,B) \swarrow & & \searrow \theta(t_2)=F(T,A) \\
(\Theta(r_1)) & & (\Theta(r_2))
\end{array}
$$

(Term classes are with respect to P_0.)

But $G(F(t_1,A))$ and $G(F(t_2,B))$ do not unify, hence there is no critical pair.

One of the basic tools in building sets of critical terms for permutative rewriting systems is a generalised notion of unification. (We use the terminology of [2] and call it P-unification.)

Definition (P-unification): Let P be an equational system. Let $u,v \in L(T,V)$, $V(u) \cap V(v)=\emptyset$; u and v are P-unifiable if there exists a substitution Θ such that $\Theta(u)=_P\Theta(v)$.

Notation: $u \nabla v$ (assuming P is understood)

In this sense $G(F(t_1,A))$ and $G(F(t_2,B))$ do unify to $G(F(F(t,A),B))$ by $\Theta(t_1)=F(t,B)$, $\Theta(t_2)=F(t,A)$.

Algorithms for P-unification are known in the cases where P consists of commutative axioms, commutative plus associative axioms, commutative plus associative plus idempotency axioms. (See [7],[8])

In general there is no most general unifier for P-unification. P-unification algorithms generate a finite "basic" set of unifiers, U, for given terms u and v in the sense that for all unifiers Θ of u and v there exists a $\mu \in U$ and a substitution λ s.t. $\Theta(w)=_P\lambda(\Theta(w))$ for all terms w. (In [2] these sets are called "complete" sets of P-unifiers.)

In the appendix we present a generalisation of Robinson's unification algorithm (see [6] p.32) such that the new algorithm decides whether or not two words $u,v \in L(T,V)$ are P_0-unifiable, and if so gives us a basic set of unifiers.

One might have hoped to use the notion of P-unification to define the set of critical pairs CP of a permutative rewriting system R in the following way: Take any pair d_1,d_2 of rules of R, d_i: $l_i \rightarrow r_i$. If s is a subterm of some $l_2' \in (l_2)$, $l_2'=\ldots s\ldots$, and $l_1 \nabla s$ by Θ, then $(\Theta(l_2))$ is a critical term class. Applying d_1 and d_2 to it results in a critical pair.

Unfortunately this is still much too naive, as can be seen by considering the pair of rules U(T,A) -> C, U(T,B) -> D, where T,A,B do not contain variables, and P consists of the commutative and associative axioms for U.

In order to get a "large enough" set of critical terms by considering pairs of rewriting rules Peterson and Stickel [2] define "P-compatible enlargements" of rewriting systems:

Definition: Let R be a rewriting system, P an equational system. R^e is a proper enlargement of R if

- $R \subseteq R^e$

- whenever $l^e \to r^e \in R^e$ then there exists $l \to r \in R$ s.t. l is a subterm of l^e: $l^e = l_1 l l_2$, and l_1, l_2 do not contain any constants (nullary function symbols), and $(l_1 r l_2)_P \overset{=}{\underset{R}{>}}{}^* (r^e)_P$. l^e as above is called a variable extension of l.

R^e is a P-compatible enlargement of R if it is a proper enlargement of R and satisfies the following condition: Whenever $t \underset{P,R}{\overset{=}{>}} s$ there exists $l \to r \in R^e$ s.t. some subterm of t is a P-instance of l: $t = t_1 u t_2$, $u \in L(T,V)$, $u =_P \Theta(l)$ for some substitution Θ, and $(s)_P \overset{=}{\underset{R}{>}}{}^* (t_1 \Theta(r) t_2)_P$.

Definition: Let u and v be terms s.t. $u \underset{P,R}{\overset{=}{>}} v$. Then p is the minimal permuted subterm of u with respect to $u \underset{P,R}{\overset{=}{>}} v$ if p is the leftmost smallest subterm of u s.t. $u = u_1 p u_2$, $u' = u_1 p' u_2$, $p =_P p'$, and $u_1 p' u_2 \overset{=}{\underset{R}{>}} v' \in (v)_P$. u' is minimally permuted with respect to u and $u \underset{P,R}{\overset{=}{>}} v$.

The main idea of P-compatible enlargements is to keep the minimal permuted subterm at any immediate reduction from a term u inside the subterm of u to which the reduction is applied.

Theorem (Peterson and Stickel): Let R be a finitely terminating rewriting system, P an equational theory, and R^e a P-compatible enlargement of R. Let

(1) $l_1 \to r_1$, $l_2 \to r_2 \in R^e$; assume that variables are renamed such that both rules have no variables in common.

(2) u be a subterm of l_1: $l_1 = v_1 u v_2$, $u \notin V$.

(3) Θ be a P-unifier out of a basic set of P-unifiers of u and l_2.

Suppose that whenever (1), (2), and (3) hold, there exists $w \in L(T,V)$ such that $(\Theta(r_1)) \overset{=}{\underset{R}{>}}{}^* (w)$ and $(\Theta(v_1 r_2 v_2)) \overset{=}{\underset{R}{>}}{}^* (w)$.

Then R has the UTP.

See [2] for the proof.

In [2] a method is presented to construct a P-compatible enlargement for a given R in case P consists of associative and commutative axioms only.

It is easy to see that P_0-compatible enlargements of rewriting systems can in general be infinite: Permutations of arbitrarily large terms of the form $F(\ldots F(u,v_1),\ldots),v_n)$ might be necessary to apply a rule, and if the left hand side of a rule does not contain a variable of sort S then it can only have instances of fixed size (disregarding the size of the terms of sorts different from S).

Example 2: Let $R = \{ F(A,B) \to C \}$. If R^e is a P_0-compatible enlargement of R, then for all $F(\ldots F(A,x_1),\ldots),x_n),B)$, $n \in N$, $x_i \in V$, there must be a rule $l \to r$ in R^e such that $\Theta(l) =_{P_0} F(\ldots F(A,x_1),\ldots),x_n),B)$ for some Θ, and $(F(\ldots F(C,x_1),\ldots),x_n)) \Rightarrow^* (\Theta(r))$. Also, because R^e must be a proper enlargement of R, F(A,B) must be a subterm of l, and l must be of the form $F(\ldots F(A,B),x_1),\ldots),x_n)$.

The fact that F has only one argument of sort S makes it impossible that l be of the form $l_1 x l_2$, $x \in V$, such that $\Theta(l) = F(\ldots F(A,x_1),\ldots),x_n),B)$ for some Θ and that all the x_i, $1 \leq i \leq n$, occur in $\Theta(x)$.

This problem does not arise if we use the functions U: SxS -> S and I: EL -> S and P as indicated before: In that case we get $R = \{U(A,I(B)) \to C\}$, and a P-compatible enlargement of R is $R^e = \{U(U(A,I(B)),t) \to U(C,t)\} \cup R$, t is a variable of sort S not occurring in A,B, or C.
Now $U(\ldots U(A,I(x_1)),\ldots),I(x_n)),I(B))$ is an instance of $U(U(A,t),I(B))$ (which is in $(U(U(A,I(B)),t))_P$) by a substitution Θ s.t. $\Theta(t) = U(\ldots U(A,I(x_1)),\ldots),I(x_n))$.

It is this freedom of being able to extend a term such that the extended term contains a new variable of type S that allows finite P-compatible enlargements of rewriting systems for commutative and associative theories P. It is equivalent to being able to quantify over all terms u s.t. an instance of a given l occurs in some $u' \in (u)$.

2.3. Critical Classes of P_0-Terms

In the sequel all term-classes will be with respect to P_0. We shall write $\underline{=}$ for $=_{P_0}$.

We shall now develop some sufficient conditions to ensure confluence of rewriting systems with respect to P_0.

Part of the strategy is to construct a set of critical pairs CP, as described above, by superpositions of all pairs $<l_1', l_2'>$, $l_i' \in (l_i)$, $l_i \to r_i \in R$, using P_0-unification. The additional concepts fall into two categories:

(1.) Deal with "critical rules" with left hand sides of the form ...F(t,E)..., $t \in V$.

(2.) Generalise superposition to "merging" to deal with cases where one needs to consider finite "extensions" of rules to generate critical pairs. (Unlike in the case of commutative and associative theories, [2], no overall bound can be given on the size of extensions for all R, rather it has to be determined for each pair of rules.)

Example 3: Consider the single rewriting rule $F(t,A) \to G(t)$, $t \in V$.

It does not on its own give rise to any critical pairs. But

(F(F(T,x),A)) => (G(F(T,x))) by $\Theta(t) = F(T,x)$ and

(F(F(T,x),A)) => (F(G(T),x)) by $\Psi(t) = T$, and in general

$(F(F(\ldots F(T,x_1),\ldots),x_n),A)) \Rightarrow (G(F(\ldots F(T,x_1),\ldots),x_n)))$,

$(F(F(\ldots F(T,x_1),\ldots),x_n),A)) \Rightarrow (F(G(F(\ldots F(T,x_1),\ldots),x_{n-1})),x_n))$,

....

$(F(F(\ldots F(T,x_1),\ldots),x_n),A)) \Rightarrow (F(\ldots F(G(T),x_1),\ldots),x_n))$.

Example 4: Let R = { G(F(t,A),B) -> C, F(T,A) -> D }, $t \in V$.

P_0-unification of F(T,A) with F(t,A) leads to the critical pair <(C), (G(D,B))>.

So consider $R' = R \cup \{ C \to Z, G(D,B) \to Z \}$. Then

$(G(F(F(T,A),x),B)) \underset{R'}{\Rightarrow} (C)$ and

$(G(F(F(T,A),x),B)) \underset{R'}{\Rightarrow} (G(F(D,x),B))$ for any $x \in EL$, and for $x \neq A$ there is no (u) such that $(C) \underset{R'}{\Rightarrow}^* (u)$ and $(G(F(D,x),B)) \underset{R'}{\Rightarrow}^* (u)$.

2.3.1. Subterms and P-Unification

Example 4 highlights an important property of P_0-unification and P-unification in general:

Fact: Let Θ be a substitution, $u,v \in L(T,V)$, and let $\Theta(u)$ be a subterm of some $s \equiv \Theta(v)$.

Then Θ does not necessarily P-unify u with any subterm of any $r \in (v)$.

Illustration: Let v = G(F(t,A),B), u = F(T,A); assume there are no variables in T,A, and B. Let $\Theta(t) = F(T,x)$. Then

$$\begin{array}{lll} \Theta(v) & = G(F(F(T,x),A),B)\ , & \Theta(u) = F(T,A) = u \\ \quad ||| & & \\ \quad s & = G(F(\underbrace{F(T,A)}_{=\Theta(u)},x),B) & \end{array}$$

$(v)=\{v\}$ and for all subterms w of v we get $\Theta(w)\not\equiv\Theta(u)$; in particular $\Theta(F(t,A))=F(F(T,x),A)\not\equiv\Theta(u)$.

This implies that even if the P-instances of two left hand sides of rewriting rules "overlap" in a term (thus leading to a fork in a derivation) there might be no critical pair by P-unification involving the two rules.

This pathological situation arises because permutations according to P can destroy subterms of a term. In the sequel we characterise the circumstances in which this can happen.

Definition (relaxed substitution): Allow different occurrences of the same variable in a term to be replaced by different but equivalent terms. More formally:
For Θ: V -> L(T,V) let $\tilde{\Theta}$: V -> L(T,V)/P_0 be defined by $\tilde{\Theta}(x)=(\Theta(x))$.
Extend $\tilde{\Theta}$ to a function from L(T,V) to FS(L(T,V)), the set of finite subsets of L(T,V):
If $w\in L(T,V)$ and $w = G(w_1,\ldots,w_n)$ then
$\tilde{\Theta}(w) = \{\ G(v_1,\ldots,v_n)\ |\ v_i\in\tilde{\Theta}(w_i),\ 1\leqslant i\leqslant n\ \}$.
Notice that $\tilde{\Theta}(w)\subseteq(\Theta(w))$, and for $w\in L(T)$: $\tilde{\Theta}(w)=\{w\}$.

In the following lemmas let $u,v\in L(T,V)$, and Θ be a substitution.

Lemma 1: Let $\Theta(u)$ be a subterm of some $s\in(\Theta(v))$;
(i) $u=G(\ldots)$, $G\neq F$;
(ii) v does not contain a subterm of the form $F(t,\ldots)$, $t\in V$.
If one of (i) or (ii) holds then there is a subterm w of some $r\in(v)$ such that $w\nabla u$ by Θ.

Lemma 2: If $\Theta(u)$ is a subterm of some $s\in\bigcup_{v'\trianglelefteq v}\tilde{\Theta}(v')$ then $u\nabla w$ by Θ for some subterm w of $r\in(v)$.

Lemma 3: Let $\Theta(u)$ be a subterm of some $s\in(\Theta(v))$ and $\Theta(u)\not\equiv\Theta(w)$ for any subterm w of any $r\in(v)$.
Then Θ is "too big": There exists a variable t of sort S in v s.t.
$v = \ldots F(\ldots F(t,A_1),\ldots),A_k)\ldots$
$\Theta(t) = F(\ldots F(T,E_1),\ldots),E_l)$, $\Theta(u)$ is a subterm of some

$s \in (F(\ldots F(F(\ldots F(T,E_1),\ldots),E_l),\Theta(A_1)),\ldots),\Theta(A_k)))$
and <u>at least one of the E_i does not occur in $\Theta(u)$</u>.

The proofs of these lemmas are fairly straightforward and tedious case distinctions of the various forms u and v can take.

We shall now try to deal with the situations characterised by lemma 3.

2.3.2. Dealing with Critical Rules

Examples 3 and 4 illustrate that there are two different types of critical rules:

(C_1) Rules of the form $F(\ldots F(t,E_1),\ldots),E_n) \to r$.
These rules are critical because some (u) might derive to different (v_1), (v_2) by the same rule, using different substitutions for t (as in example 3).

(C_2) Rules of the form $G(u_1F(\ldots F(t,E_1),\ldots),E_n)u_2) \to r_1$. These rules are not critical by themselves but only in conjunction with other rules of the form $F(\ldots F(T,A_1),\ldots),A_m) \to r_2$ s.t. T L(T,V) and $F(\ldots F(t,E_1),\ldots),E_k) \nabla F(\ldots F(T,A_1),\ldots),A_k)$, $1 \leqslant k \leqslant \min(n,m)$, by a basic set of P_0-unifiers U.

Notice that we assume the E_i and A_i suitably permuted s.t. the P_0-unifiable arguments are on the first k positions.
(Rules of this type are not critical on their own because, unlike in the case of C_1, arguments of F cannot permute "across G and out of the left hand side".)

In case of C_1 we have to show that

(α) $\forall n \in N: \forall (x_1,\ldots,x_n) \in V^n:$
$(\Theta(F(\ldots F(r,x_1),\ldots),x_n)) \downarrow (\Psi(r))$ where $\Psi(t) = F(\ldots F(t,x_1),\ldots),x_n)$
$\Theta(t) = t$ and $\Psi(x) = \Theta(x)$ for $x \neq t$.

and analogously in case of C_2:

(β) $\forall l \in N: \forall (x_1,\ldots,x_l) \in V^l: \forall \Theta \in U:$
$(\Theta(G(u_1F(\ldots F(\ldots F(r_2,E_{k+1}),\ldots),E_n),x_1),\ldots),x_l)u_2))) \downarrow (\Psi(r_1))$
where $\Psi(t) = F(\ldots F(F(\ldots F(T,x_1),\ldots),x_l),\Theta(A_{k+1})),\ldots),\Theta(A_m))$,
$\Psi(x) = \Theta(x)$ for $x \neq t$.

So it seems that one has to consider an unbounded number of critical pairs, thus losing the decidability of the UTP, unless we find decidable necessary and sufficient conditions that ensure that and hold.

Let us take a closer look at critical rules searching for such conditions.

__Lemma 4__: Let R be a permutative rewriting system with FTP, and suppose R contains a critical rule of type C_1,
$F(...F(t,E_1),...),E_n) \to r$, such that $\underline{t \notin V(r)}$.
If R is confluent then there exists a terminal (u) such that
$(r) =>c (u)$ and $(F(u,x)) =>^* (u)$ for $x \in V$, $x \notin V(u)$.

__Proof__: Since R is confluent there exists a (u_1) such that

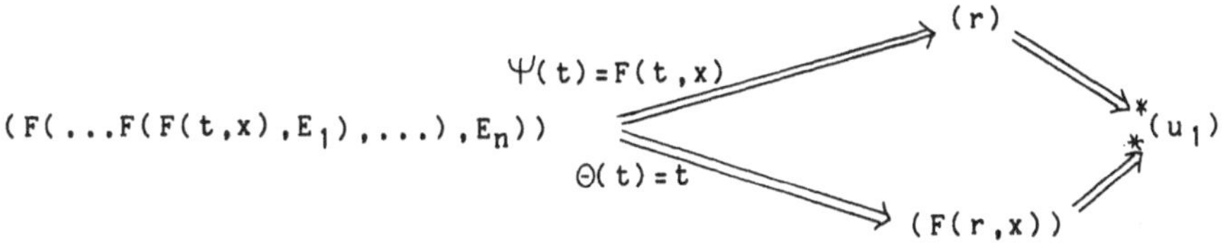

and for all $x \in V$, $x \neq t$, $\Psi(x) = \Theta(x) = x$.
But since $t \notin V(r)$ we also have $(F(r,x)) =>^* (F(u_1,x))$. Thus we get the following situation:

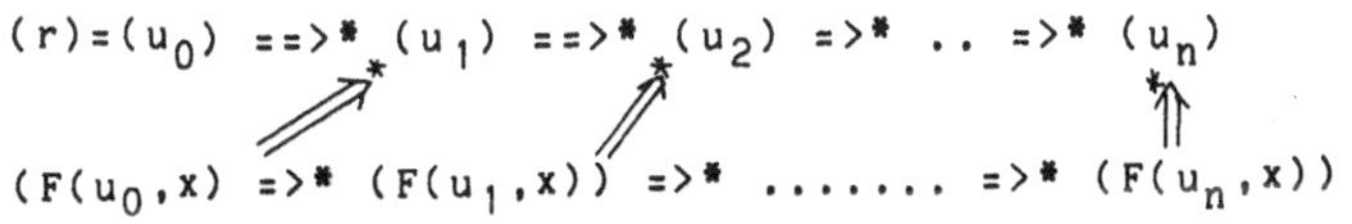

$(r) = (u_0) ==>^* (u_1) ==>^* (u_2) =>^* .. =>^* (u_n)$ (u_n) terminal, $n \geqslant 0$

$(F(u_0,x)) =>^* (F(u_1,x)) =>^* .... =>^* (F(u_n,x))$

Since R has the FTP there are only finitely many (u_i) with $(r) =>^* (u_i)$, i.e. for some $n \in N$ (u_n) is terminal. From the above two diagrams it is obvious that R is not confluent unless $(F(u_n,x)) =>^* (u_n)$. (Notice that $(u_n) =>^* (F(u_n,x))$ is not possible for an R with FTP.)
Q.e.d.

Lemma 4 can in fact be generalised without effort to

__Corollary 1__: Let R be as in lemma 4 but $t \in V(r)$.   If $(r) =>^* (u)$ and $t \notin V(u)$ then R is confluent only if $(F(u,x)) =>^* (u)$ for $x \in V$, $x \notin V(u)$.

__Lemma 5__: Let $d: l \to r$ be a rule of type C_1, i.e.
          $l = F(...F(t,E_1),...),E_n)$, $t \notin V(E_i)$ for $1 \leqslant i \leqslant n$, and let
(i) $r = t$ or
(ii) $r = F(...F(t,A_1),...),A_m)$, $t \notin V(A_i)$ for $1 \leqslant i \leqslant m$,
then d does not cause any nontrivial divergence on its own:
For all $(u) \in L(T,V)/P_0$: If $(u) \underset{d}{=}> (v_1)$ and $(u) \underset{d}{=}> (v_2)$ then
either $\exists(w): (v_1) \underset{d}{=}> (w)$ and $(v_2) \underset{d}{=}> (w)$ (the divergence is trivial)
or $(v_1) = (v_2)$.

Proof: (i) Case 1: d is applied to two non overlapping subterms of some $u' \in (u)$: This case is trivial.

Case 2: $(u) \ni u' = u_1\Theta(F(\ldots F(t,E_1),\ldots),E_n))u_2$ and
$(\Theta(t)) \ni u_3\Psi(F(\ldots F(t,E_1),\ldots E_n))u_4$.
Then (v_1) and (v_2) derive to $(u_1u_3\Psi(t)u_4u_2)$.

Case 3: The two applications of d "overlap" in (u):

$(u) \ni u' = \ldots F(\ldots F(\Theta(1),B_1),\ldots),B_j)\ldots,$ where
$\Theta(t) = F(\ldots F(T,C_1),\ldots),C_k)$

$(u) \ni u'' = \ldots F(\ldots F(\Psi(1),C_1),\ldots),C_k)\ldots,$ where
$\Psi(t) = F(\ldots F(T,B_1),\ldots),B_j)$

Then $v_1 = \ldots F(\ldots F(F(\ldots F(T,C_1),\ldots)C_k),B_1)\ldots),B_j)\ldots,$
$|||$
$v_2 = \ldots F(\ldots F(F(\ldots F(T,B_1),\ldots),B_j),C_1),\ldots),C_k)\ldots$

(ii) This case goes through analogously since a term like $r = F(\ldots F(t,A_1),\ldots),A_m)$ allows free permutations of its EL-type arguments with an enclosing term of the form $F(\ldots F(r,\ldots),\ldots)$ and with any instantiated term $\Theta(t) = F(\ldots F(\ldots)\ldots)$.
Q.e.d.

Corollary 1 and lemma 5 provide necessary and sufficient conditions to check unique termination of axiomatic specifications of the type SET mentionned above, using P_0 with F=INSERT, which partly motivated this investigation. They should also provide tools to deal with a number of similar data types.

In the sequel we introduce some conditions that are sufficient to "defuse" critical rules of type C_1 which are not of the forms described above.

Notation: To indicate that we assume t to be a variable occurring in a term r we write r[t], and to indicate substitution of t in r by u we simply write r[u] (in analogy to conventional functional notation like f(x)=x*x and f(3)=3*3). If t does not occur in r then r[t]=r[u]=r.

Remember that critical rules of type C_1 allow a term class of the form $(F(\ldots F(F(\ldots F(t,E_1),\ldots),E_n),x_1),\ldots),x_m))$, $x_i \in V$, to derive to any element of the set

$$CF(x_1,\ldots,x_m) = \{\ (R[F(\ldots F(t,x_1),\ldots),x_m)]),\ (F(R[F(\ldots F(t,x_2),\ldots),x_m)],x_1)),\ \ldots\ldots,\ (F(\ldots F(r[t],x_1),\ldots),x_m))\ \}$$

To ensure confluence we have to ensure that for all m and all $x_1,\dots,x_m \in V$ all elements of $CF(x_1,\dots,x_m)$ derive to one single $(w) \in L(T,V)/P_0$

Each of the following two conditions is sufficient to guarantee this:

(CS_1) $\exists (u[t])$ s.t. $(r[t]) \Rightarrow^* (u[t])$, and for $x \in V$, $x \notin V(u)$,
either $(F(u[t],x)) \Rightarrow^* (u[F(t,x)])$
or $(u[F(t,x)]) \Rightarrow^* (F(u[t],x))$

In that case all elements of $CF(x_1,\dots,x_m)$ derive either to $(u[F(\dots F(t,x_1),\dots),x_m)])$ or to $(F(\dots F(u[t],x_1),\dots),x_m))$.

(CS_2) $\exists (u[t])$ s.t. $(r[t]) \Rightarrow^* (u[t])$, and for $x \in V$, $x \notin V(u)$,
$(u[F(t,x)]) \Rightarrow^* (u[t])$ <u>and</u>
$(F(u[t],x)) \Rightarrow^* (u[t])$.

In that case all elements of $CF(x_1,\dots,x_m)$ derive to $(u[t])$.

Let us turn to critical rules of type C_2 now. For these rules CP contains in general critical pairs of the form
$\langle(\Theta(\dots G(u_1F(\dots F(r_2,E_{\pi k+1}),\dots),E_{\pi n})u_2)\dots)),\ (\Theta(r_1))\rangle$
where $F(\dots F(T,A_1),\dots),A_m) \triangledown F(\dots F(t,E_{\ 1}),\dots),E_{\ k})$ for some basic θ, π a permutation of $(1,\dots,n)$.

Apart from these we get, in analogy to the case of C_1, critical pairs of the following form:
$\langle(\Theta(\dots G(u_1F(\dots F(u,x_1),\dots),x_j)u_2)\dots)),\ (\Theta(\lambda_j(r_1)))\rangle$ where $x_1,\dots,x_j \in V$, $u = F(\dots F(r_2,E_{\pi k+1}),\dots),E_{\pi n})$, and $\lambda_j(t) = F(\dots F(t,x_1),\dots),x_j)$. $(\Theta(\lambda_j(r_1))$, not $\lambda_j(\Theta(r_1))$!) Let us call these "<u>critical pairs of type C_2</u>".

Each of the following conditions is sufficient to ensure that for all j: $(\Theta(\dots G(u_1F(\dots F(u,x_1),\dots),x_j)u_2)\dots)) \downarrow (\Theta(\lambda_j(r_1)))$.

Let $r_1'[t] = \Theta'(r_1[t])$ where $\Theta'(t)=t$ and $\Theta'(x)=\Theta(x)$ for $x \neq t$.

(CS_3) $(\Theta(\dots G(u_1t_1u_2)\dots)) \downarrow (r_1'[t_2])$ where $t_1,t_2 \in V$ do not occur in the terms under consideration.

(CS_4) $(r_1'[F(t,x)]) \Rightarrow^* (r_1'[t])$ <u>and</u>
$(\dots G(u_1F(u,x)u_2)\dots) \Rightarrow^* (\dots G(u_1uu_2)\dots)$, for $x \in V$.

2.3.3. Merging of Terms

One of the cases not yet considered is illustrated in the next example:

Example 5: Consider rules F(A,B) -> D and F(A,C) -> E, $A \notin V$. These rules are not critical, and there are no critical pairs for them resulting from superposition using P_0-unification. Yet (F(F(A,B),C)) => (F(D,C)) and (F(F(A,B),C)) => (F(E,B)).

In this example term classes can be critical because of arguments at any "distance" from A, the argument of sort S: $F(\ldots F(A,B),x_1),\ldots),x_n),C)$ is critical for any n and $x_1,\ldots,x_n \in V$. But fortunately, if (F(D,C)) =>* (u) and (F(E,B)) =>* (u) then

$$(F(\ldots F(A,B),x_1),\ldots),x_n),C))$$

$$\swarrow \qquad\qquad \searrow$$

$$(F(\ldots F(D,x_1),\ldots),x_n),C)) \qquad\qquad (F(\ldots F(E,B),x_1),\ldots),x_n))$$

$$\searrow \qquad\qquad \swarrow$$

$$(F(\ldots F(u,x_1),\ldots),x_n))$$

Thus, all the ingredients that make these term classes critical are already contained in F(A,B) and F(A,C).

To deal with this particular problem we introduce the concept of merging of terms.

Definition (merging): Consider two terms $u = F(\ldots F(T_1,A_1),\ldots),A_m)$ and $v = F(\ldots F(T_2,B_1),\ldots),B_n)$.
First suppose $T_1,T_2 \notin V$.
Start an inside-out unifying process, building up a set M of merged terms:
Starting point: If $T_1 \nabla T_2$ by a basic set of unifiers U_0 then take $\Theta(F(\ldots F(F(\ldots F(T_1,A_1),\ldots),A_m),B_1),\ldots),B_n))$ into M for all $\Theta \in U_0$.
If $T_1 \not\nabla T_2$ then u and v cannot be merged.

Next try, for $i=1,\ldots,\min(m,n)$, to P_0-unify $F(\ldots F(T_1,A_1),\ldots),A_i)$ and $F(\ldots F(T_2,B_1),\ldots),B_i)$, and if this is possible by a basic set of P_0-unifiers U_i, then take
$\Theta(F(\ldots F(F(\ldots F(T_1,A_1),\ldots),A_m),B_{i+1}),\ldots),B_n))$ into M for all $\Theta \in U_i$.

(Notice that, if T_1 and T_2 are not both variables, all substitutions in any U_i are extensions of substitutions in U_0: $\forall \Theta \in U_i\ \exists \Psi \in U_0$ such that $\Theta|_{dom_\Psi} = \Psi$.)

If for some i, $1 \leq i \leq \min(m,n)$, the corresponding terms cannot be P_0-unified, then start the process all over again with a permutation of u and v until all possible pairs (u',v'), $u' \in (u)$, $v' \in (v)$, are exhausted.

In fact, we also restart the process with further permutations of u and v if P_0-unification has been successful for $i=1,\dots,\min(m,n)$ and $m\neq n$.

Next suppose one of the T_i, T_1 say, is a variable.
In this case go through the inside-out unification trials just as above starting with $\Theta(T_1)=T_2$, then with $\Theta(T_1)=F(T_2,B_1)$, etc. until $\Theta(T_1)=F(\dots F(T_2,B_1),\dots),B_{n-1})$.

As before, go through the whole process for all pairs of permutations of u and v.

Notice that this time merging is successful in each step: If one starts with $\Theta(T_1)=F(\dots F(T_2,B_1),\dots),B_i)$ then there is at least one merged term from this trial:
$F(\dots F(F(\dots F(F(\dots F(T_2,B_1),\dots),B_i),B_{i+1}),\dots),B_n),A_1),\dots),A_m)$.

We do not consider merging of terms where T_1 and T_2 are both varibles. In that case P_0-unification covers all the critical cases.

In example 5 F(A,B) and F(A,C) merge to F(F(A,B),C).

Example 6: Consider the following two rules:

F(F(t,i),i) -> F(t,i) (t,i variables)

F(F(A,B),C) -> F(A,D)

Produce the following critical term classes by merging:

(F(F(F(A,B),C),B)) $\Theta(t) = A$, $\Theta(i) = B$

(F(F(F(A,B),C),C)) $\Theta(t) = F(A,B)$, $\Theta(i) = C$

3. A Unique Termination Theorem

Definition: The set of critical pairs CP of a permutative rewriting system R can now be defined as follows:

$CP:=\emptyset$;

for any pair (l_1,l_2), d_i: $l_i \to r_i \in R$, $i=1,2$, do the following:

(i) If l_1 P_0-unifies with any subterm s of an $l_2' \in (l_2)$, $l_2'=w_1 s w_2$, and U is the basic set of P_0-unifiers of l_1 and s then set
$CP:= CP \cup \{<(\Theta(r_2)),(\Theta(w_1 r_1 w_2))> \mid \Theta\in U\}$;

(ii) If $l_1= F(\dots F(T_1,A_1),\dots),A_k)$, $l_2= F(\dots F(T_2,B_1),\dots),B_1)$ and not both T_1 and T_2 are variables then let M be the set of terms merged from l_1 and l_2 and for all $u\in M$ set
$CP:= CP\cup\{<(v_1),(v_2)>\}$ where $(u) \underset{\{d_1\}}{\overset{=}{\Rightarrow}} (v_1)$ and $(u) \underset{\{d_2\}}{\overset{=}{\Rightarrow}} (v_2)$.

Critical pairs lemma: Let R be a permutative rewriting system with FTP. Let the set of critical pairs of R, CP, be constructed as described above.

Then R is locally confluent if the following conditions are satisfied:

(i) $\forall \langle (u),(v)\rangle \in CP$: $(u)\downarrow(v)$

(ii) For any critical rule $l \to r$ of type C_1 s.t. $r \neq t$ and $r \neq F(\ldots F(t,A_1),\ldots),A_m)$, for some $t \in V$, condition CS_1 or CS_2 holds.

(iii) For any critical pair of type C_2 condition CS_3 or CS_4 holds.

<u>Sketch of the Proof</u>: Let $(u) \Rightarrow (v_1)$ and $(u) \Rightarrow (v_2)$ for some $u, v_1, v_2 \in L(T,V)$, i.e. $u' \underset{\{l_1 \to r_1\}}{\Rightarrow} v_1'$, $u'' \underset{\{l_2 \to r_2\}}{\Rightarrow} v_2'$ for $u', u'' \in (u)$ minimally permuted, $v_1' \in (v_1)$, $v_2' \in (v_2)$.

Distinguish three cases of how the minimal permuted subterms and the left hand sides of the rewriting rules "overlap" in u:

(1.) There exists $u' \in (u)$ s.t. $u' \Rightarrow v_1'$ and $u' \Rightarrow v_2'$.
In this case the proof is just like in the "classical" critical pairs case described in [1].

(2.) $u' = u_1\Theta(l_1)u_2$, $u'' = u_1\underline{w_1\Theta(l_2)w_2}u_2$, i.e. $w_1\Theta(l_1)w_2 \equiv \Theta(l_2)$.
Using lemmas (1) to (3) one can show that either there are relevant critical pairs in CP or that (CS_1) to (CS_4) guarantee confluence.

(3.) $u' = u_1pu_2$, $u'' = u_1qu_2$ and $\Theta(l_1)$ and $\Theta(l_2)$ are proper subterms of p, resp. q.
Then l_1 and l_2 must be of the form $F(\ldots)$ since otherwise the permutations enclosing $\Theta(l_1)$ and $\Theta(l_2)$ are irrelevant for the application of the two rules. There is a relevant critical pair in CP from merging l_1 and l_2.

<u>Corollary 2</u>: Let R be a permutative rewriting system with FTP. Let CP be the set of critical pairs of R. If R does not contain any critical rules of type C_2, and all the critical rules of type C_1 are as described in corollary 1 and lemma 5, then R is locally confluent if and only if for all $\langle (a),(b)\rangle \in CP$: $(a)\downarrow(b)$.

<u>Theorem</u>: Let R be as in the critical pairs lemma. Then R is confluent if conditions (i) to (iii) of that lemma are satisfied.

This follows from the diamond lemma and the critical pairs lemma.

Appendix:

In the sequel we shall generalise Robinson's unification algorithm (see [6] p.32) such that the new algorithm decides whether or not two words $u,v \in L(T,V)$ are P_0-unifiable, and if so gives us a basic set of P_0-unifiers.

Assume there is an ordering on variables and function symbols, inducing a lexicographical ordering on terms.

Definition: Let $u,v \in L(T,V)$. The disagreement pair of u and v is the pair $\langle p,q \rangle$ of those subexpressions of u and v which begin at the first symbol position at which u and v do not have the same symbol; p comes before q in the lexicographic ordering of terms.

$\langle p',q' \rangle$ is the P_0-disagreement pair of u and v if p' and q' are the largest "directly enclosing F-terms" of p and q in u and v: p' and q' are the largest subterms of u and v (or v and u respectively) s.t. $p'=F(\ldots F(T_1,A_1),\ldots),A_k)$, $q'=F(\ldots F(T_2,B_1),\ldots),B_l)$, $k,l \geq 1$, with $T_1 \neq F(\ldots) \neq T_2$ and: $p=T_1$ or $q=T_2$ or for some i: $A_i=apb$, $B_i=aqc$ and F does not occur in a.

E.g. the disagreement pair of $F(T_1,G(F(T_2,A)))$ and $F(T_1,G(F(T_3,B)))$ is $\langle T_1,T_2 \rangle$, the P_0-disagreement pair is $\langle F(T_2,A),F(T_2,B) \rangle$.

The P_0-Unification Algorithm:

Given $u,v \in L(T,V)$; collect P_0-unifiers for u and v in U; keep candidates for P_0-unifiers in P.

```
U:=∅; P:={ε} where ε(x)=x for all x∈V;
cycle
  for all Θ∈P do: [if Θ(u)≡Θ(v) then U:=U∪{Θ}; P:=P-{Θ}]
  if P=∅ then stop;
  P':=∅;
  for all Θ∈P do: [ extend(Θ,p,q) where <p,q> is the P0-disagreement
                    pair of Θ(u) and Θ(v) ];
  P:=P'
repeat the cycle
where extend is the following procedure:
if p∈V start if p∉V(q) then P':=P'∪{μ∘Θ} where μ(p)=q and
                                  μ(x)=x for x≠p;
           return (Θ has been extended or rejected)
      finish
if q∈V start ... (analogously) finish;
```

```
if p=G(...), q=H(...), G,H∈𝔉, G≠H, then return (reject Θ);
if p=F(...F(T1,A1)...),Ak), q=F(...F(T2,B1),...),Bl), l,k≥1 start
  for all <p',q'> s.t. p'≡p and q'≡q do the following:
     distinguish three cases:
    (1.) T1,T2∉V: if k≠l then return (Θ is no longer a candidate);
         now k=l: extend(Θ,p",q") where <p",q"> is the (ordinary)
         disagreement pair of Θ(u) and Θ(v);
         continue with the next pair <p',q'>;
    (2.) T1∈V, T2∉V: if l<k then return (reject Θ);
         if T1∉V(F(...F(T2,B1),...),Bl-k)) then P':=P'∪{μ•Θ}
            where μ(T1)=F(...F(T2,B1),...),Bl-k) and μ(x)=x for x≠T1;
         continue with the next <p',q'>
    analogously for T2∈V, T1∉V.
    (3.) T1,T2∈V:
       if k<l start
         if T1∉V(F(...F(T2,B1),...),Bl-k)) then P':=P'∪{μ•Θ}
            where μ(T1)=F(...F(t,B1),...),Bl-k), μ(x)=x for x≠T1,
            and t is a new variable;
         continue with a new <p',q'>
       finish
       if k>l start ... (analogously) finish
       now k=l:
       if T1≠T2 start
         P':=P'∪{μ∘Θ} where μ(T1)=T2, μ(x)=x for x≠T1;
         (notice that T1<T2 in the lexicographic ordering)
         if μ(p)≠ μ(q) start (now μ(Ai)≠μ(Bi) for some i)
           if T1∉V(Bi) and T2∉V(Ai) start (now Ai≠Bi)
             if Ai≠Bj and Bi≠Aj for j>i then P':=P'∪{μ•Θ}
               where μ(T1)=F(t3,Bi) and μ(T2)=F(t4,Ai) and t3,t4
               are new variables;
           finish
         finish
         continue with the next <p',q'>
       finish
       now T1=T2: then Ai≠Bi for some i and, by the way we defined
       P0-disagreement pairs, also Ai≢Bi;
       if Ai≠Bj and Bi≠Aj for j>i then P':=P'∪{μ∘Θ} where
         μ(T1)=F(F(t,Ai),Bi), and t is a new variable.
       continue with the next <p',q'>;
finish case p=F(...F(T1,A1),...),Ak), q=F(...F(T2,B1),...),Bl)
This ends the procedure extend.
```

Proposition: The set of P_0-unifiers generated by the algorithm above is basic.

References:

[1] D.E.Knuth & P.B.Bendix: Simple Word Problems in Universal Algebras in Computational Problems in Abstract Algebra
Ed. J.Leech, Pergamon Press 1970, pp.263-297

[2] G.E. Peterson & M.E. Stickel: Complete Sets of Reductions for Equational Theories
Unpublished

[3] G. Huet: Confluent Reductions: Abstract Properties and Applications to Term Rewriting Systems
IRIA-LABORIA, Domaine de Voluceau, F-78150 Rocquencourt France.
Preliminary version in 18th IEEE Symposium on Foundations of Computer Science, Oct 1977

[4] D.S.Lankford & A.M.Ballantyne: Decision Procedures for Simple Equational Theories with a Commutative Axiom: Complete Sets of Commutative Reductions
Automatic Theorem Proving Project, Depts. Math. and Comp. Science, University of Texas at Austin; Report #ATP-35

[5] D.S.Lankford & A.M.Ballantyne: Decision Procedures for Simple Equational Theories with Commutative-Associative Axioms: Complete Sets of Commutative-Associative Reductions
As [2], Report #ATP-39

[6] J.A.Robinson: A Machine-Oriented Logic Based on the Resolution Principle. JACM Vol.12, No.1; January 1965; pp.23-41

[7] M.E. Stickel: A Complete Unification Algorithm for Associative-Commutative Functions
Proceedings of IJCAI, Tblisi Georgia, USSR

[8] M. Livesey, J. Sieckmann: Unification of A+C-Terms (Bags) and A+C+I-Terms (Sets)
Interner Bericht Nr.5/76, Institut für Informatik I, Universität Karlsruhe

HOW TO PROVE ALGEBRAIC INDUCTIVE HYPOTHESES WITHOUT INDUCTION

with Applications to the Correctness of Data Type Implementation

by J. A. Goguen
SRI International, 333 Ravenswood Ave., Menlo Park, CA 94025[1]
Computer Science Dept., UCLA, Los Angeles, CA 90024[2]

ABSTRACT

This paper proves the correctness of algebraic methods for deciding the equivalence of expressions by applying rewrite rules, and for proving inductive equational hypotheses without using induction; it also shows that the equations true in the initial algebra are just those provable by structural induction. The major results generalize, simplify and rigorize Musser's method for proving inductive hypotheses with the Knuth-Bendix algorithm; our approach uses a very general result, that (under certain conditions) an equation is true iff it is consistent. Finally, we show how these results can be extended to proving the correctness of an implementation of one data abstraction by another.

1. INTRODUCTION

This paper is concerned with methods for automatically deciding in a given algebraic theory, whether or not two terms are deductively equivalent, and whether or not a given equation can be proved inductively. We make extensive use of the connection between algebra and rewrite rules, and we have generally preferred to use the notation and terminology of universal algebra rather than that of automatic deduction, in order to emphasize the very general nature of our results, as well their relations to other fields. The restriction to equational theories seems to be a natural one, in view of the fact that all the data structures one normally uses in theorem proving can be defined equationally [Goguen, Thatcher & Wagner 78].

Our algebraic justification for the use of rewrite rules to decide term equality seems to be new, although experts will probably not be surprised. Our correctness proof for the inductive methods is based on a very general theorem (having nothing to do with rewrite rules) that under certain conditions, an equation is true iff it is consistent. We then relate this to the Knuth-Bendix [Knuth & Bendix 70] algorithm, thus generalizing, simplifying and rigorizing some work of Musser [Musser 80a].

One could say that our approach makes available for rewrite rules the powerful techniques of many-sorted universal algebra and the algebraic approach to abstract data types; or one could say it provides a rewrite-rule-based operational interpretation for algebraic semantics, thus supporting the execution of test cases

---

[1]supported in part by NSF Grant MCS-7816783.

[2]On leave; supported in part by NSF Grant No. MCS-7818918.

and automatic deduction.

It is well known that there are two different kinds of semantics for a given collection of equations; consequently there are two different kinds of theorem proving. For what we shall call *varietal semantics*, we wonder if some new equations can be proved by equational reasoning from some given equations; speaking semantically, we are asking whether the new equations hold in the *variety*, i.e., in every model, of the original axioms. In the second case, which we shall call *initial algebra semantics*, the axioms are taken to define a particular standard model, and we wonder if some other equations hold in that particular model. Alternatively, the first case is deciding the equality of terms, while the second is proving inductive assertions.

The algebraic approach to data abstraction [Zilles 74], [Goguen, Thatcher, Wagner & Wright 75], [Guttag 75] has stimulated a number of languages and automatic systems, including CLEAR [Burstall & Goguen 77], OBJ-O [Goguen 77], OBJ-T [Goguen & Tardo 79], DTVS [Guttag, Horowitz & Musser 78], AFFIRM [Musser 80b] and [Wand 77]. The work reported in this paper arose partly as an attempt to prove correctness of some algorithms in OBJ and partly as an attempt to clarify and generalize Musser's [Musser 80a] ideas so that they could be used to compute the inductive closure of a theory (as needed in CLEAR) and for proofs of correctness of implementations.

I would like to thank Drs. Thatcher, Wagner and Wright of IBM Watson Research Center for their earlier collaboration in the area of algebraic semantics, and Prof. R. M. Burstall of the University of Edinburgh for his continuing collaboration and inspiration, both personally and professionally. In addition, I would like to thank Drs. Boyer, Moore, Shostak, and especially Huet[3], of SRI, International and Dr. Oppen of Stanford University, for their comments on, and interest in, the topics of this paper, and I would like to thank Prof. Ehrig and Dr. Kreowski of the Technical University of Berlin for their suggestions on the material of Section 6. Last but not least, I would like to thank J. Tardo and D. Smallberg for their implementations of the OBJ and kb systems referred to in this paper; this practical work provided much of the inspiration for the theoretical results.

2. BACKGROUND

This section briefly introduces some material from rewrite rules and algebra.

Definition 1. Let T be a set, and let R be a relation on T. Thinking of aRb as meaning that a can be rewritten to b, we also write $a \to b$. R^+, R^*, $\equiv_R$ respectively denote the transitive, transitive-reflexive , and transitive-reflexive-symmetric closures of R; if R is clear from context, we may also write $a \to^+ b$, $a \to^* b$, $a \equiv b$,

---

[3]on leave from INRIA

respectively. t is (R-) <u>reduced</u> iff there is no t' such that tR^+t'. t' is the (R-) <u>reduced form of</u> t iff tR^*t' and t' is R-reduced. Notice that $\equiv$ is an equivalence relation, so that it makes sense to speak of $\equiv$-equivalence classes.

We shall say that R is <u>unique terminating</u> iff each $\equiv$-class contains at most one reduced form, and R is <u>finite terminating</u> iff there is no infinite sequence

$$t_1 \to t_2 \to \ldots \to t_n \to \ldots$$

R is <u>globally finite</u> iff $\{b \mid aR^*b\}$ is finite for each a in T. R is <u>confluent</u> iff aR^*b and aR^*c imply there is some d such that bR^*d and cR^*d; and R is <u>locally confluent</u> iff aRb and aRc imply there is some d such that bR^*d and cR^*d. □

<u>Proposition 1.</u> If R is confluent, then R is unique terminating; moreover, if $t\equiv_R t'$ and t' is R-reduced, then tR^*t'. If R is confluent and finite terminating, then each $\equiv_R$-equivalence class contains a unique R-reduced element. □

The second assertion (c.f. [Wand 77]) says that, in looking for the reduced form of t, if R is confluent, we never need to apply converses of rewrites: if t' is reduced and $t\equiv t'$, then actually tR^*t'.

Letting R be a relation on a set T as above, we define a new relation ~ on T as follows: $t\sim t'$ iff tR^*t' and $t'R^*t$; noting that ~ is an equivalence relation, let us denote the ~-equivalence class of t by [t]. Now define R/~ on T/~ by [t] R/~ [t'] iff there are u,u' in T such that $u\sim t$, $u'\sim t'$, uRu' and $[t] \neq [t']$.

<u>Proposition 2.</u> If a relation R on a set T is finite terminating and locally confluent, then it is confluent. Moreover, if R is globally finite and R/~ is locally confluent, then R is confluent. □

The first assertion is the well known "diamond lemma" and the second is a new generalization proved in the notes from which this paper was drawn [Goguen 80] (actually, global finiteness generalizes finite termination only under the additional assumption of local finiteness).

We now turn to algebra, developed in its many-sorted form because of the importance of treating more than one sort of data in applications.

<u>Definition 2.</u> A (many-sorted) <u>signature</u> is a pair $\langle S,\Sigma\rangle$, where S is a set whose elements are called <u>sorts</u>, and Σ is an $(S \times S)$-indexed family of sets,

$$\Sigma = \langle \Sigma_{w,s} \mid w \in S^*,\ s \in S\rangle,$$

where an operation symbol σ in $\Sigma_{w,s}$ has <u>arity</u> w, <u>sort</u> s, and <u>rank</u> $\langle w,s\rangle$. Usually we write Σ for $\langle S,\Sigma\rangle$, leaving the sort set implicit.

Call a sort s <u>void</u> in Σ iff Σ contains no constants of sort s, and every non-constant operation symbol of sort s has at least one void sort in its arity. Call Σ <u>sensible</u> iff whenever σ has rank $\langle s1\ldots sn,s\rangle$ and s is non-void, then s1,...,sn are non-void.
□

Some (but not all) of the results of this paper require a sensible signature; we shall simply assume that all signatures are sensible. (I thank G. Huet for pointing out the need for this notion.)

Notice that our formulation of signature permits a given symbol to have more than one rank, i.e., to be "overloaded". Thus, we can have + in $\Sigma_{\mathbf{nat\ nat,\ nat}}$ and + in $\Sigma_{\mathbf{bool\ bool,bool}}$. Note that an operation of rank $\langle\lambda,s\rangle$, where λ is the empty string of sorts, is just a constant of sort s; e.g., $0 \in \Sigma_{\lambda,\mathbf{nat}}$ and true $\in \Sigma_{\lambda,\mathbf{bool}}$.

Definition 3. A Σ-algebra A consists of an S-indexed family $\langle A_s \mid s \in S\rangle$ of sets (A_s is called the carrier of sort s), plus a function A_σ: $A_w \to A_s$ for each σ in $\Sigma_{w,s}$, where $A_w = A_{s1} \times \ldots \times A_{sn}$ if w = s1...sn, and $A_w = \{.\}$ (some one-pointed set) if w = λ. Hereafter, we write σ for A_σ when A is clear. □

An s-sorted Σ-(ground) term is a sequence of symbols from Σ plus (and), constructed as follows, where $T_{\Sigma,s}$ denotes the set of all Σ-terms of sort s:

(t0) $\sigma \in T_{\Sigma,s}$ if $\sigma \in \Sigma_{\lambda,s}$

(t1) σ(t1...tn) $\in T_{\Sigma,s}$ if
$\sigma \in \Sigma_{w,s}$, where w = s1...sn and
ti $\in T_{\Sigma,si}$.

T_Σ can be made into a Σ-algebra by defining the function $T_{\Sigma,\sigma}$: $T_{\Sigma,w} \to T_{\Sigma,s}$ for $\sigma \in \Sigma_{w,s}$, as follows:

(T0) $(T_\Sigma)_\sigma = \sigma$ for w = λ;

(T1) $(T_\Sigma)_\sigma$(t1,...,tn) = σ(t1...tn)
for w = s1...sn and ti $\in T_{\Sigma,si}$.

Notice that a sort s is void in Σ iff $T_{\Sigma,s}$ is void.

Definition 4. A Σ-homomorphism h from a Σ-algebra A to a Σ-algebra B is an S-indexed family of functions h_s: $A_s \to B_s$ such that for each σ in $\Sigma_{w,s}$, the diagram

$$\begin{array}{ccc} A_w & \xrightarrow{A_\sigma} & A_s \\ {\scriptstyle h_w}\downarrow & & \downarrow{\scriptstyle h_s} \\ B_w & \xrightarrow[B_\sigma]{} & B_s \end{array}$$

commutes, where h_w(a1,...,an) = $\langle h_{s1}$(a1),...,h_{sn}(an)$\rangle$, where w = s1...sn and a = $\langle$a1,...,an$\rangle \in A_w$. We will write h: A $\to$ B. A Σ-homomorphism h is a Σ-isomorphism iff each h_s is bijective. A Σ-algebra T is initial iff for any other Σ-algebra A, there is a unique Σ-homomorphism h: T$\to$ A. □

Proposition 3. T_Σ is an initial Σ-algebra. □

Concretely, T_Σ is the Herbrand Universe under the standard interpretation. Initiality characterizes this object "abstractly," that is, uniquely up to isomorphism, using the notion of homomorphism. It is therefore possible to prove properties of this object with "diagram chasing" rather than induction [Goguen, Thatcher, Wagner & Wright 75].

Proposition 4. Any two initial Σ-algebras are isomorphic; and any Σ-algebra isomorphic to an initial Σ-algebra is initial. □

For example, the non-negative integers are characterized independently of representation, as being initial among Σ-algebras, where Σ contains the constant 0 and the unary successor operator s; both the base 10 representations and the binary representations form Σ-algebras, and these algebras are Σ-isomorphic. The integers modulo 2 are a Σ-algebra which is not initial.

It is necessary to use equations in order to define more complex structures, and even such simple operations as + on the integers. We proceed as follows: let X be an S-indexed family of sets, $\langle X_s \mid s \in S\rangle$; think of $x \in X_s$ as a "variable symbol" of sort s. Define a new signature $\Sigma(X)$ by $\Sigma(X)_{\lambda,s} = \Sigma_{\lambda,s} \cup X_s$ and $\Sigma(X)_{w,s} = \Sigma_{w,s}$ for w not λ. $\Sigma(X)$ is Σ with variable symbols adjoined as constants, and we will call its elements "expressions." Note that these include all ground terms. Next, a Σ-equation of sort s is a pair of elements of $T_{\Sigma(X),s}$ (for some X). It is convenient to write this as a pair, (t1 = t2) for t1,t2 in $T_{\Sigma(X),s}$. We shall say that this equation holds in (or is satisfied by) a Σ-algebra A iff for every assignment α: X → A of values in A to variables (this means, $\alpha_s: X_s \to A_s$ for s in S), $\alpha(t1) = \alpha(t2)$ in A, where $\alpha(t)$ denotes the result of evaluating t in A with values for the variables supplied by α.

Definition 5. A presentation P is a triple $\langle S,\Sigma,E\rangle$, where S is a set (of sorts), Σ is an S-sorted signature, and E is an S-indexed family of Σ-equations, E_s containing the equations of sort s. The presentation P is satisfied by a Σ-algebra A iff each equation in E holds in A; we also say that A is a P-algebra. □

We now construct an initial algebra satisfying a presentation, thus generalizing Theorem 3.

Definition 6. A Σ-congruence $\equiv$ on a Σ-algebra A is an S-indexed family $\langle \equiv_s \mid s \in S\rangle$ of equivalence relations, $\equiv_s$ on A_s, such that for each σ in $\Sigma_{w,s}$ and ai,bi in A_{si}, for i=1,...,n where w = s1...sn,

$$ai \equiv_{si} bi \text{ for } i = 1,\dots,n \text{ implies}$$
$$\sigma(a1,\dots,an) \equiv_s \sigma(b1,\dots,bn).$$

(This is called the substitution property of $\equiv$.) Given a Σ-congruence $\equiv$ on a Σ-algebra A, we define a Σ-algebra $A/\equiv$, called the quotient of A by $\equiv$, as follows: $(A/\equiv)_s = A_s/\equiv_s$, the set theoretic quotient of A_s by the equivalence relation $\equiv_s$, and

$$(A/\equiv)_\sigma([a1],\dots,[an]) = [\sigma(a1,\dots,an)],$$

for σ in $\Sigma_{w,s}$ with w = s1...sn, where [a] denotes the $\equiv$-equivalence class of a. (The substitution property guarantees that this is well defined.) □

Any S-indexed relation R on a Σ-algebra A (this means that $R_s \subseteq A_s \times A_s$) is contained in a least Σ-congruence on A, called the Σ-congruence generated by R. Now, given a presentation $P = \langle S,\Sigma,E\rangle$, define a relation R_P on T_Σ as follows:

$$R_{P,s} = \{\langle t,t'\rangle \mid t = \alpha(L) \text{ and } t' = \alpha(R), \text{ for some substitution } \alpha, \text{ where } (L = R) \in E\}.$$

Next, let $\equiv_P$ be the Σ-congruence generated by R_P. Finally, define T_P (also written $T_{\Sigma,E}$) to be the quotient $T_\Sigma/\equiv_P$.

Theorem 5. There is one and only one Σ-homomorphism from T_P to any other given Σ-algebra satisfying P; i.e., T_P is an initial P-algebra. □

Any Σ-homomorphism $h: A \to B$ has a kernel, denoted ker(h), which is a Σ-congruence relation on A defined for a,a' in A_s by

$a \ker(h)_s a'$ iff $h_s(a) = h_s(a')$.

Then we have

Theorem 6. (Homomorphism Theorem) For $h: A \to B$ a Σ-homomorphism, A/ker(h) is isomorphic to a sub-Σ-algebra of B; if h is surjective (i.e., each component h_s is surjective), then A/ker(h) is Σ-isomorphic to B. □

We conclude this section by generalizing term rewriting systems to many sorts.

Definition 7. An S-sorted rewrite (rule) system is an S-indexed set $T = \langle T_s \mid s \in S\rangle$ and an S-indexed relation $R = \langle R_s \mid s \in S\rangle$. We write $a \to b$ if aR_sb for some s in S. Then R is confluent, finite terminating, etc. iff each R_s is. □

Let Σ be an S-sorted signature, let T be T_Σ, and let E be a family of Σ-equations. Now let each $e = (l = r)$ in E_s define a relation R_e on $T_{\Sigma,s}$ as follows: $\langle t,t'\rangle \in R$ iff $t = \alpha(l)$ and $t' = \alpha(r)$ for some assignment α of values in T_Σ to the variables occurring in l and r. Next let $Q_s = \bigcup\{R_e \mid e \in E_s\}$, with $Q = \langle Q_s \mid s \in S\rangle$. Now define tR_st' iff there are t1 and t2 such that $t1Q_st2$, and t and t' have the form $\alpha(u)$ and $\alpha'(u)$ respectively, where α substitutes t1 for a variable x in u, α' substitutes t2 for x, and u is in $T_{\Sigma(\{x\}),s}$ (i.e., u has only the one variable x). Then T_Σ with R is the Σ-(term) rewriting system defined by the presentation $P = \langle S,\Sigma,E\rangle$. If helpful, we write R_E and $R_{E,s}$ instead of R and R_s. We shall say that E, or P, is confluent, finite terminating, etc. iff R_E on T_E is.

3. REWRITE RULES AND ALGEBRAS

This section discusses relationships between algebraic semantics and the corresponding operational semantics which regards the equations as rewrite rules. First, we show that, under certain conditions, the reduced ground terms form an initial algebra; in fact, it is a canonical term algebra in the sense of [Goguen, Thatcher & Wagner 78]. We also justify deciding the truth of an equation by applying rewrite rules to both sides, and seeing if they reduce to the same expression. We begin with a general set-theoretic result.

Proposition 7. Let R be a relation on a set A such that each a $\in$ A has an R-reduced form, denoted [[a]]. Let E be the equivalence relation on A generated by R. Then bR[[a]] iff bE[[a]] iff [[b]] = [[a]] iff aEb. □

This result justifies using reduction to decide equivalence (for either ground terms or for expressions with variables, for suitable choices of A), but it does **not** justify the use of rewrite rules to represent data types (with their operations on data), because the criterion for adequacy (following [Goguen, Thatcher, Wagner & Wright 75]) is that one gets an **algebra** (not just a set) initial among all algebras satisfying the defining equations.

Assume now that we are given a family S of sorts, an S-sorted signature Σ, and a family E of Σ-rewrite rules such that the family R_E of substitution instances is confluent and finite terminating. Let $C \subseteq T_\Sigma$ (this means that $C_s \subseteq T_{P,s}$ for each s in S). For σ in $\Sigma_{s1...sn,\ s}$ and ci in C_{si}, define

$C_\sigma(c1,...,cn)$ = $[[\sigma(c1...cn)]]$,

where [[c]] denotes the unique R_E-reduced form of c. If this always lies in C, then C can be given the structure of a Σ-algebra with the above definition for the operations, and we shall say that C (with R_E) represents a Σ-algebra A iff A and C are Σ-isomorphic as Σ-algebras.

Theorem 8. Let E be a confluent and finite terminating Σ-rewrite rule system, and let C be the family of all R_E-reduced Σ-terms. Then C (with R_E) represents the Σ-algebra $T_{\Sigma,E}$. In particular, C is an initial $\langle S,\Sigma,E\rangle$-algebra. □

A simple proof using the homomorphism theorem is in the full version of this paper [Goguen 80]. The notes from which this paper was derived [Goguen 79] generalize this result from finite termination to global finiteness. This permits one to decide term equality in a much larger class of equational theories; for example, if a commutative rule, such as (x*y = y*x) is included, the resulting system is not finite terminating, but it is globally finite. We begin as follows, omitting all proofs.

Proposition 9. If E is a confluent and globally finite Σ-rewrite rule system, let $\underline{\equiv}>$ denote the relation $R_E/^\sim$ on $T_\Sigma/^\sim$ (where $^\sim$ is the relation defined just before Theorem 2 above). Then $\underline{\equiv}>$ on T_Σ is confluent and finite terminating. □

Now let C_E denote the set of all cycles of E, where a cycle of E is a $\underline{\equiv}>$-reduced element of T_Σ. We can make C_E into a Σ-algebra as follows: letting σ in $\Sigma_{s1...s1,s}$ and ci in $C_{E,si}$ for i = 1,...,n; and letting [[[t]]] denote the $\underline{\equiv}>$- reduced form of the $^\sim$-equivalence class [t] of t inT_Σ, we define

$(C_E)_\sigma(c1,...,cn)$ = $[[[\sigma(t1,...,tn)]]]$

for any choice of ti in ci. This definition can easily be shown to be independent of the choice of the ti. Then C (with $\underline{\equiv}>$) represents T_P, because

Theorem 10. If E is a confluent and globally finite rewrite rule system, then C_E is an initial Σ-algebra. □

Theorem 8 can also be sharpened in another way. First, we need

Definition 8. Let C be a Σ-algebra of Σ-terms (thus, $C_s \subseteq T_{\Sigma,E,s}$ for each s in S). Then C is a canonical (Σ-) term algebra iff for $\sigma \in \Sigma_{s1...sn,s}$ $\sigma(t1...tn) \in C_s$ implies that ti $\in C_{si}$ for i = 1,...,n; and that $C_\sigma(t1,...,tn) = \sigma(t1...tn)$. □

Theorem 11. Under the assumptions of Theorem 8, C is a canonical term algebra. □

The preceding covers ground terms, but not expressions which contain variables. The situation for these is actually much the same, but the relevant structure is not simply a many-sorted algebra, but rather what is called an "algebraic theory" [Lawvere 63] or a "clone." Limitations of space, and our desire not to require category theory preclude detailed discussion. But we assert that a result analogous to Theorem 8 holds: the algebraic theory whose morphisms are reduced expressions (using the rewrite rules) is a free theory on the corresponding equational presentation.

4. STRUCTURAL INDUCTION

The main result of this section says an equation is true in the initial algebra of an equational theory iff it is provable by structural induction [Burstall 69]. To minimize effort in actually carrying out such proofs, it is worthwhile to consider subsignatures which are still large enough to provide canonical terms for the initial algebra.

Definition 9 and Theorem 12. $\Omega \subseteq \Sigma$ is a constructor signature for a presentation P = <S,Σ,E> iff one of the following equivalent conditions holds:

(C1) the composition $T_\Omega \to T_\Sigma \to T_P$ is surjective, where the first map is the inclusion;

(C2) each equivalence class in T_P contains an element of T_Ω;

(C3) there is a canonical term algebra for P contained in T_Ω.

We shall say that Ω is a minimal constructor signature for P iff it does not properly contain any other constructor signature. We shall say an operation symbol ω is a constructor iff it is contained in every constructor signature. We shall call Ω a signature of constructors iff it is a constructor signature which consists only of constructors; such a signature is unique for P, if it exists, and it is minimal.

Proof. (C1) and (C2) are clearly equivalent, and (C3) immediately implies (C2). To show that (C1) implies (C3), let $\equiv$ be the Ω-congruence on T_Ω induced by the composition regarded as an Ω-homomorphism; then $T_\Omega/\equiv$ is Ω-isomorphic to T_P regarded as an Ω-algebra, by the Homomorphism Theorem. We now apply the basic existence theorem (Theorem 8, p 144) of [Goguen, Thatcher & Wagner 78] to get a canonical Ω-term algebra for T_Ω, which is therefore isomorphic to T_P. For the last assertions, note that the signature consisting of all constructors is certainly minimal, if it is a constructor signature, because every constructor signature must contain it. □

Notice that, for a given presentation, minimal constructor signatures are not unique; moreover, there does not necessarily exist a signature of constructors.

For each s in S, let Q_s be a predicate defined on $T_{P,s}$, and let Q^*_s denote the corresponding predicate defined on $T_{\Sigma,s}$ by $Q^*_s(t)$ iff $Q_s([t])$, where [t] denotes the P-equivalence class of t. Note that $Q_s([t])$ iff $Q^*_s([[t]])$, where [[t]] is any representative of [t] in T_Ω.

Theorem 13. Let P = <S,Σ,E> be a presentation, and let Ω be a constructor signature for P. Then Q is true on T_P (i.e., Q([t]) is true for each [t] in T_P) iff:

(I0) $Q^*_s(\omega)$ is true for each constant ω in $\Omega_{\lambda,s}$; and
(I1) for each ω in $\Omega_{s1,\ldots,sn,s}$ with n > 0, if $Q^*_{si}(ti)$ for each ti in $T_{\Omega,si}$, then $Q^*_s(\omega(t1 \ldots tn))$.

Proof. Clearly, if Q is true on T_P then (I0) and (I1) are true of Q*.

For the converse, we assume that Q* satisfies (I0) and (I1). Then for each t in T_Σ we must show that $Q_s([t])$ is true. Let C be a canonical term algebra for T_P containing only terms constructed from the signature Ω (this exists by condition (C3) above), and let [[t]] denote the canonical representative in C for [t] in T_P.

It will suffice to show that Q*([[t]]) is true. We do this by structural induction on C. If [[t]] is a constant ω, then Q*([[t]]) is true by (I0). Otherwise, [[t]] has the form ω(t1...tn), where t1,...,tn are terms in C for which we may assume that Q* is true. Then Q*([[t]]) is true by (I1). □

Results and definitions in this section are related to [Nourani 79a], to a conjecture in [Nourani 79b], and to work of [Aubin 76].

5. PROVING INDUCTIVE HYPOTHESES WITHOUT INDUCTION

There are many applications in which one wants to prove that some equations hold for all values ranging over some data types. This section provides a rigorous and general foundation for approaching such problems using the Knuth-Bendix algorithm, as in [Musser 80a]. The results given here as a foundation for such proofs generalize those in previous versions of this paper [Goguen 79] and in [Huet & Oppen 80], in that they do not require equationally defined equalities (or equivalently, inequalities) on each sort, but only on sort bool, plus the possibility of distinguishing elements of other sorts. For example, the usual equational specification of stack does not give an equality for sort stack; but two stacks t and t' can be distinguished by an appropriate expression; e.g., if the third element of t, top(pop(pop(t))), is 0, which is different from the third element of t', then u(t) = zero?(top(pop(pop(t)))) = true,

while u(t') = false.

Definition 10. A presentation P = <S,Σ,E> is s-taut for a sort s iff:

(st1) S contains sort **bool** with constants true and false, and an operation symbol == of rank <**bool bool**,**bool**>; and == is an equationally defined equality, in the sense that, for all ground terms t,t' of sort **bool**, (t==t') = true if $t \equiv_P t'$ and (t==t') = false iff not($t \equiv_P t'$);

(st2) for all t,t' in $T_{P,s}$ with not($t \equiv_P t'$), there is a **bool**-valued expression u with a single variable of sort s such that (u(t)==u(t')) $\equiv_{P,\mathbf{bool}}$ false.

Call P taut iff P is s-taut for all s in S, and call P consistent if not(true$\equiv_P$false). □

These notions appear related to those of sufficient completeness and consistency advanced in [Guttag 75]. Tautness is a generalization of the condition that each sort have an equationally defined equality (this concept is defined by the obvious generalization of condition (st1) to sorts other than **bool**).

Theorem 14. Let P = <S,Σ,E> be a consistent specification, and let E' be a set of Σ-equations such that, if E' contains an equation of sort s, then P is s-taut. Then the following are equivalent, where P' = <S,Σ,E∪E'>:

(i) each equation in E' holds in T_P;

(ii) $T_P = T_{P'}$;

(iii) P' is consistent.

Proof. We see that (i) and (ii) are equivalent, because, if e holds in T_P, then the congruence which e generates on T_Σ is contained in that generated by P.

(i) => (iii): If P' is inconsistent, then so is P, contradicting our assumption.

(iii) => (i): Assume that P' is consistent but some equation e does not hold in T_P. Then there is a substitution α such that it is not true that $\alpha(L) \equiv_P \alpha(R)$. By condition (st2), there is an expression u of sort **bool** such that

$$u(\alpha(L))==u(\alpha(R)) \equiv_P \text{false}$$

and $\equiv_P \subseteq \equiv_{P'}$ implies

$$u(\alpha(L))==u(\alpha(R)) \equiv_{P'} \text{false}.$$

But since (L = R) is in P', $u(\alpha(L)) \equiv_{P'} u(\alpha(R))$, and therefore $u(\alpha(L)==u(\alpha(R)) \equiv_{P'}$ true), by condition (st1). This contradicts the consistency of P'. Therefore, there is no such substitution α, and e holds in T_P. □

The motivation for the word "taut" is given by Theorem 15, but first we need

Definition 11. A Σ-algebra A is reachable iff the unique Σ-homomorphism T_Σ -> A is surjective. □

Intuitively, A is reachable iff each element in A is named by a Σ-ground term.

Theorem 15. If P is a taut specification and A is a reachable consistent P-algebra, then A is an initial P-algebra.

Proof. Let P = $\langle S,\Sigma,E\rangle$. First, notice that if there is a consistent P-algebra, then P is consistent. Next, note that A is a reachable P-algebra iff A is Σ-isomorphic to a quotient of T_P iff A is Σ-isomorphic to $T_{P'}$, where P' = $\langle S,\Sigma,E\cup E'\rangle$ for some set E' of Σ-equations. P' is consistent because A is, so by Theorem 14, E' holds in T_P; thus T_P is isomorphic to $T_{P'}$, and therefore to A. □

Note that $T_{P,\mathbf{bool}}$ may properly contain {[true], [false]}; the extra elements might be useful as error messages, for example. We now introduce the Knuth-Bendix [Knuth & Bendix 70] algorithm; there is no need to give details of the algorithm, because all we need are certain general properties, which are summarized in the following

Proposition 16. Given a family E of Σ-equations, the Knuth-Bendix algorithm, if it halts, gives a set KB(E) of Σ-equations, such that:

(kb1) KB(E) is consistent iff E is consistent;

(kb2) KB(E) is locally confluent;

Moreover, even if the algorithm doesn't halt:

(kb3) an equation e follows by equational deduction from E iff it follows from $KB_n(E)$, the result of n iterations of the algorithm; and

(kb4) there is an n such that $KB_n(E)$ contains (true = false) or else (false = true) iff E is inconsistent.

□

There is a great advantage to breaking the discussion into two pieces, one of which gives a general algebraic result, and the other of which gives some properties of the Knuth-Bendix algorithm: any algorithm having the properties (kb1) - (kb4) can be used to do inductionless theorem proving. For example, the Knuth-Bendix algorithm enriched with commutative, or associative, or commutative-and-associative axioms for some operations, has the necessary properties, and can therefore be so used. The main result now follows easily from Theorem 15 and Pròposition 16.

Theorem 17. Let P = $\langle S,\Sigma,E\rangle$ be a consistent presentation and let E' be a family of Σ-equations such that, if E' contains an equation of sort s, then P is s-taut. Let KB be an algorithm satisfying conditions (kb1) - (kb4). Then

(1) if KB(E∪E') exists, is finite terminating, and is consistent, then each equation in E' is true in T_P;

(2) there is an n such that KB_n(E∪E') contains (true = false) or (false = true) iff some equation in E' does not hold in T_P.

□

This exposition eliminates the complicated and confusing (to this author) hypotheses regarding "hierarchical specifications" in [Musser 80a], and also avoids the confusion between an equationally defined equality operation, and the congruence relation defined by the equations (this latter condition may not in fact be decidable). Moreover, our approach is very general. Of course, Musser should be given credit for being first to realize that proofs using the Knuth-Bendix algorithm are possible and desirable.

A number of examples based on the above results have been run using a program called "kb" written by David Smallberg at UCLA in the language C. Jean-Marie Hullot has a similar system; and of course, there is the original AFFIRM system [Guttag, Horowitz & Musser 78]. Here is a proof using kb that the reverse of the reverse of a list is the original list. We define an equality relation on each sort, which more than satisfies tautness. Equations are broken into groups called "objects," and delimited by pairs of the form OBJ...END. The kb prompt character is ".". The command "in" reads a file, and the command "kb" requests that the Knuth-Bendix algorithm be run. If ">kb" returns with a ">", this means that no new rules were added, so that the current set (possibly with some deletions) is locally confluent. Of course, one still has to prove finite termination in order to get confluence, but that is pretty obvious in this case. If there had been an inconsistency, it would have shown up as T = F, and we would know that the hypothesis was wrong. Finally, an example is "run" to show how the structure works. Hopefully, the rest of the transcript is close enough to the discussions above that no further explanation is required.

Date: 15 Oct 1979 1513-PDT

```
> in revp

revp:
OBJ
  SORTS bool
  OPS
       T : -> bool
       F : -> bool
       & : bool,bool -> bool
      == : bool,bool -> bool
  VARS
  EQNS
       (bool : &(T,T) = T)
       (bool : &(T,F) = F)
       (bool : &(F,T) = F)
       (bool : &(F,F) = F)
       (bool : ==(T,T) = T)
       (bool : ==(T,F) = F)
```

```
        (bool : ==(F,T) = F)
        (bool : ==(F,F) = T)
 END

 OBJ
   SORTS
     nat,bool
   OPS
     0 : -> nat
     s : nat -> nat
     == : nat,nat -> bool
   VARS
     n,m : nat
   EQNS
     (bool : ==(n,n) = T)
     (bool : ==(0,s(n)) = F)
     (bool : ==(s(n),0) = F)
     (bool : ==(s(n),s(m)) = ==(n,m))
 END

 OBJ
   SORTS nat
   OPS
     1 : -> nat
     2 : -> nat
     3 : -> nat
 EQNS
     (nat : 1 = s(0))
     (nat : 2 = s(1))
     (nat : 3 = s(2))
 END

 OBJ
   SORTS
     atom,bool,nat
   OPS
     a : nat -> atom
     nil : -> atom
     == : atom,atom -> bool
   VARS
     n,m : nat
   EQNS
     (bool : ==(nil,nil) = T)
     (bool : ==(nil,a(n)) = F)
     (bool : ==(a(n),nil) = F)
     (bool : ==(a(n),a(m)) = ==(n,m))
 END

 OBJ
   SORTS
     list,bool,atom
   OPS
     nil : -> list
    c : atom,list -> list
     == : list,list -> bool
   VARS
     a,a' : atom ; l,l' : list
   EQNS
     (bool : ==(nil,c(a,l)) = F)
     (bool : ==(c(a,l),nil) = F)
     (bool : ==(l,l) = T)
     (bool : ==(c(a,l),c(a',l')) = &(==(a,a'),==(l,l')))
 END
```

```
OBJ
  SORTS
    list,atom,nat,bool
  OPS
    * : list,list -> list
  VARS
    a : atom ; l,l' : list
  EQNS
    (list : *(nil,l) = l)
    (list : *(c(a,l),l') = c(a,*(l,l')))
END

OBJ
  SORTS bool,nat,atom,list
  OPS
    r : list -> list
  VARS
    a : atom ; l : list
  EQNS
    (list : r(nil) = nil)
    (list : r(c(a,l)) = *(r(l),c(a,nil)))
    (list : r(r(l)) = l)
END

> kb
New rule # 29 -- which way should it go?
(1)  c(atom.1,list.2)      (2)  r(*(r(list.2),c(atom.1,nil)))
      ? 2
New rule # 30 -- which way should it go?
(1)  &(==(atom.1,atom.1),T)      (2)  T      ? 1
New rule # 31 -- which way should it go?
(1)  c(atom.1,r(list.2))      (2)  r(*(list.2,c(atom.1,nil)))
      ? 2
>
> show rules
                13. nat 1 = s(0)
                14. nat 2 = s(s(0))
                15. nat 3 = s(s(s(0)))
                26. list r(nil) = nil
                 1. bool &(T,T) = T
                 2. bool &(T,F) = F
                 3. bool &(F,T) = F
                 4. bool &(F,F) = F
                 5. bool ==(T,T) = T
                 6. bool ==(T,F) = F
                 7. bool ==(F,T) = F
                 8. bool ==(F,F) = T
                16. bool ==(nil,nil) = T
                24. list *(nil,list.1) = list.1
                 9. bool ==(nat.1,nat.1) = T
                22. bool ==(list.1,list.1) = T
                28. list r(r(list.1)) = list.1
                10. bool ==(0,s(nat.1)) = F
                11. bool ==(s(nat.1),0) = F
                17. bool ==(nil,a(nat.1)) = F
                18. bool ==(a(nat.1),nil) = F
  22 on  23:    30. bool &(==(atom.1,atom.1),T) = T
                27. list r(c(atom.1,list.2)) =
                         *(r(list.2),c(atom.1,nil))
                20. bool ==(nil,c(atom.1,list.2)) = F
                21. bool ==(c(atom.1,list.2),nil) = F
                25. list *(c(atom.1,list.2),list.3) =
```

```
                 c(atom.1,*(list.2,list.3))
            12. bool ==(s(nat.1),s(nat.2)) =
                 ==(nat.1,nat.2)
            19. bool ==(a(nat.1),a(nat.2)) =
                 ==(nat.1,nat.2)
 28 on  29:  31. list r(*(list.2,c(atom.1,nil))) =
                 c(atom.1,r(list.2))
            23. bool ==(c(atom.1,list.2),c(atom.3,list.4)) =
                 &(==(atom.1,atom.3),==(list.2,list.4))
Deleted rules:
 27 on  28:  29. list r(*(r(list.2),c(atom.1,nil))) =
                 c(atom.1,list.2)
> quit
```

6. THE CORRECTNESS OF IMPLEMENTATIONS

The results in Section 5 can be extended for use in proving that one algebra is a subalgebra of another; this is useful in verifying the correctness of implementations.

Definition 12. Let $\langle S,\Sigma\rangle$ and $\langle S',\Sigma'\rangle$ be signatures with $S \subseteq S'$ and $\Sigma \subseteq \Sigma'$. If A is a Σ'-algebra, let $U_\Sigma(A)$ denote the Σ-algebra which results from A by 'forgetting' those sorts and operations which are in $\langle S',\Sigma'\rangle$ but not in $\langle S,\Sigma\rangle$. Then we say that an Σ'-algebra A' is an extrusion of a Σ-algebra A iff

$$A \subseteq U_\Sigma(A')$$

(as Σ-algebras). Further, A' is an extension of A iff = replaces $\subseteq$ in the above, and A' is an enrichment of A iff A' is an extension of A and S = S'. □

Now the results:

Theorem 18. Let $P = \langle S,\Sigma,E\rangle$ be a consistent presentation, let S' be disjoint from S, let Σ' be an $(S\cup S')$-sorted signature disjoint from Σ, and let E' be a set of $(\Sigma\cup\Sigma')$-equations. Then $P' = \langle S\cup S', \Sigma\cup\Sigma', E\cup E'\rangle$ consistent and taut implies $T_{P'}$ is an extrusion of T_P. □

Theorem 19. Under the assumptions of the previous theorem, with KB an algorithm satisfying conditions (kb1) - (kb4):

(1) if KB($E\cup E'$) is finite terminating, consistent and taut, then each equation in E' holds in T_P and $T_{P'}$ is a consistent extrusion of T_P;

(2) some equation in E' fails in T_P iff there is an n such that $KB_n(E\cup E')$ contains (true = false) or (false = true); thus, $T_{P'}$ is not a consistent extrusion of T_P iff there is an n such that $KB_n(E\cup E')$ contains (true = false) or (false = true). □

These results can be used to prove the correctness of an implementation of one abstraction by another as follows: First, let $P = \langle S,\Sigma,E\rangle$, the "base abstraction," define the operations with which the implementation will be done, plus (possibly) additional equations to define an equivalence relation on the data representations to be used in the implementation. This presentation should be taut and consistent. Next, any new operations needed by the implementation should be added to the base, and checked for consistency. Finally, the definition of the abstraction to be implemented should be given, together with equations defining its operations in terms of the base. Then the whole thing should be checked for consistency. If it is consistent, then the type to be implemented is an extrusion of the base abstraction, and this expresses correctness of the implementation. An example of this technique run by the author in Smallberg's system is given in the Appendix of the full version of this paper [Goguen 80]. Note that this technique can also be extended to commutative, associative, etc. systems. Finally, it should be noted that constructors can be used to simplify proofs of this kind [Nourani 79a].

The notion of implementation used in the above discussion is essentially that of [Goguen, Thatcher & Wagner 78]: a subalgebra of a quotient. This notion is simpler and more restrictive than some others (e.g., [Ehrig 79] or [Ehrich 78]), but it appears to be sufficient for practical purposes.

REFERENCES

[Aubin 76]
Aubin, R.
Mechanizing Structural Induction.
PhD thesis, University of Edinburgh, 1976.

[Burstall & Goguen 77]
Burstall, R. M. and Goguen, J. A.
Putting Theories together to Make Specifications.
Proc. 5th Int. Joint Confr. on Artificial Intelligence , 1977.

[Burstall 69]
Burstall, R. M.
Proving Properties of Programs by Structural Induction.
Computer Journal , 1969.

[Ehrich 78]
Ehrich, H.-D.
On the Theory of Specification, Implementation and Parameterization of Abstract Data Types.
Technical Report, Forschungsbericht, Dortmund, 1978.

[Ehrig 79]
Ehrig, H., Kreowski, H.-J. and Padawitz, P.
Algebraic Implementation of Abstract Data Types.
Technical Report, Technical University of Berlin, 1979.

[Goguen & Tardo 79]
Goguen, J. A. and Tardo, J.
An Introduction to OBJ-T.
In Specification of Reliable Software. IEEE, 1979.

[Goguen, Thatcher & Wagner 78]
Goguen, J. A., Thatcher, J. W. and Wagner, E.
An Initial Algebra Approach to the Specification, Correctness and Implementation of Abstract Data Types.
In R. Yeh, editor, Current Trends in Programming Methodology, . Prentice-Hall, 1978.
also published as IBM T.J.Watson Research Center Report, 1876.

[Goguen, Thatcher, Wagner & Wright 75]
Goguen, J. A., Thatcher, J. W., Wagner, E. and Wright, J. B.
Abstract Data Types as Initial Algebras and the Correctness of Data Representations.
In Computer Graphics, Pattern Recognition and Data Structure. IEEE, Beverley Hills, CA, 1975.

[Goguen 77]
Goguen, J. A.
Abstract Errors for Abstract Data Types.
In _Working Confr. on Formal Description of Programming Concepts_. IFIP, 1977.
also published by North-Holland, 1979, editor P. Neuhold.

[Goguen 79]
Goguen, J. A.
Proving Inductive Hypotheses without Induction and Evaluating Expressions with Non-terminating Rewrite Rules.
1979.
class notes at UCLA, and draft paper at SRI, 1980.

[Goguen 80]
Goguen, J. A.
How to Prove Inductive Hypotheses without Induction.
Technical Report, SRI International, 1980.

[Guttag, Horowitz & Musser 78]
Guttag, J. V., Horowitz, E. and Musser, D. R.
Abstract Data Types and Software Validation.
Communications of the ACM , 1978.

[Guttag 75]
Guttag, J.V.
The Specification and Application to Programming of Abstract Data Types.
PhD thesis, Univ. of Toronto, 1975.

[Huet & Oppen 80]
Huet, G. and Oppen, D.
Equations and Rewrite Rules: A Survey.
1980.

[Knuth & Bendix 70]
Knuth, D. and Bendix, P.
Simple Word Problems in Universal Algebra.
In J. Leech, editor, _Computational Problems in Abstract Algebra_, . Pergamon Press, 1970.

[Lawvere 63]
Lawvere, F. W.
Functorial Semantics of Algebraic Theories.
(_Proc. Nat. Acad. Sciences_) , 1963.

[Musser 80a]
Musser, D.
On Proving Inductive Properties of Abstract Data Types.
1980.
to appear in 7th ACM Symp. on Principles of Programming Languages.

[Musser 80b]
Musser, D.
Abstract Data Type Specification in the AFFIRM System.
IEEE Trans. Software Eng. , 1980.
to appear.

[Nourani 79a]
Nourani, F.
Constructive Extension and Implementation of Abstract Data Types and Algorithms.
PhD thesis, UCLA, Dept. of Computer Science, 1979.

[Nourani 79b]
Nourani, N.
Inductive Extensions of Equational Theories of Data Types (Working Outline).
Technical Report, University of Michigan, Dept. of Elec. Eng. and Computer Science, 1979.
unpublished memorandum, November 1979.

[Wand 77]
Wand, M.
Algebraic Theories and Tree Rewriting Systems.
Technical Report 66, Computer Science Dept., Indiana Univ., 1977.

[Zilles 74]
Zilles, S.
Abstract Specification of Data Types.
Technical Report 119, Computation Structures Group, MIT, 1974.

A COMPLETE, NONREDUNDANT ALGORITHM FOR REVERSED SKOLEMIZATION*

by

P. T. Cox
University of Auckland, New Zealand

and

T. Pietrzykowski
University of Waterloo,
Acadia University, Canada

ABSTRACT

An algorithm is presented which, for an arbitrary literal containing Skolem functions, outputs a set of closed quantified literals with the following properties. If a and b are formulae we define $a \supset b$ iff $\{sk(a), dsk(b)\}$ is unifiable where sk denotes Skolemization and dsk denotes the dual operation, where the roles of $\forall$ and $\exists$ are reversed. If d is an arbitrary literal and X is the output, then:

(i) Soundness: if $x \in X$ then $x \supset d$

(ii) Completeness: if $a \supset d$ then $\exists x \in X$ such that $a \supset x$

(iii) Nonredundancy: if $x,y \in X$ then $x \not\supset y$ and $y \not\supset x$.

1. INTRODUCTION

We consider the problem of reversing Skolemization and present an algorithm which assigns to a literal one or more closed literals where here, as in the rest of this paper, "closed literal" means a closed formula whose matrix is a literal. In the simplest case, if the input literal is the result of skolemizing a closed literal then by applying our algorithm, skolemizing and applying the algorithm again we will produce the original closed literal.

In the general case, however, the situation is more complex; for example if the input literal is one deduced by a mechanical question answering system. The ability to quantify such literals is especially important when the system attempts to form a hypothesis. This can be formalized as follows. Given a set of wffs K (knowledge base) and a conjunction of closed literals E (effect) we define a conjunction of closed literals C as a cause of E -under K iff E can be deduced from K and C combined together. Among all possible causes we are only interested in such which do not imply other causes (minimality) and have no causes under K other than themselves (basicness). For example let:

$$K = \{\forall x\forall y\forall u\forall v(P(x,y,u,v) \supset (R(x,y) \vee Q(u,v))), \quad \forall x\exists y(R(x,y) \supset T(y)), \forall u\exists v(Q(u,v) \supset T(v))\}$$

and $$E = \exists xT(x).$$

---

* This work was supported by NSERC Grants: A3025 and A5267.

To find causes of E under K we skolemize K and negated E, combine them together and produce all resolvents using negated E as a set of support. However, since we are looking only for the basic causes we select such resolvents which cannot be resolved any more (dead ends). In the example above there is only one dead end: $\neg P(x,\phi(x),u,\psi(u))$ where ϕ and ψ are distinct skolem functions. Finally in order to find all minimal and basic causes of E we apply QUANTIFY to $P(x,\phi(x),u,\psi(u))$ which produces output consisting of two closed literals: $\exists x \forall y \exists u \forall v M$ and $\exists u \forall v \exists x \forall y M$ where $M = P(x,y,u,v)$. More detailed discussion of mechanical hypothesis formation can be found in [3]. For such applications, the output of our algorithm must have properties of completeness and implicational independence. By "completeness" we mean that if some closed literal A implies the input literal B in a general sense to be defined later, then there is an output C from our algorithm such that A implies C. By "implicational independence" we mean that no output implies another different output.

2. PRELIMINARIES

In this section we review standard concepts and notation, as well as introduce some specific definitions.

We shall use the word expression to refer to literals, terms and variables, where a variable is not a term.

Any term beginning with a Skolem function is called a Skolem term.

A quantifier string is a string of the form $Q_1x_1 \ldots Q_nx_n$ $(n \geq 0)$ where Q_i is either $\exists$ or $\forall$ $(1 \leq i \leq n)$ and $x_1,\ldots,x_n$ are distinct variables.

We use the word "formula" with its standard meaning in mathematical logic.

If $s = pm$ is a formula such that p is a quantifier string and m contains no quantifiers then we define:

$$\text{prefix}(s) = p$$
$$\text{matrix}(s) = m$$

If a is any string, the head of a is the leftmost symbol of a.

If $a = x(t_1,\ldots,t_n)$, $b = x(s_1,\ldots,s_n)$ are expressions, then expressions t and s are said to be vis-a-vis in a and b iff for some i $(1 \leq i \leq n)$ either $s = s_i$ and $t = t_i$, or t and s are vis-a-vis in t_i and s_i.

If m is a literal, p is a quantifier string and v is a variable which does not occur in p, we define:

$$sk(m) = m$$
$$sk(p\forall vm) = sk(pm)$$
$$sk(p\exists vm) = sk(pm\theta)$$

where $\theta = \{v \leftarrow f(u_1,\ldots,u_r)\}$, f is a new Skolem function and $u_1,\ldots,u_r$ are all the variables immediately preceded by $\forall$ in p. We also define a function dsk with the same range as sk by replacing in the above definition "sk" by "dsk", "$\forall$" by "$\exists$", and "$\exists$" by "$\forall$". Clearly, sk is Skolemization and dsk is the dual operation (see [4]).

If m is a formula and b is an occurrence of an expression in m, then b is called a <u>top-level occurrence (in m)</u> iff it is not a proper subexpression of a Skolem subterm (of m).

If m is a formula and b is an expression with a top-level occurrence in m, then we will say that b is <u>top-level (in m)</u>.

We will abbreviate the phrase "top-level Skolem" to TS.

If x is an expression we wrote $x[t]$ to indicate that an expression t occurs in x.

We assume that the reader is familiar with standard definitions of such concepts as "substitution", "variant", "unification". Substitutions will be denoted by lower case Greek letters. We will call a substitution that transforms a literal into one of its variants a <u>renaming</u>. We abbreviate the phrase "most general unifier" to mgu.

3. THE ALGORITHMS

In this section we describe two algorithms which together produce the required set of closed literals. The first of these algorithms, PREPROCESS, is unnecessary in the case where the input literal has no unifiable TS subterms.

If σ is a substitution, m is a literal, t is a top-level subexpression of m and x is a top-level variable of m, we say that <u>σ disturbs t in m with x</u> iff x occurs in t but not in $t\sigma$, or x occurs in $t\sigma$ but x not in t. We will omit "in m" and "with x" when m is understood from context, and x is irrelevant. We say that <u>σ disturbs m</u> iff σ disturbs some top-level subexpression of m.

Let d be a literal.

```
algorithm PREPROCESS(d)
W ← {d}
R ← ϕ
while  W ≠ ϕ
   do ⎧delete  e  from  W;
      ⎪if   every unifier of every pair of distinct TS subterms of  e  disturbs  e,
      ⎨     and  R  contains no variant of  e
      ⎪then R ← R ∪ {e}
      ⎩W ← W ∪ {eσ | t  and  s  are distinct TS subterms of  e  with mgu  σ}
PREPROCESS ← R
stop
```

If m is a formula, we define:

$$\mathcal{W}(m) = \{\{v \mid v \text{ is a free top-level variable of } m \text{ and occurs in } t\} \mid t \text{ is a TS subterm of } m\}$$

We then define free(m) as the set of lower bounds of the set $\mathcal{W}(m)$ with the partial ordering $\subseteq$. For example, if $m = \exists wP(\alpha(x,w),\beta(y,z,u),\gamma(x,\delta(y),z),x,y,z,w)$, where all the function symbols are Skolem, then $\mathcal{W}(m) = \{\{x\},\{y,z\},\{x,y,z\}\}$ and $\text{free}(m) = \{\{x\},\{y,z\}\}$.

If m is a formula, we define:

$$\text{ground}(m) = \{t \mid t \text{ is a TS subterm of } m\text{, and contains no free top-level variables of } m\}.$$

If X is a set, we denote by $\vec{X}$ an arbitrary but fixed ordering of X.

If $\vec{X} = (x_1,\ldots,x_m)$ we define:

$$\forall(\vec{X}) = \forall x_1 \forall x_2 \ldots \forall x_n \qquad (n \geq 0)$$

$$\exists(\vec{X}) = \exists x_1 \exists x_2 \ldots \exists x_n \qquad (n \geq 0)$$

If X is a set of variables and D is a set of variables or terms such that $|X| = |D|$, and if $\vec{X} = (x_1,\ldots,x_n)$ and $\vec{D} = (g_1,\ldots,g_n)$, then we denote the substitution $\{x_1 \leftarrow g_1,\ldots,x_n \leftarrow g_n\}$ by $\vec{X} \leftarrow \vec{D}$.

If m is a formula or expression, and a and b are expressions, then repl(a,b,m) is the formula or expression obtained by replacing all top-level occurrences of a in m by b. We extend this definition to ordered sets of expressions as follows:

$$\text{repl}((a_1,\ldots,a_n),(b_1,\ldots,b_n),m)$$
$$= \text{repl}((a_1,\ldots,a_{n-1}),(b_1,\ldots,b_{n-1}),\text{repl}(a_n,b_n,m))$$

Let us note that if u is a free, top-level variable of m then $\text{repl}(u,b,m) = m\{u \leftarrow b\}$.

Let d be a literal.

```
algorithm QUANTIFY(d)
Q ← φ
S ← {d}
while S ≠ φ
   do  { delete  s  from  S;
       G ← ground(s)
       if    ( for every mgu  σ  of every pair of distinct terms in  G
             ( either  σ  disturbs a top-level variable in  s
       (#)   ( or  for some  v  and  y,  where  v  is the new variable
             (     corresponding to some  TS  subterm  t  of  d,  ∀v
             (     occurs in prefix(s) to the left of  ∃y  and  σ
             (     disturbs  t  with  y.
       then  ( F ← free(s)
             ( H ← {v | v  is a free variable in  s,  and does not occur in
             (       any  TS  subterm of  s}
             ( p ← prefix(s)∀(V⃗)                 (see * below)
             ( m ← repl(G,V,matrix(s))
             ( if  F = φ
             ( then  Q ← Q ∪ {p∃(H⃗)m}
             ( else  ( while  F ≠ φ
             (       (    do  ( delete  F  from  F:
             (       (        ( S ← S ∪ {p∃(F⃗)m}
QUANTIFY ← Q
stop
```

(* V is a set of variables which do not occur in s, and $|V| = |G|$.)

In order to illustrate how QUANTIFY works let $d = P(x,\phi(x),u,\psi(u))$ where ϕ, ψ are Skolem functions. In the following there are presented values of s, S and Q at the end of the main do-loop, in consecutive iterations.

s	S	Q
$P(x,\phi(x),u,\psi(u))$	$\{\exists xP(x,\phi(x),u,\psi(u)),$ $\exists uP(x,\phi(x),u,\psi(u))\}$	$\emptyset$
$\exists xP(x,\phi(x),u,\psi(u))$	$\{\exists x\forall yP(x,y,u,\psi(u)),$ $\exists uP(x,\phi(x),u,\psi(u))\}$	$\emptyset$
$\exists x\forall yP(x,y,u,\psi(u))$	$\{\exists x\forall y\exists uP(x,y,u,\psi(u)),$ $\exists uP(x,\phi(x),u,\psi(u))\}$	$\emptyset$
$\exists x\forall y\exists uP(x,y,u,\psi(u))$	$\{\exists uP(x,\phi(x),u,\psi(u))\}$	$\{\exists x\forall y\exists u\forall vP(x,y,u,v)\}$
$\exists uP(x,\phi(x),u,\psi(u))$	$\{\exists u\forall vP(x,\phi(x),u,v)\}$	the same
$\exists u\forall vP(x,\phi(x),u,v)$	$\{\exists u\forall v\exists xP(x,\phi(x),u,v)\}$	the same
$\exists u\forall v\exists xP(x,\phi(x),u,v)$	$\emptyset$	$\{\exists x\forall y\exists u\forall vP(x,y,u,v),$ $\exists u\forall v\exists x\forall yP(x,y,u,v)\}$

Let us note that (#) condition has been always satisfied. Otherwise some s would be simply deleted.

More complicated examples which illustrate these algorithms are presented in section 5 where they may be more fully appreciated in the light of the results presented in section 4.

4. CORRECTNESS AND IMPLICATIONAL INDEPENDENCE

Here we present some properties of the algorithms considered independently, such as their termination; and some properties of the algorithms combined. However, because of the space limitation only some proofs or their part will be presented. The full version of the proof, the reader can find in [2].

4.1 Termination

4.1.1 Lemma

For any literal d PREPROCESS(d) and QUANTIFY(d) halts.

Proof We define non-negative, integer functions F and G for PREPROCESS and QUANTIFY respectively. The domain of these functions are set of sets of partially quantified literals (in the case of PREPROCESS they are not quantified). Let D be a set of partially quantified literals then:

$$F(D) = \sum_{x\in D} f(x)$$

$$G(D) = \sum_{x\in D} g(x)$$

where $f(x) = (n!)^2$ and $g(x) = m!$ where n = number of TS subterms of x and m = n + number of free variables of x. It can be verified that both functions strictly decrease from D' to D" where D' and D" are the value of W (or S) at the beginning and at the end of the major loop of PREPROCESS (or QUANTIFY). In view of nonnegativity of F and G the algorithms must halt. □

4.2 Soundness

If a and b are formulae whose matrices are literals we write $a \supset b$ iff {sk(a),dsk(b)} are unifiable. In the case when a and b are closed, our definition coincides with the standard definition of $\supset$.

4.2.1 Theorem:

For any literal d if $c \in$ PREPROCESS(d) and $b \in$ QUANTIFY(c) then $b \supset d$.

Proof It suffices to show that: $c \supset d$ and $b \supset c$. The first implication follows immediately from the definition of PREPROCESS. The second, involving QUANTIFY, is based on the proof that for all $x \in S \cup Q$ {sk(x),c} has a unifier σ without free variables of x occurring in it.

At the beginning of the first execution of the major loop $S \cup Q = \{c\}$, so the result clearly holds.

Let S', Q' and S", Q" be the values of S and Q at the beginning and end respectively, of some execution of the major loop. Assume the result holds for $S' \cup Q'$. Now suppose $x \in S'' \cup Q''$ then either $x \in S' \cup Q'$ in which case the result holds, by the above assumption; or b is introduced during the current execution of the loop. Let s be the element of S' deleted, then:

$$x = p'\forall(\vec{V})\exists(\vec{F})\text{repl}(\vec{G},\vec{V},m') \quad \text{and} \quad y = p'\forall(\vec{V})\text{repl}(\vec{G},\vec{V},m')$$

where p' = prefix(s), m' = matrix(s), and V, F and G are as defined in the algorithm.

Let σ denote a unifier of {sk(s),c} with required properties. Denoting $\gamma = \{\vec{Y} \leftarrow \vec{G}\} \circ \sigma$ we verify that γ unifies {sk(y),c} and has no free variables of x occurring in it. Denoting m" = matrix(y) and $\delta = \{\vec{F} \leftarrow \vec{T}\} \circ \gamma$ where T is a set of Skolem terms introduced by proceedings from sk(y) to sk(x), we can prove that δ is a unifier of {sk(x),c} with no free variables of x in it. The details of the verification are presented in [2]. □

4.3 Completness

If d and d' are literals such that $d' \in$ PREPROCESS(d), we shall denote by $P_{d'}$, the partition of the set of all TS subterms of d such that $s,t \in X \in P_{d'}$, iff s' = t', where s' and t' are TS subterms of d' which are vis-a-vis s and t respectively.

4.3.1 Lemma: Completeness of PREPROCESS

If c is a closed literal such that $c \supset d$, then there is a literal $b \in$ PREPROCESS(d) such that {sk(c),b} has a unifier ξ with the property that

$t\xi \neq s\xi$ for all pairs of distinct TS subterms t and s of b.

Proof Let μ be an mgu of $\{sk(c),d\}$; denote $sk(c)\mu$ by e; let η be an mgu for P_e; and let $d' = d\eta$. It is easy to prove that $sk(c)$ and d' are unifiable; let their mgu be σ. It is easy to show that at some time during the execution of PREPROCESS(d), $d' \in W \cup R$: to show this, we can select pairs of terms which belong to the same class of P_e; these terms are obviously unifiable, and will be unified during some execution of the loop. By continuing this process we can construct a sequence of literals such that each is in $W \cup R$, and the last in the sequence is d'. If $d' \in R$, then $b = d'$ is obviously the required literal.

If $d' \notin$ PREPROCESS(d), we construct a sequence $d_1, d_2, \ldots, d_n$ $(n \geq 2)$ where $d_1 = d'$, and $d_{i+1} = d_i\theta_i$ where θ_i $(1 \leq i \leq n-1)$ is an mgu of some pair of distinct TS subterms of d_i which does not disturb d_i, and d_n has no pair of distinct TS subterms with a unifier θ_1 that does not disturb d_1. Also since the number of distinct unifiable TS subterms is reduced at each extension of the sequence, the construction must terminate. Obviously $d_n \in$ PREPROCESS(d). It remains to prove that d_n and $sk(c)$ are unifiable. This, rather lengthy demonstration the reader can find in [2].

4.3.2 Theorem: Completeness

If a is a closed literal, and d is a literal such that $a \supset d$, there exists c and b such that $c \in$ PREPROCESS(d), $b \in$ QUANTIFY(c), and $a \supset b$.

The rather long and involved proof of the above theorem is presented in [2].

4.4 Implicational Independence

4.4.1 Lemma: Implicational Independence of QUANTIFY

If d is a literal, $e,f \in$ QUANTIFY(d), and $e \neq f$, then $e \not\supset f$ and $f \not\supset e$.

Proof First let us assume that matrix(e) = matrix(f): this assumption is justified by noting that no two formulae in QUANTIFY(d) are variants of each other, and that each TS subterm can always be replaced by the same new variable. Let us also assume that the variables of e and f are ordered by a relation $<$ in an arbitrary but fixed manner, and that blocks of quantifiers of the same type in the prefixes of e and f are arranged according to this ordering: that is, if Qu_1Qu_2 occurs in prefix(e) or prefix(f), where Q is $\forall$ or $\exists$, then $u_1 < u_2$.

Since $e \neq f$, prefix(e) $\neq$ prefix(f). Let p be the longest quantifier string which is the left part of both prefixes (note that p could be empty), then:

$$\text{prefix}(e) = pQ'u \ldots$$
$$\text{prefix}(f) = pQ''v \ldots,$$

where $Q'u \neq Q''v$.

Now, if $Q' = Q'' = \forall$, then $u \neq v$. Suppose u and v were introduced to replace TS subterms s and t, respectively; then all top level variables occurring in s and t must occur in p. Therefore, $Q''v$ must occur to the right of $Q'u$ in prefix(e), and is introduced by QUANTIFY at the same time as $Q'u$; hence $u < v$. Similarly, by considering prefix(f), we find that $v < u$. Consequently, not both Q' and Q'' are $\forall$. Now suppose that $Q' = \forall$ and $Q'' = \exists$ (or vice-versa); then there is a TS subterm s with all its top-level variables occurring in p: QUANTIFY always removes all such terms before existentially quantifying any further top-level variables. This contradicts the supposition that $Q'' = \exists$. The only remaining possibility is that $Q' = Q'' = \exists$, which implies $u \neq v$. Let U be the set of variables existentially quantified at the same time as u in the production of e: we define v analgously for $v \Delta f$.

Clearly $U \neq V$, since if $U = V$ we can conclude $u < v$ (from prefix(e)) and $v < u$ (from prefix(f)); also $U \not\subset V$ since $U \subset V$ implies that V cannot be chosen as a minimal set of free variables to be existentially quantified. Consequently, there exists variables $x \in U \backslash V$ and $y \in V \backslash U$, and TS subterms s and t such that x occurs in s not t, and y occurs in t not s, and:

$$\text{prefix(e)} = p \ldots \exists x \ldots \forall w \ldots \exists y \ldots \forall z \ldots$$

$$\text{prefix(f)} = p \ldots \exists y \ldots \forall z \ldots \exists x \ldots \forall w \ldots$$

where w and z are new variables corresponding to s and t respectively. Hence in order to unify {sk(e),dsk(f)} it is necessary to unify $\{\{\alpha,x\}, \{w,\gamma[y]\}, \{\beta[w],y\}, \{z,\delta[x,y]\}\}$, where α and β are Skolem terms introduced by the application of sk to e, and γ and δ are Skolem terms introduced by the application of dsk to f. Therefore, {sk(e),dsk(f)} is not unifiable; by symmetry, neither is {dsk(e),sk(f)}.

4.4.2 <u>Lemma</u>:

If $d' \in$ PREPROCESS(d) and Θ is an mgu of $P_{d'}$, then $d' = d\Theta$.

The proof follows easily from the definition of PREPROCESS, and is left to the reader.

4.4.3 <u>Corollary</u>

If $d', d'' \in$ PREPROCESS(d) and $P_{d'} = P_{d''}$, then $d' = d''$.

<u>Proof</u> If $P_{d'} = P_{d''}$ then $d' = d\sigma'$ and $d'' = d\sigma''$ where σ' and σ'' are mgus of $P_{d'}$. Hence either d'' and d' are variants, contradicting the definition of PREPROCESS, or $d' = d''$. □

If $d', d'' \in$ PREPROCESS(d), we shall say that $d' > d''$ iff for each $X \in P_{d'}$ there exists $Y \in P_{d''}$ such that $X \subseteq Y$.

4.4.4 <u>Lemma</u>:

If $d', d'' \in$ PREPROCESS(d), $d' > d''$, σ' is an mgu of $P_{d'}$, and Θ is an mgu of $P_{d''}\sigma'$ then $d'' = d'\Theta$. Again, the proof, based on the definition of PREPROCESS and lemma 4.4.2, is left to the reader.

4.4.5 Theorem: Implicational Independence

Let d, d', d" be literals such that d', d" $\in$ PREPROCESS(d) and d' $\neq$ d"; and suppose g' $\in$ QUANTIFY(d'), g" $\in$ QUANTIFY(d"), then

(i) $g'' \not> g'$

(ii) $g' \not> g''$

Proof We consider two cases as follows:

(a) Suppose $d' \not> d''$ and $d'' \not> d'$. From the definition of $>$, it follows that d' has TS subterms s_1', t_1', s_2', t_2' which are vis-a-vis s_1'', t_1'', s_2'', t_2'' in d", such that $s_1'' = t_1''$ and $s_2' = t_2'$, but $s_1' \neq t_1'$ and $s_2'' \neq t_2''$. Consequently, $\forall x_1''$, $\forall y_1'$ and $\forall x_2'$ occur in prefix(g") where x_1', y_1'', x_2', x_1'', x_2'', y_2'' are the new variables corresponding to s_1', t_1', s_2' (= t_2'), s_1', (= t_1''), s_2'', t_2'' respectively. Therefore, to unify {dsk(g'),sk(g")} it is necessary to unify $\{\{\alpha_1',x_1''\},\{\beta_1'',s_1''\}\}$ where α_1', β_1' are distinct Skolem terms introduced by the application of dsk to g'; this is clearly impossible. Hence (i) is proved, and (ii) is similarly proved by considering the new terms in dsk(g") which are vis-a-vis x_2 in sk(g').

(b) Suppose $d' > d''$. Let Θ be a substitution as defined in lemma 4.4.4 such that $d'' = d'\Theta$; by corollary 4.4.3, since $d' \neq d''$, $P_{d'} \neq P_{d''}$ so Θ must unify at least one pair of distinct TS subterms of d'. Therefore, vis-a-vis these distinct terms of d' are identical terms of d", so by reasoning identical to that used in case (a), (i) holds.

Since d' $\in$ PREPROCESS(d) but has distinct unifiable TS subterms (since Θ unifies at least two of them), Θ must disturb d'. There are two ways this can happen:

either (A) Θ disturbs a top-level variable u of d'

or (B) Θ disturbs a top-level subterm of d'.

In case (A), u in d' occurs vis-a-vis a new Skolem term α in sk(g'). Suppose u occurs vis-a-vis an expression t in d". If t is a term, then it occurs vis-a-vis a term in dsk(g") with a head different from that of α, so (ii) clearly holds. If t is a variable, say v then it occurs in d" vis-a-vis both u and w in d', where $w \neq u$, since Θ disturbs u. Then v occurs in dsk(g") vis-a-vis two different Skolem terms in sk(g'). Then v occurs in dsk(g") vis-a-vis two different Skolem terms in sk(g'). Again it is clear that (ii) holds.

In case (B) we will show that g' has a TS subterm s' such that:

(1) some top-level variable y' occurs in $s'' = s'\Theta$ but not in s', and

(2) $\forall x'$ occurs to the left of $\exists y'$ in prefix(g'), where x' is the new variable corresponding to s'.

Let t' and r' be two distinct TS subterms of d' which are unified by Θ. We have two cases to consider:

(a) Suppose t' and r' are not replaced by new variables during the same execution of the major loop of QUANTIFY. Then some top-level variable y' occurs in t" but not in r' and y' is still free. Clearly $\forall x'$ occurs to the left of $\exists y'$ in prefix(g'), where x' is the new variable corresponding to r'.

(b) If t' and r' are replaced by new variables in the same execution of the major loop, then suppose that for all TS subterms s and all top-level variables y and d', if Θ disturbs s with y, $\forall x$ occurs to the right of $\exists y$ in prefix(g') where x is the new variable corresponding to s. In the case when Θ is an mgu of t' and r', processing of the formula by QUANTIFY will be terminated because of condition (#), contradicting the fact that g' $\in$ QUANTIFY(d'). The general case, when Θ is not an mgu of r' and t', is left to the reader.

Now for g $\in$ QUANTIFY(d"), it follows from (1) that $\exists y'$ occurs to the left of $\forall x"$ in prefix(g'), where x' is the new variable corresponding to s". From this fact, and (2) it follows that to unify {sk(g'),dsk(g")} it is necessary to unify $\{\{x',\alpha[y']\},\{y',\beta[x']\}\}$ where α and β are Skolem terms introduced by sk and dsk. This proves case (ii). □

5. EXAMPLES AND FINAL REMARKS

First we illustrate QUANTIFY.

5.1 Example

Let d be the literal $P(\alpha(x,y),\beta(x,z),\gamma(y,z),f(x,y),z)$ where α, β and γ are Skolem functions and f is not. Then:

$$\text{QUANTIFY}(d) = \{\ \exists x\exists y\forall a\exists z\forall b\forall c\ m,\ \ \exists y\exists z\forall c\exists x\forall a\forall b\ m,\ \ \exists x\exists z\forall b\exists y\forall a\forall c\ m\ \}$$

$$\text{where } m = P(a,b,c,f(x,y),z)$$

The reader should note that for every b $\in$ QUANTIFY(d), dsk(b) $\neq$ d.

The next example illustrates that expressions which are not top-level have no influence on the output of QUANTIFY; and that the names of Skolem functions are unimportant.

5.2 Example

Let $d_1 = P(f(\alpha),\gamma(g(z)),\beta(x),x)$ and $d_2 = P(f(\beta(z)),\delta(a),\gamma(h(x)),x)$ where α, β, γ, δ are Skolem functions, and f is not. Then:

$$\text{QUANTIFY}(d_1) = \text{QUANTIFY}(d_2) = \{\forall y\forall w\exists x\forall v P(f(y),w,v,x)\}.$$

In the preceding examples, no TS subterms are unifiable, so PREPROCESS would have no effect. The next example shows that PREPROCESS is required for completeness.

5.3 Example

Let $d = P(\alpha(z),\alpha(x),x)$, where α is a Skolem function, then:

$$\text{QUANTIFY}(d) = \{\forall y\exists x\forall v P(y,v,x)\} = \{b\}.$$

Consider the closed literal $c = \forall y P(y,y,a)$. Clearly $c \supset d$, but $c \not\supset b$. This is because QUANTIFY distinguishes between Skolem terms which are not identical but are unifiable. However:

$$\text{PREPROCESS}(d) = \{d,e\}$$
where $e = P(\alpha(x),\alpha(x),x)$.

Then $\text{QUANTIFY}(e) = \{\exists x\forall y P(y,y,x)\} = \{f\}$, and $c \supset f$.

Now we provide an example to illustrate how PREPROCESS avoids redundancy.

5.4 Example

Let $d = P(\alpha(x,z),\alpha(x,x),x)$ then $\text{PREPROCESS}(d) = \{P(\alpha(x,x),\alpha(x,x),x)\} = \{b\}$. Note that unlike example 5.3, d is not in PREPROCESS(d), since the unification does not cause a disturbance; and that $d \supset b$.

Our final example illustrates the need for the condition (#) in QUANTIFY.

5.5 Example

Let $d = P(\alpha(x),\gamma(y,v),\beta(z,y),\beta(z,x),x,z,v)$ then:

$$\text{PREPROCESS} = \{d,e\}$$
where $e = P(\alpha(x),\gamma(x,v),\beta(z,x),\beta(z,x),x,z,v)$.

Suppose condition (#) is removed from QUANTIFY, then this modified algorithm produces from d the following:

$$\{\exists x\forall a\exists z\forall b_1\forall b_2\exists v\forall c\ m\ (= d_1),$$
$$\exists x\forall a\exists v\forall c\exists z\forall b_1\forall b_2\ m\ (= d_2),$$
$$\exists z\forall b_1\exists v\forall c\exists x\forall a\forall b_2\ m,$$
$$\exists z\forall b_1\exists x\forall b_2\forall a\exists v\forall c\ m,$$
$$\exists v\forall c\exists x\forall a\exists z\forall b_1\forall b_2\ m,$$
$$\exists v\forall c\exists z\forall b_1\exists x\forall b_2\forall a\ m\}$$

where $m = P(a,c,b_1,b_2,x,z,v)$.

From e, QUANTIFY produces:

$$\{\exists x\forall a\exists v\forall c\exists z\forall b\ m'\ (= e_1),$$
$$\exists x\forall a\exists z\forall b\exists v\forall c\ m'\ (= e_2)\}$$

where $m' = P(a,c,b,b,x,z,v)$.

Clearly $d_1 \supset e_2$ and $d_2 \supset e_1$. However, condition (#) restricts QUANTIFY such that d_1 and d_2 are not produced, since at the point where the two unifiable terms have no free variables remaining, the term $\gamma(y,v)$ which is disturbed by the mgu with x is either still in the matrix (in d_1) or its corresponding new variable c occurs to the right of $\exists x$ in the prefix (in d_2).

BIBLIOGRAPHY

[1] Bledsoe, W.W., and Ballantyne, A.M., "Unskolemizing", Mathematics Dept. Memo ATP-41, University of Texas, July 1978.

[2] Cox, P.T. and Pietrzykowski, T., On reverse Skolemization, Research Report CS-80-01, Department of Computer Science, University of Waterloo, 1980.

[3] Pietrzykowski, T., Mechanical Hypothesis Formation, Research Report CS-78-33, Department of Computer Science, University of Waterloo, 1978.

[4] Skolem, T., Über die mathematische Logik, Norsk mathematisk Tidskrift, 10, pp. 125-142, 1928.

Vol. 49: Interactive Systems. Proceedings 1976. Edited by A. Blaser and C. Hackl. VI, 380 pages. 1976.

Vol. 50: A. C. Hartmann, A Concurrent Pascal Compiler for Minicomputers. VI, 119 pages. 1977.

Vol. 51: B. S. Garbow, Matrix Eigensystem Routines – Eispack Guide Extension. VIII, 343 pages. 1977.

Vol. 52: Automata, Languages and Programming. Fourth Colloquium, University of Turku, July 1977. Edited by A. Salomaa and M. Steinby. X, 569 pages. 1977.

Vol. 53: Mathematical Foundations of Computer Science. Proceedings 1977. Edited by J. Gruska. XII, 608 pages. 1977.

Vol. 54: Design and Implementation of Programming Languages. Proceedings 1976. Edited by J. H. Williams and D. A. Fisher. X, 496 pages. 1977.

Vol. 55: A. Gerbier, Mes premières constructions de programmes. XII, 256 pages. 1977.

Vol. 56: Fundamentals of Computation Theory. Proceedings 1977. Edited by M. Karpiński. XII, 542 pages. 1977.

Vol. 57: Portability of Numerical Software. Proceedings 1976. Edited by W. Cowell. VIII, 539 pages. 1977.

Vol. 58: M. J. O'Donnell, Computing in Systems Described by Equations. XIV, 111 pages. 1977.

Vol. 59: E. Hill, Jr., A Comparative Study of Very Large Data Bases. X, 140 pages. 1978.

Vol. 60: Operating Systems, An Advanced Course. Edited by R. Bayer, R. M. Graham, and G. Seegmüller. X, 593 pages. 1978.

Vol. 61: The Vienna Development Method: The Meta-Language. Edited by D. Bjørner and C. B. Jones. XVIII, 382 pages. 1978.

Vol. 62: Automata, Languages and Programming. Proceedings 1978. Edited by G. Ausiello and C. Böhm. VIII, 508 pages. 1978.

Vol. 63: Natural Language Communication with Computers. Edited by Leonard Bolc. VI, 292 pages. 1978.

Vol. 64: Mathematical Foundations of Computer Science. Proceedings 1978. Edited by J. Winkowski. X, 551 pages. 1978.

Vol. 65: Information Systems Methodology, Proceedings, 1978. Edited by G. Bracchi and P. C. Lockemann. XII, 696 pages. 1978.

Vol. 66: N. D. Jones and S. S. Muchnick, TEMPO: A Unified Treatment of Binding Time and Parameter Passing Concepts in Programming Languages. IX, 118 pages. 1978.

Vol. 67: Theoretical Computer Science, 4th GI Conference, Aachen, March 1979. Edited by K. Weihrauch. VII, 324 pages. 1979.

Vol. 68: D. Harel, First-Order Dynamic Logic. X, 133 pages. 1979.

Vol. 69: Program Construction. International Summer School. Edited by F. L. Bauer and M. Broy. VII, 651 pages. 1979.

Vol. 70: Semantics of Concurrent Computation. Proceedings 1979. Edited by G. Kahn. VI, 368 pages. 1979.

Vol. 71: Automata, Languages and Programming. Proceedings 1979. Edited by H. A. Maurer. IX, 684 pages. 1979.

Vol. 72: Symbolic and Algebraic Computation. Proceedings 1979. Edited by E. W. Ng. XV, 557 pages. 1979.

Vol. 73: Graph-Grammars and Their Application to Computer Science and Biology. Proceedings 1978. Edited by V. Claus, H. Ehrig and G. Rozenberg. VII, 477 pages. 1979.

Vol. 74: Mathematical Foundations of Computer Science. Proceedings 1979. Edited by J. Bečvář. IX, 580 pages. 1979.

Vol. 75: Mathematical Studies of Information Processing. Proceedings 1978. Edited by E. K. Blum, M. Paul and S. Takasu. VIII, 629 pages. 1979.

Vol. 76: Codes for Boundary-Value Problems in Ordinary Differential Equations. Proceedings 1978. Edited by B. Childs et al. VIII, 388 pages. 1979.

Vol. 77: G. V. Bochmann, Architecture of Distributed Computer Systems. VIII, 238 pages. 1979.

Vol. 78: M. Gordon, R. Milner and C. Wadsworth, Edinburgh LCF. VIII, 159 pages. 1979.

Vol. 79: Language Design and Programming Methodology. Proceedings, 1979. Edited by J. Tobias. IX, 255 pages. 1980.

Vol. 80: Pictorial Information Systems. Edited by S. K. Chang and K. S. Fu. IX, 445 pages. 1980.

Vol. 81: Data Base Techniques for Pictorial Applications. Proceedings, 1979. Edited by A. Blaser. XI, 599 pages. 1980.

Vol. 82: J. G. Sanderson, A Relational Theory of Computing. VI, 147 pages. 1980.

Vol. 83: International Symposium Programming. Proceedings, 1980. Edited by B. Robinet. VII, 341 pages. 1980.

Vol. 84: Net Theory and Applications. Proceedings, 1979. Edited by W. Brauer. XIII, 537 Seiten. 1980.

Vol. 85: Automata, Languages and Programming. Proceedings, 1980. Edited by J. de Bakker and J. van Leeuwen. VIII, 671 pages. 1980.

Vol. 86: Abstract Software Specifications. Proceedings, 1979. Edited by D. Bjørner. XIII, 567 pages. 1980

Vol. 87: 5th Conference on Automated Deduction. Proceedings, 1980. Edited by W. Bibel and R. Kowalski. VII, 385 pages. 1980.

This series reports new developments in computer science research and teaching – quickly, informally and at a high level. The type of material considered for publication includes:

1. Preliminary drafts of original papers and monographs
2. Lectures on a new field or presentations of a new angle in a classical field
3. Seminar work-outs
4. Reports of meetings, provided they are
   a) of exceptional interest and
   b) devoted to a single topic.

Texts which are out of print but still in demand may also be considered if they fall within these categories.

The timeliness of a manuscript is more important than its form, which may be unfinished or tentative. Thus, in some instances, proofs may be merely outlined and results presented which have been or will later be published elsewhere. If possible, a subject index should be included. Publication of Lecture Notes is intended as a service to the international computer science community, in that a commercial publisher, Springer-Verlag, can offer a wide distribution of documents which would otherwise have a restricted readership. Once published and copyrighted, they can be documented in the scientific literature.

Manuscripts

Manuscripts should be no less than 100 and preferably no more than 500 pages in length.
They are reproduced by a photographic process and therefore must be typed with extreme care. Symbols not on the typewriter should be inserted by hand in indelible black ink. Corrections to the typescript should be made by pasting in the new text or painting out errors with white correction fluid. Authors receive 75 free copies and are free to use the material in other publications. The typescript is reduced slightly in size during reproduction; best results will not be obtained unless the text on any one page is kept within the overall limit of 18 x 26.5 cm (7 x 10½ inches). On request, the publisher will supply special paper with the typing area outlined.
Manuscripts should be sent to Prof. G. Goos, Institut für Informatik, Universität Karlsruhe, Zirkel 2, 7500 Karlsruhe/Germany, Prof. J. Hartmanis, Cornell University, Dept. of Computer-Science, Ithaca, NY/USA 14850, or directly to Springer-Verlag Heidelberg.

Springer-Verlag, Heidelberger Platz 3, D-1000 Berlin 33
Springer-Verlag, Neuenheimer Landstraße 28–30, D-6900 Heidelberg 1
Springer-Verlag, 175 Fifth Avenue, New York, NY 10010/USA

ISBN 3 540-**10009-1**
ISBN 0-387-**10009-1**